Barriers and Challenges in Computational Fluid Dynamics

ICASE/LaRC Interdisciplinary Series in Science and Engineering

Managing Editor:

MANUEL D. SALAS
ICASE, NASA Langley Research Center, Hampton, Virginia, U.S.A.

Volume 6

Barriers and Challenges in Computational Fluid Dynamics

edited by

V. Venkatakrishnan
Boeing Commercial Airplane Group,
Seattle, Washington, U.S.A.

Manuel D. Salas
ICASE,
Hampton, Virginia, U.S.A.

and

Sukumar R. Chakravarthy
Metacomp Technologies Inc.,
Westlake Village, California, USA

SPRINGER-SCIENCE+BUSINESS MEDIA, B.V.

A C.I.P. Catalogue record for this book is available from the Library of Congress.

ISBN 978-0-7923-4855-9   ISBN 978-94-011-5169-6 (eBook)
DOI 10.1007/978-94-011-5169-6

The cover illustration shows the ejection of water into air from a nozzle driven by a 10 microsecond impulse in the shape of 1/2 of a cosine wave. The domain is 60 x 240 microns and has been covered by a uniform square grid of 32 x 128 cells. This computation was conducted by Igor D. Aleinov and E.G. Puckett in the Department of Mathematics at the University of California, Davis using a high-order accurate projection method with a volume-of-fluid interface tracking algorithm, a Cartesian grid model of the geometry and a generalized continuum surface force model of surface tension.

Printed on acid-free paper

TABLE OF CONTENTS

PREFACE

This volume contains the proceedings of the ICASE/LaRC workshop, "Barriers and Challenges in Computational Fluid Dynamics," conducted by the Institute for Computer Applications in Science and Engineering and NASA Langley Research Center during August 5-7, 1996. This workshop and the proceedings are sequels to the ICASE/LaRC workshop held in August 1991 entitled, "Algorithmic Trends for Computational Fluid Dynamics (CFD) in the 90's" and the bound proceedings with the same title. Significant developments have taken place since the first workshop and this book brings together the key developments in a single unified volume.

As a new millennium approaches, Computational Fluid Dynamics (CFD) is undergoing an important transition. It is playing an integral role in many inter-disciplinary applications. In addition, CFD techniques are increasingly being used in nontraditional areas such as material sciences and manufacturing technologies, and others. New algorithmic issues arise in such contexts and need to be addressed. At this time, therefore, it is particularly important to review the challenges that face the field and identify the barriers to be overcome. In addition, techniques deemed inappropriate in CFD may well be the methods of choice when applied to new disciplines. The objectives of the workshop were two-fold:

- Identify the barriers and challenges facing CFD and propose strategies to overcome them
- Identify new application areas that can exploit CFD techniques and propose new algorithms to deal with emerging disciplines.

The present volume is comprised of five chapters, with an invited paper leading off each chapter. The first chapter "Accuracy" begins with an invited paper by M.B. Giles entitled "Progress in Applied Numerical Analysis for Computational Fluid Dynamics." This paper discusses the role of numerical analysis in helping to understand the issues of accuracy of one-dimensional shock capturing, stability of aerothermal coupling, accuracy of aeroelastic coupling and the stability of Navier-Stokes computations on unstructured grids. The rest of the chapter consists of contributions by B. Engquist and B. Sjögreen, J.W. Goodrich, M.H. Carpenter and J. Casper, and R. Lowrie et al. The second chapter "Boundary Conditions and Stiffness Issues" leads off with an invited paper by W.G. Habashi and co-workers entitled "Anisotropic Mesh Adaptation: A Step Towards a Mesh-Independent and User-Independent CFD." This paper presents the details of a mesh adaptation strategy and demonstrates convergence of the strategy for a variety of inviscid and viscous flows. This paper is followed by contributions from S.V. Tsynkov, and M. Arora and P.L. Roe. The third chapter "Discontinuities" begins with an invited paper by Osher and coworkers entitled "Numerical Methods For A One-Dimensional Interface Separating Compressible and Incompressible Flows." This paper presents a method in one dimension for tracking the interface between a droplet governed by the incompressible Navier-Stokes equations and the surrounding compressible, chemically reacting, multi-species gas also governed by the Navier-Stokes equations. This paper is followed by contributions from R.L. Löhner, G. Puckett, and D. Jacqmin. The fourth chapter "Other Applications" leads off with an invited paper by P.L. Roe entitled "Compounded of Many Simples: Reflections on the Role of Model Problems in CFD." This paper addresses two topics, the first concerning the choice of suitable model problems for a differential system, and the second topic describing a methodology that simultaneously optimizes the grid and the discrete scheme by minimizing a desired objective function. This paper is followed by contributions from R.K. Agarwal, W. Dai and P.R. Woodward, T. Linde and P.L. Roe, and N. Aslan and T. Kammash. The final chapter "Convergence" begins with a paper by B. van Leer entitled "Local Preconditioning: Who Needs It?" This paper reviews the progress made in the area of local preconditioning for the Euler and Navier-Stokes equations to equilibrate the eigenvalues and improve convergence as well as accuracy at low Mach numbers. This paper is followed by contributions from E. Turkel and P.L. Roe, and D. Sidilkover.

The editors would like to thank all the participants for their contributions and cooperation in making the workshop a success. The efforts of Ms. Emily Todd in organizing the workshop and collecting the papers, and the editorial assistance of Mrs. Shannon Verstynen are deeply appreciated.

S.R. Chakravarthy
M.D. Salas
V. Venkatakrishnan

PROGRESS IN APPLIED NUMERICAL ANALYSIS
FOR COMPUTATIONAL FLUID DYNAMICS

M.B. GILES
Oxford University Computing Laboratory
Wolfson Building, Parks Road, Oxford OX1 3QD, UK

1. Introduction

The last 20 years have seen phenomenal progress in the development and application of CFD algorithms, advancing from 1D to 3D calculations, from steady to unsteady flows, from potential flow modelling to the Reynolds-averaged Navier-Stokes equations, from single-block structured grids to unstructured and hybrid grids, and from pure CFD applications to a wide variety of multi-disciplinary applications.

Much of this progress has been built upon a relatively small base of numerical analysis theory. The numerical stability of constant coefficient finite difference equations on infinite structured grids is determined using Fourier analysis. This also gives a necessary, and usually sufficient, local condition for stability when the coefficients of the finite difference equation vary smoothly. For unstructured grids, the CFL theorem has been the mainstay, giving a condition for stability which is necessary, and usually within a constant factor of $2 - 5$ of being sufficient.

Truncation error (or modified equation) analysis is the basis for determining the order of accuracy of algorithms on structured grids, but for finite volume methods on unstructured grids this theory is inadequate. Many finite element methods have their own distinct mathematical theory, but the accuracy that is achieved in actual computations is often much better than the error bounds predicted by theory. With such a large discrepancy, it is not obvious that the numerical analysis provides a good basis for designing improved discretisations.

Perhaps one of the best examples of the relative strengths and weaknesses of engineering computations and numerical analysis has been in grid adaptation. Numerical analysis theory exists for some very simple applications, such as the Laplace equation. However, most developments in adap-

1

V. Venkatakrishnan et al (eds.), Barriers and Challenges in Computational Fluid Dynamics, 1-25.
© 1998 *Kluwer Academic Publishers.*

tive 3D computations using the Euler and Navier-Stokes equations have been with *ad hoc* adaptation criteria based on a combination of a very good understanding of fluid dynamics, knowledge of the truncation errors in flux evaluations, and a considerable amount of numerical experimentation.

My conjecture is that in the next 10 years there will continue to be great progress in the development and application of CFD algorithms, much of it in multi-disciplinary and design applications. However, I think algorithmic developments in 'core' areas of CFD, for example improving the accuracy of a discretisation, or the effectiveness of grid adaptation, will depend increasingly on more detailed numerical analysis of the accuracy and stability of existing algorithms. In doing so, the numerical analysis will have to cope with the following aspects of engineering computations:

- systems of equations
- nonlinearity
- irregular and unstructured grids
- boundary conditions
- high Reynolds number viscous flow
- multidisciplinary applications

This paper cannot attempt to survey the range of old and new theory in numerical analysis which can be applied to address these issues. Instead, it presents a number of recent analyses performed by the author:

- accuracy of quasi-1D shock capturing (Giles, 1996)
- stability of aerothermal coupling (Giles, 1997b)
- accuracy of aeroelastic coupling (Giles, 1997c)
- stability of N-S computations on unstructured grids (Giles, 1997)

Each is motivated by an engineering application and involves the selection of a relevant model problem. Together, they illustrate the application of a selection of the numerical analysis theory which is able to treat some of the difficulties listed above.

2. Accuracy of quasi-1D shock capturing

This analysis was motivated by the question of how best to adapt grids for 2D and 3D transonic flow computations in which there are shocks. Ideally, the criterion will lead to the adaptation of only those cells in which large flow gradients generate large numerical errors. Unfortunately, at present there is no complete theory of *a posteriori* error estimation for the discretisation of nonlinear p.d.e.'s, on which to base rigorous adaptation criteria and so they have instead been developed based on a combination of model linear p.d.e.'s, engineering intuition and practical experience (e.g. (Löhner *et al.*, 1986; Mavriplis & Jameson, 1987; Peraire *et al.*, 1987; Dannenhöffer, 1988; Weatherill *et al.*, 1993)).

One typical adaptation parameter that is used is $A_p = h|\delta p|$, where h is some measure of the cell length, and δp is a first difference of the pressure field. At shocks, δp is independent of the cell size and so the shock cells are adapted repeatedly until h is sufficiently small that A_p falls below the adaptation threshold. Away from shocks, $A_p \approx h^2|\nabla p|$, and so the adapted grid resolution is related to the flow gradient, as desired.

In designing adaptation criteria such as this which will generate a large number of adapted cells at shocks and so obtain very thin discrete shocks, it is implicitly assumed that the shock would otherwise cause substantial numerical errors. The lift on a wing is one of the most important engineering quantities obtained from a solution of the Euler equations. For such a calculation it appears, intuitively, that since the shock is 'smeared' over one or two cells there must be an error in the lift prediction of order $h_s \Delta p$ where h_s is the cell size at the shock and Δp is the jump in pressure across the shock. This appears to be the basis for the particular adaptation criterion above, but other adaptation criteria also lead to very substantial refinement of shock cells and so the belief in a significant first order error at shocks seems widespread although not stated.

2.1. ANALYSIS

The model problem which was selected is the discretisation of transonic inviscid flow in a quasi-1D diverging duct. The steady quasi-1D Euler equations in conservative form are

$$\frac{d}{dx}(AF) - \frac{dA}{dx}P = 0, \tag{1}$$

where U is the state vector and F and P are the usual flux vectors given by

$$U = \begin{pmatrix} \rho \\ \rho u \\ \rho E \end{pmatrix}, \quad F = \begin{pmatrix} \rho u \\ \rho u^2 + p \\ \rho u H \end{pmatrix}, \quad P = \begin{pmatrix} 0 \\ p \\ 0 \end{pmatrix}. \tag{2}$$

$A(x)$ is the cross-sectional area of the duct which, for convenience, is assumed to be locally constant at the two ends.

At the supersonic inflow at $x=0$, the entire state vector $U(0)$ is specified. At the subsonic outflow at $x=1$, the static pressure is specified. Integration of Equation (1) over the domain gives

$$[AF]_0^1 = \int_0^1 \frac{dA}{dx} P \, dx = \begin{pmatrix} 0 \\ D \\ 0 \end{pmatrix}, \tag{3}$$

where the 'drag' D (the force exerted by the sidewall on the fluid) is defined as

$$D = \int_0^1 p \frac{dA}{dx} \, dx. \tag{4}$$

The first and third components of Equation (3) together with the one outflow boundary condition totally specify the three components of $U(1)$ given that $U(0)$ has already been specified. The second component of Equation (3) then defines D uniquely as a function of the boundary conditions independent of the precise variation of $p(x)$ or $A(x)$ between the end points. This is the key in determining the accuracy with which the discretisation approximates the quantity

$$\int_0^1 p \, dx$$

which represents the lift in 2D and 3D Euler calculations for lifting bodies.

The full details of the numerical analysis are presented in Reference (Giles, 1996), but the outline approach is as follows. The analysis considers steady discrete equations of the following conservative form

$$A_{j+1/2}F_{j+1/2} - A_{j-1/2}F_{j-1/2} - \left(A_{j+1/2} - A_{j-1/2}\right) P_j = 0, \tag{5}$$

on a computational grid with uniform mesh spacing h.

The three components of the discrete solution at the inflow are specified as boundary conditions. Because the discretisation is conservative, mass and energy conservation together with the specification of the static pressure at the outflow fully determine the three components of the discrete solution at the outflow as well. Momentum conservation then implies that D_h, the discrete equivalent of the 'drag', is obtained exactly.

The drag integral can be split into two pieces, a 'shock' piece from a region of width $O(h)$ spanning the shock, and a 'smooth' piece from the regions on either side of the shock in which the flow is smooth.

In the smooth flow regions, the solution error is $O(h^m)$ where m is the order of the truncation error. The corresponding errors in the 'smooth' pieces of the discrete drag and lift are also $O(h^m)$. Since the combined drag integral is exact, the error in the shock piece of the discrete drag must be equal and opposite, and so is $O(h^m)$. Reference (Giles, 1996) presents an asymptotic analysis which shows that, as a consequence, the error in the 'shock' piece of the discrete lift integral is $O(h^2)$. Provided $m \geq 2$, this means that the total error in the discrete lift integral is also $O(h^2)$.

Writing the total lift error as

$$L_h - L = C_h h^2, \tag{6}$$

there is nothing in the analysis to suggest that C_h should asymptote to a constant as $h \to 0$. The proven second order accuracy only requires that C_h be bounded, leaving the possibility that C_h may depend on the location of the shock within the shock cell (e.g. whether the shock is at a grid node or halfway between two nodes).

2.2. NUMERICAL RESULTS

Numerical results were obtained using a discretisation in which the numerical smoothing is a blend of second and fourth difference terms. The duct geometry and boundary conditions were chosen so that the peak Mach number on the upstream side of the shock was 1.3. The steady-state discrete solutions were obtained by a fully-converged Runge-Kutta time-marching procedure. Figure 1 shows the Mach number distribution for the solution near the shock using a uniform grid of 64 points.

To investigate the effect of mesh resolution, a sequence of grids was used, with the number of grid points ranging from 64 to 192. For each grid, the influence of the shock position relative to the grid nodes was investigated by performing a number of calculations with the grid displaced by an amount δx in the range $0 \leq \delta x \leq h$. Figure 2a) shows the errors in the computed lift. The 'error bar' indicates the range of values obtained depending on the position of the shock relative to the grid. Figure 2b) plots the magnitude of these error bars $L_h^{max} - L_h^{min}$. Note that in both figures the quantities are plotted against h^2, not h. The linear behaviour in Figure 2b) corresponds to the second order 'shock' component of the error, as predicted by the numerical analysis, with C_h being a function of the shock position. Figure 2a) also shows an almost linear behaviour for small values of h, but for larger values the error increases more rapidly, due to the 'smooth' component of the error which is $O(h^3)$. Figure 2a) also illustrates the possibility for non-monotonic convergence as h is refined; for sufficiently small values of h there are some points within the error bar which show an overprediction of the lift, while others show an underprediction.

The result that the lift is determined with second order accuracy for the model quasi-1D problem is surprising and counter-intuitive. If one performs

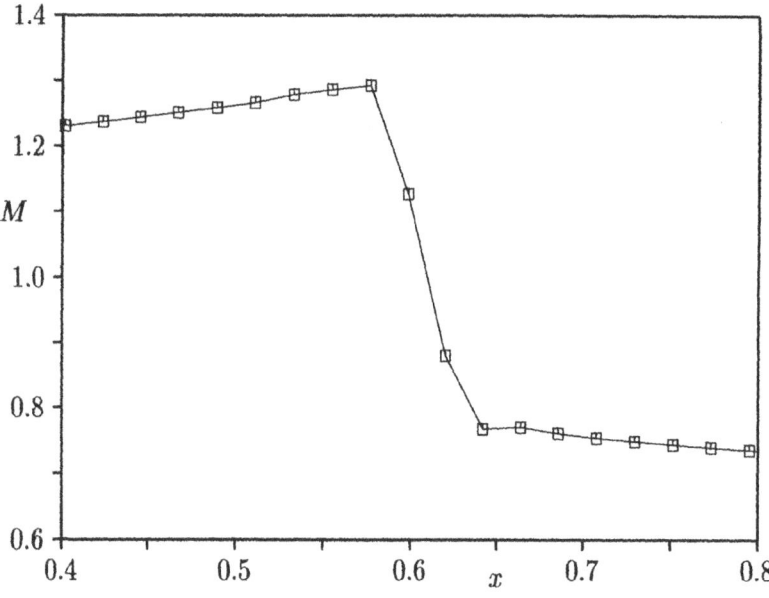

Figure 1. Mach number distribution near the shock

numerical integration of the analytic solution at the discrete grid points using the trapezoidal method, the integration error is $O(h)$ since the value of the trapezoidal integral will be independent of the precise location of the shock within the shock cell. A similar argument applies to the use of any other numerical integration scheme. Since the asymptotic analysis and numerical evidence show that the discrete lift is $O(h^2)$, there must be an equal but opposite error which is also $O(h)$. This can only be due to a $O(1)$ difference between the analytic solution and the discrete solution at the grid points near the shock.

2.3. RELEVANCE TO 2D/3D APPLICATIONS

There is obviously a question about the relevance of the quasi-1D model problem to the 2D and 3D computations which are of real engineering interest. Unpublished grid refinement studies by Jameson show a variety of behaviour for different test cases. Almost all show convergence in lift and drag to be faster than first order. A substantial fraction, but not the majority, show clear second order convergence with the error proportional to h^2. The majority show very rapid convergence which does not appear to be proportional to h^m for any value of m; in many cases the convergence is not even monotonic. These results are consistent with the quasi-1D analysis. However, extending the rigorous numerical analysis from the quasi-1D duct problem to a 2D airfoil problem may prove to be very difficult.

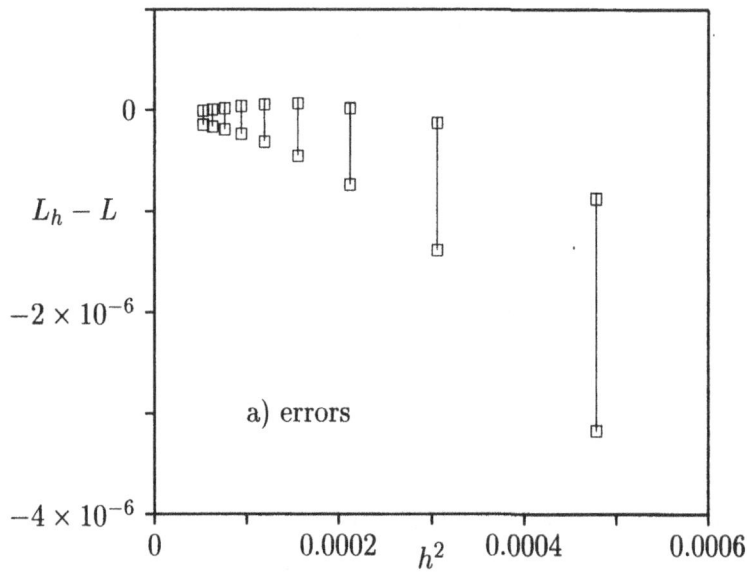

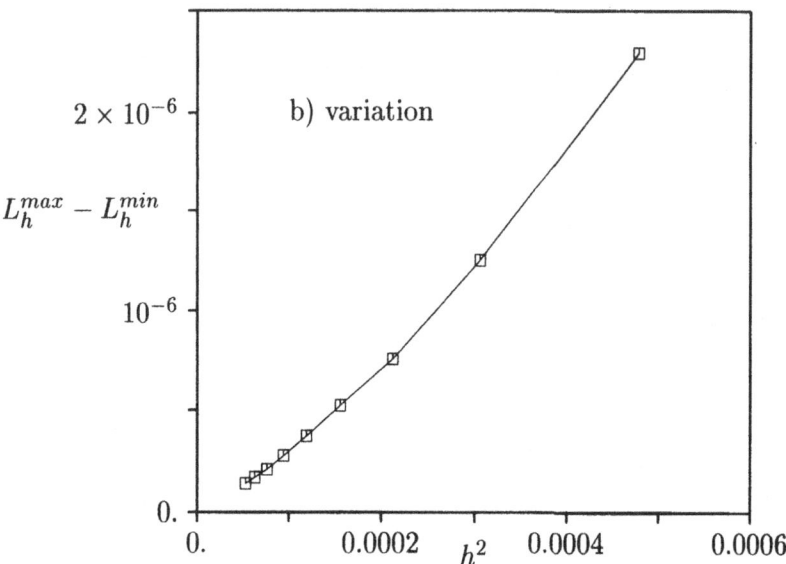

Figure 2. Errors and variation in computed lift

3. Stability of aerothermal analysis

This research was motivated by interest in numerical procedures for combined aerothermal analysis, coupling a thermal diffusion analysis of the heat flux in a solid turbine blade with a Navier-Stokes computation of the surrounding fluid.

One approach to the numerical approximation of this situation would be the use of a single consistent fully-coupled discretisation modelling both the solid and the fluid, plus the boundary conditions at the interfaces (Moore et al., 1989).

However, in practice, a simpler approach is to link two separate codes modelling the solid and fluid, exchanging information at the interface between the two (Amano et al., 1994; Chew et al., 1994; Heselhaus & Vogel, 1995; Bohn et al., 1995). Both CFD codes and thermal analysis codes usually have the capability to specify either the temperature or the heat flux at boundaries. A natural choice therefore for coupling these codes is to specify the surface temperature at the interface in one code, taking the value from the other code, and specify the boundary heat flux in the second code, taking its value from the first code (Amano et al., 1994; Chew et al., 1994). A concern was whether there is any possibility that the coupling procedure could introduce a spurious numerical instability. Therefore, the numerical stability of a model 1D problem was analysed.

3.1. MODEL PROBLEM

As indicated in Figure 3, the 1D model problem has a solid in the region $x < 0$, and a fluid in $x > 0$. In the solid, the evolution of the unsteady temperature is governed by the diffusion equation

$$c_- \frac{\partial T}{\partial t} = -\frac{\partial q}{\partial x}, \qquad q = -k_- \frac{\partial T}{\partial x}, \tag{7}$$

in which $T(x, t)$ is the temperature, $q(x, t)$ is the heat flux, and c_- and k_- are the heat capacity and conductivity, respectively, which are taken to be

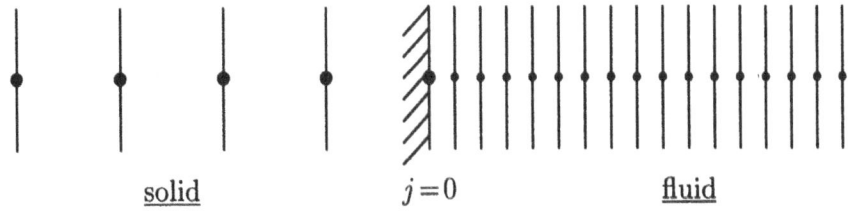

solid $j = 0$ fluid

Figure 3.   1D geometry for aerothermal analysis

uniform.

In the fluid, the convection velocity is neglected, and so the Navier-Stokes equations reduce to a thermal diffusion equation of the same form, but with uniform heat capacity c_+ and conductivity k_+.

At $x=0$, the interface conditions are that T and q must be continuous. The boundary conditions as $x \to \pm\infty$ are that $q \to 0$.

3.2. STABILITY ANALYSIS

Using a computational grid with uniform spacing Δx_+ for the fluid, and uniform spacing Δx_- for the solid, explicit Forward Euler central space differencing of the diffusion equation gives the algorithm

$$\frac{c_\pm \Delta x_\pm}{\Delta t}(T_j^{(n+1)} - T_j^{(n)}) = \frac{k_\pm}{\Delta x_\pm}(T_{j+1}^{(n)} - 2T_j^{(n)} + T_{j-1}^{(n)}), \qquad (8)$$

on either side of the interface, i.e. for $j \neq 0$.

This can be re-expressed as

$$(T_j^{(n+1)} - T_j^{(n)}) = d_\pm(T_{j+1}^{(n)} - 2T_j^{(n)} + T_{j-1}^{(n)}), \quad j \neq 0, \qquad (9)$$

where

$$d_\pm \equiv \frac{k_\pm \Delta t}{c_\pm \Delta x_\pm^2}. \qquad (10)$$

Standard Fourier analysis shows that this is stable provided $d_\pm < \frac{1}{2}$.

At the interface, we choose to enforce continuity of temperature and heat flux by using the solid surface temperature as the boundary condition for the fluid, and using the fluid surface heat flux as the boundary condition for the solid. To be precise, the calculation of $T_1^{(n+1)}$ in the fluid uses the temperature $T_0^{(n)}$ from the solid, and the temperature $T_0^{(n+1)}$ in the solid is calculated from

$$\frac{c_- \Delta x_-}{2\Delta t}(T_0^{(n+1)} - T_0^{(n)}) = -q_w - \frac{k_-}{\Delta x_-}(T_0^{(n)} - T_{-1}^{(n)}), \qquad (11)$$

with the fluid surface heat flux being evaluated by a first order one-sided difference (Heselhaus & Vogel, 1995),

$$q_w = -\frac{k_+}{\Delta x_+}(T_1^{(n)} - T_0^{(n)}). \qquad (12)$$

It is more convenient to consolidate these last two equations into the following equation,

$$T_0^{(n+1)} = T_0^{(n)} - 2d_-\left(T_0^{(n)} - T_{-1}^{(n)}\right) + 2rd_+\left(T_1^{(n)} - T_0^{(n)}\right), \qquad (13)$$

in which r is the ratio of the thermal capacities of the computational cells on either side of the interface,

$$r = \frac{c_+ \Delta x_+}{c_- \Delta x_-}. \tag{14}$$

The interface stability analysis uses the well-established theory of Godunov and Ryabenkii (Godunov & Ryabenkii, 1964; Richtmyer & Morton, 1967), in which the task is to investigate the existence of separable normal modes of the form

$$T_j^{(n)} = z^n f_j. \tag{15}$$

The discretisation is unstable if the difference equation admits such solutions which satisfy the far-field boundary conditions, $f_j \to 0$ as $j \to \pm\infty$, and have $|z| > 1$, giving exponential growth in time. The form of the solution is very similar to that of Fourier modes, except that the amplitude of the spatial oscillation decays exponentially away from the interface.

For this application the normal mode must be of the form

$$T_j^{(n)} = \begin{cases} z^n \kappa_-^j, & j \le 0 \\ z^n \kappa_+^j, & j \ge 0 \end{cases} \tag{16}$$

The difference equations, Equation (9) and Equation (13), are satisfied provided the three variables z, κ_-, κ_+ satisfy the following equations.

$$\begin{aligned} z &= 1 + d_-(\kappa_- - 2 + \kappa_-^{-1}) \\ z &= 1 + 2d_-(\kappa_-^{-1} - 1) + 2rd_+(\kappa_+ - 1) \\ z &= 1 + d_+(\kappa_+ - 2 + \kappa_+^{-1}) \end{aligned} \tag{17}$$

Solving the first of these equations to obtain κ_-^{-1} gives

$$\kappa_-^{-1} = 1 - \frac{1-z}{2d_-}\left(1 \pm \sqrt{1 - \frac{4d_-}{1-z}}\right). \tag{18}$$

To satisfy the far-field boundary conditions as $j \to -\infty$ it is necessary to choose the negative square root when the argument is real and positive; when it is complex, the choice of root is defined by the requirement that $|\kappa_-^{-1}| < 1$. Solving the third of the equations similarly to obtain κ_+, and substituting these into the second equation gives the following nonlinear equation for z.

$$\sqrt{1 - \frac{4d_-}{1-z}} - r\left(1 - \sqrt{1 - \frac{4d_+}{1-z}}\right) = 0 \tag{19}$$

There is no simple closed form solution to this, giving z as an explicit function of the parameters d_-, d_+, r, but analysis of this equation reveals that $|z| < 1$ if, and only if,

$$r < \frac{\sqrt{1 - 2d_-}}{1 - \sqrt{1 - 2d_+}}. \tag{20}$$

The full details are presented in (Giles, 1997b).

This analysis is supported by the numerical results presented in Figure 4. The computations use the finite domain $-2000 \le j \le 2000$, initial conditions $T_j^0 = -1$ for $j < 0$ and $T_j^0 = 1$ for $j \ge 0$ and boundary conditions $T_{-2000}^{(n)} = -1$, $T_{2000}^{(n)} = 1$. d_- and d_+ are both taken to be $\frac{3}{8}$, for which the analysis above predicts the coupled system to be stable only for $r < 1$.

Figure 4 shows two sets of results with $T_j^{(n)}$ plotted every 25 iterations. In a), $r = 0.99$ and the solution appears to be stable, with a slowly decaying interface transient, while in b), $r = 1.01$ and the solution is clearly unstable.

Reference (Giles, 1997b) also analyses the stability of two other algorithms. One is a hybrid algorithm in which the fluid is discretised using the same explicit algorithm, but the solid is discretised using an implicit algorithm. The other uses an implicit discretisation for both the solid and the fluid, but an explicit updating of the boundary conditions for each. For both of these algorithms, the analysis reveals that the stability depends on the parameters r, d_- and d_+, with the coupling being unstable when $r \gg 1$ and stable when $r \ll 1$.

In practice, typical values for $c_\pm$ and Δx usually result in $r \ll 1$, and so the coupled fluid/structural calculations will be stable. This is based on the assumption that the fluid takes its surface temperature from the solid, and the solid takes its surface heat flux from the fluid. If the roles are reversed, specifying the heat flux into the fluid and the surface temperature of the solid, then the above analysis remains valid with the fluid in $x < 0$ and the solid in $x > 0$. In this case, $r \gg 1$, and so the coupling would be unstable.

3.3. CONCLUSIONS

The stability analysis shows the viability of a loosely-coupled approach to computing the temperature and heat flux in coupled fluid/structure interactions. The key point to achieving numerical stability is the use of Dirichlet boundary conditions for the fluid calculation and Neumann boundary conditions for the structural calculation. Although the analysis is performed for a 1D model diffusion equation, this conclusion should remain valid for the real engineering calculations in which the 3D diffusion equation is used to model the heat flux in the structure and the 3D Navier-Stokes equations are used to model the behaviour of the fluid.

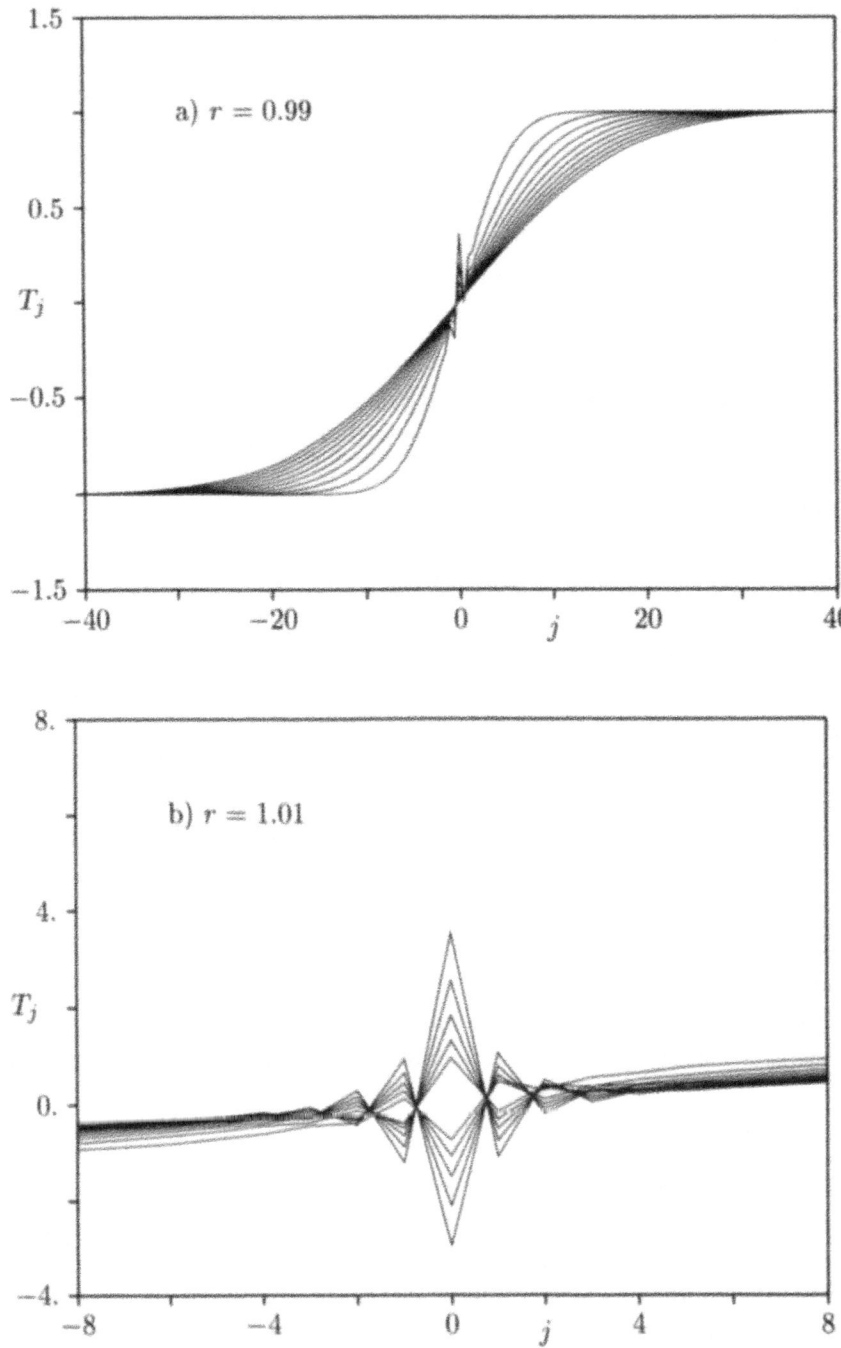

Figure 4.   Aerothermal coupling with results every 25 iterations

4. Accuracy of aeroelastic coupling

The possible flutter of aircraft wings and turbomachinery blades can now be investigated by the simultaneous solution of the coupled 3D nonlinear p.d.e.'s describing the unsteady aerodynamics and structural dynamics of the application (Guruswamy, 1990; Guruswamy, 1990; Marshall & Imregun, 1995; Nomura & Hughes, 1992). However, such calculations are computationally demanding, preventing extensive investigations of some of the underlying algorithmic issues. One issue is whether the coupling procedure may introduce a spurious Godunov–Ryabenkii numerical instability, unrelated to the real flutter instabilities which are the focus of engineering attention. Another is the accuracy of the resulting coupled analysis, particularly when there are very few timesteps per period of oscillation.

To investigate these issues, a model 1D problem was constructed, and a number of different discretisations were analysed and tested numerically (Giles, 1997c).

4.1. MODEL PROBLEM

As illustrated in Figure 5, the 1D model problem consists of a wall oscillating about $x = 0$, and a semi-infinite fluid in $x > 0$.

Neglecting all viscous effects, the fluid dynamics is modelled by the inviscid acoustic equations expressed as a coupled system of first order differential equations for the pressure, p, and velocity, u,

$$\frac{\partial}{\partial t} \begin{pmatrix} p \\ u \end{pmatrix} + \begin{pmatrix} 0 & \rho c^2 \\ \frac{1}{\rho} & 0 \end{pmatrix} \frac{\partial}{\partial x} \begin{pmatrix} p \\ u \end{pmatrix} = 0. \qquad (21)$$

Here ρ and c are the density and speed of sound, respectively, of the undisturbed fluid.

The dynamics of the wall's motion are modelled by a simple mass-spring system subject to the external unsteady aerodynamic pressure.

$$m \ddot{x}_w + m \omega_o^2 x_w = -p(0, t). \qquad (22)$$

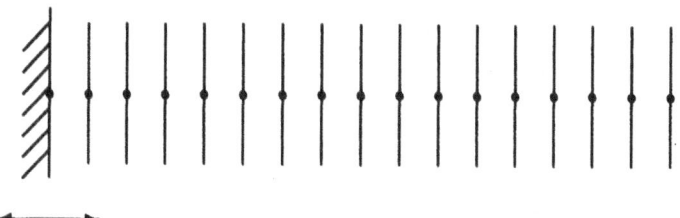

Figure 5.   1D geometry for aeroelastic analysis

Here m represents the mass per unit area and ω_o is the natural frequency of oscillation in the absence of any aerodynamic coupling.

There is also a kinematic compatibility condition, requiring that the velocity of the wall must match that of the fluid.

$$\dot{x}_w(t) = u(0, t). \tag{23}$$

In the far-field, the boundary condition is the radiation condition, that all waves should be outgoing, travelling away from the oscillating wall.

This simple model problem admits an eigenmode solution of the form

$$\begin{aligned}
x_w(t) &= X\, e^{i\omega t}, \\
p(x, t) &= P\, e^{i\omega(t - x/c)}, \\
u(x, t) &= U\, e^{i\omega(t - x/c)},
\end{aligned} \tag{24}$$

where

$$\frac{\omega}{\omega_o} = \sqrt{1 - d^2} + id \approx 1 + id - \tfrac{1}{2}d^2, \tag{25}$$

and

$$d = \frac{\rho c}{2m\omega_o}. \tag{26}$$

The positive imaginary component of ω indicates the amplitude of the wall's oscillation is decaying exponentially; this is because the wall's kinetic and potential energy is being converted into radiating acoustic energy of the fluid.

d is the non-dimensional damping factor which plays a critical role in the aeroelastic analysis. In engineering applications, it is usually in the range $0.005 - 0.02$ for turbomachinery flutter, and in the range $0.05 - 0.2$ got aircraft wing flutter.

4.2. NUMERICAL ANALYSIS

Using first order upwinding for the CFD, with either explicit or implicit time differencing, the discrete equivalent of the far-field radiation condition leads to the conclusion that the discrete characteristic variables corresponding to the incoming acoustic mode are all zero. As a consequence, the pressure and velocity at the wall node, $j = 0$, are related by

$$p_0^{(n)} = \rho c\, u_0^{(n)}. \tag{27}$$

A central difference approximation to the wall dynamics gives

$$\frac{m}{\Delta t^2}\left(x_w^{(n+1)} - 2x_w^{(n)} + x_w^{(n-1)}\right) + m\omega_o^2 x_w^{(n)} = -p_0^{(n)}. \tag{28}$$

The final discrete equation is the kinematic compatibility condition. A simple first order approximation of this is

$$\frac{1}{\Delta t}\left(x_w^{(n+1)} - x_w^{(n)}\right) = u_0^{(n)}. \tag{29}$$

An eigenmode of the form

$$\begin{aligned}
x_w^{(n)} &= X z^n, \\
p_0^{(n)} &= P z^n, \\
u_0^{(n)} &= U z^n,
\end{aligned} \tag{30}$$

is a solution if, and only if, z satisfies the equation

$$z - 2 + z^{-1} + (\omega_0 \Delta t)^2 = -2 d\, \omega_0 \Delta t (1 - z^{-1}). \tag{31}$$

It can be shown that, for $0 < d < 1$, the roots of this quadratic equation have magnitude less than unity provided

$$\omega_0 \Delta t \le \sqrt{4 + d^2} - d < 2. \tag{32}$$

Thus, there is no numerical instability provided there are more than 3 timesteps per period of natural oscillation of the wall.

To determine the accuracy of the discretisation, we let $z = e^{i\omega \Delta t}$ and performing a Taylor series expansion in both d and $\omega_0 \Delta t$ to obtain

$$\frac{\omega}{\omega_0} \approx 1 + id - \tfrac{1}{2}d^2 + \tfrac{1}{2}d\omega_0\Delta t + id^2\omega_0\Delta t + \tfrac{1}{24}(\omega_0\Delta t)^2 + O(d^4, (\omega_0\Delta t)^4). \tag{33}$$

This shows that the first order error in the coupling produces a first order error in both the real and imaginary components of the complex frequency, corresponding to the frequency and damping rate of the coupled oscillation.

The accuracy of this analysis is shown in Figure 6. Numerical calculations were performed for $\omega_o \Delta t = 0.02, 0.05, 0.1, 0.2$ (corresponding approximately to 300, 120, 60, 30 timesteps per period) and values of d in the range $0.005 - 0.1$. Each calculation was performed for 10,000 iterations, and from the results the frequency and damping rate were deduced. These are presented as solid lines in the two parts of Figure 6, while the dashed lines show the predictions from the asymptotic analysis above. The agreement is excellent over the whole parameter range studied.

For a typical flutter frequency and a timestep limited by the explicit CFL stability restriction $\frac{c\Delta t}{\Delta x} < 1$ for a typical grid resolution, $\omega_0 \Delta t$ will be in the range $10^{-3} - 10^{-2}$. In this case, the errors in both the frequency and the damping are negligible compared to other errors such as modelling approximations and uncertainty about structural damping factors. However, when using implicit methods (Jameson, 1991; Rausch *et al.*, 1993), the timestep is no longer limited by the CFL condition and $\omega_0 \Delta t$ will typically be $O(10^{-1})$. In this case the first order coupling is no longer sufficiently accurate.

Reference (Giles, 1997c) contains analyses of three alternative discretisations of the compatibility equation, all of which are second order. The best of the three is the implicit discretisation

$$\frac{1}{2\Delta t}\left(x_w^{(n+1)} - x_w^{(n-1)}\right) = u_0^{(n)}, \tag{34}$$

which can be implemented using a predictor/corrector procedure. Some alternative discretisations of the wall dynamic equations are also analysed and tested numerically. This includes the very accurate state-transition algorithm used by Rausch *et al* (Rausch *et al.*, 1993).

4.3. CONCLUSIONS

One conclusion from all of the analyses and comparisons with numerical experiments is that the asymptotic numerical analysis is very accurate in predicting the accuracy of the coupled aeroelastic damping and frequency when there are at least 30 timesteps per period and the non-dimensional damping parameter d is in the range $0.005 - 0.1$.

If an explicit CFD algorithm is used for the aerodynamic equations, then for typical flutter frequencies and aerodynamic grid resolution the number of timesteps per period will so large that any algorithm for the discretisation of the structural dynamics and the kinematic boundary condition will be sufficiently accurate provided it is at least second order accurate for the uncoupled vibration.

If, on the other hand, an implicit CFD algorithm is used for the aerodynamic equations, then it is possible that there may be as few as 30 timesteps per period. In this case it is necessary to use a discretisation which is second-order accurate for both the uncoupled and coupled systems. For turbomachinery applications with extremely low levels of structural and aerodynamic damping, it is also best to avoid the use of the many standard structural dynamics algorithms which cause spurious numerical damping of the uncoupled wall dynamics.

Although the real 3D aeroelastic applications (which can exhibit unstable flutter) are quite different to this model problem (which is always stable) it is thought these conclusions remain valid for the engineering applications of interest. Further discussion of this point is presented in Reference (Giles, 1997c).

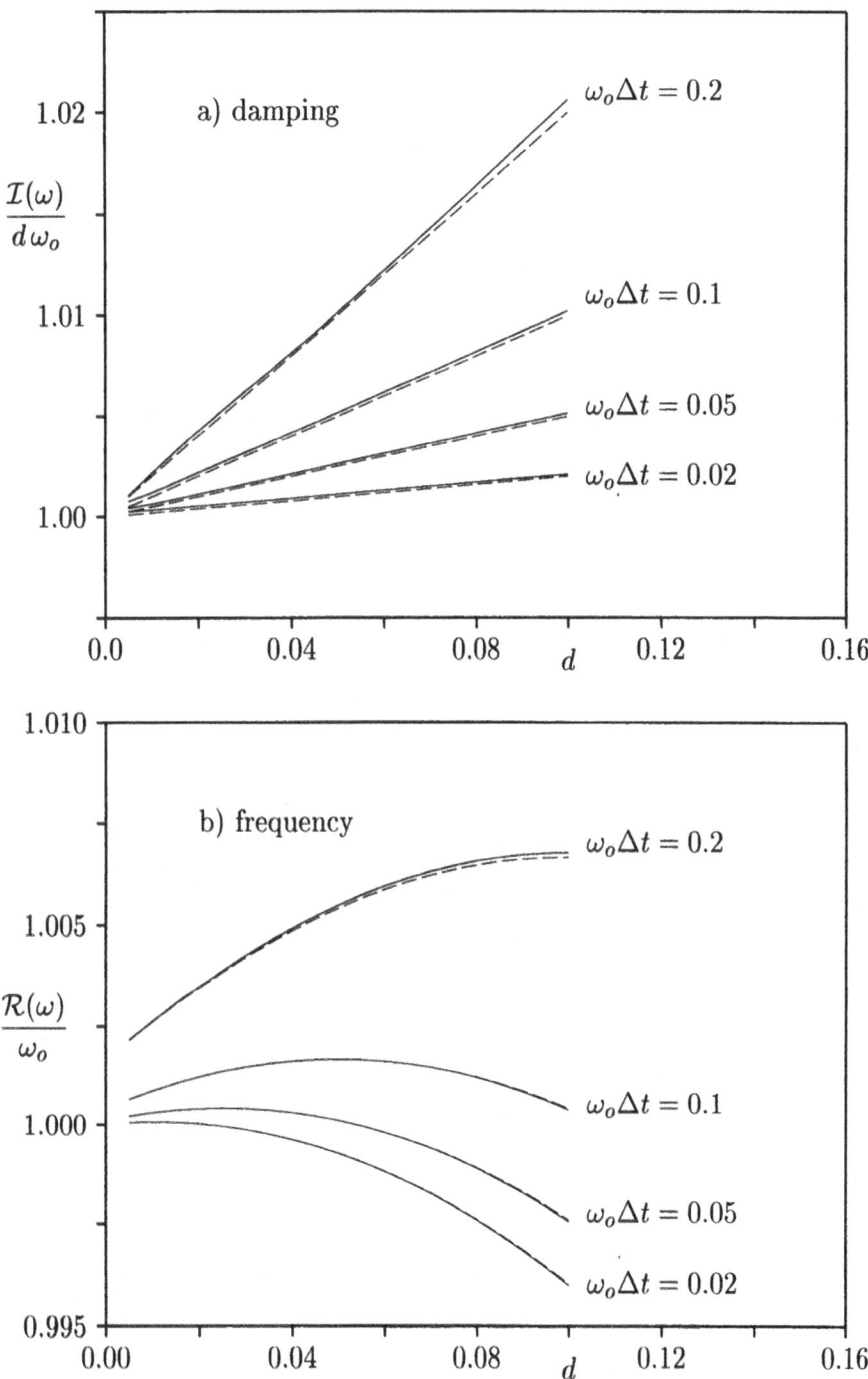

Figure 6. Aeroelastic damping and frequency using first order coupling algorithm
(solid lines – numerical computation; dashed lines – numerical analysis)

5. Stability of N-S discretisation

5.1. INTRODUCTION

Inviscid flow calculations are now being performed almost routinely on unstructured grids for complete aircraft geometries (Weatherill *et al.*, 1993; Peraire *et al.*, 1993; Rausch *et al.*, 1993; Crumpton & Giles, 1995). Many researchers are now working on the development of more accurate and more efficient Navier-Stokes discretisations, and these calculations will also become routine in the next five years.

This raises the problem of determining the timestep stability limit for explicit time-marching methods. Because the grid is unstructured, standard Fourier analysis is not applicable. The CFL theorem still applies, giving an upper bound for the maximum stable timestep and an rough estimate of the actual stability limit. However, it could be that these timestep stability limits are unnecessarily restrictive leading to a large increase in computational cost. This is likely to be particularly true for 3D computations, for which it is much harder to avoid poorly shaped computational cells.

The analysis discussed here, (Giles, 1997), uses recent theoretical developments in numerical analysis. A Galerkin spatial discretisation of the Navier-Stokes equations leads to a coupled system of semi-discrete equations which is solved using Runge-Kutta time-marching. The stability of this is analysed using the concept of *algebraic stability* developed by Spijker and others. In the case of the Euler equations, this leads to stability conditions which are equivalent to those obtained by Giles using an energy analysis method (Richtmyer & Morton, 1967; Giles, 1987).

5.2. NAVIER-STOKES DISCRETISATION

The equations which are considered are a linearised form of the Navier-Stokes equations, for perturbations from a steady-state which is uniform apart from possible variations in the viscosity and conductivity. A periodic domain is considered to avoid the complications of boundary conditions. Changing from the usual conservation variables to symmetrising variables U (Gustafsson & Sundström, 1978; Abarbanel & Gottlieb, 1981), it can be shown that the 'energy' $\iiint \|U\|^2 \, dV$ is non-increasing, and so the flow is stable.

Using a Galerkin spatial discretisation of the p.d.e. leads to a semi-discrete system of equations of the form

$$M \frac{dU}{dt} = (C+D)\,U. \tag{35}$$

The 'mass' matrix M and the diffusion matrix D are both symmetric, and positive definite and positive semi-definite, respectively. Furthermore, the convection matrix C is anti-symmetric. As a consequence of these properties, the semi-discrete 'energy' $U^T M U$ is non-increasing and so the semi-discrete solution is also stable.

5.3. STABILITY THEORY FOR RUNGE-KUTTA METHODS

Discretisation of the scalar o.d.e.

$$\frac{du}{dt} = \lambda u, \tag{36}$$

using an explicit Runge-Kutta method with timestep k yields a difference equation of the form

$$u^{(n+1)} = L(\lambda k)\, u^{(n)} \tag{37}$$

where $L(z)$ is a polynomial function of degree p

$$L(z) = \sum_{m=0}^{p} a_m z^m, \tag{38}$$

with $a_0 = a_1 = 1$, $a_p \neq 0$. Discrete solutions of this difference equation on a finite time interval $0 \leq t \leq t_0$ will converge to the analytic solution as $k \to 0$. In addition, the discretisation is said to be *absolutely stable* for a particular value of k if it does not allow exponentially growing solutions as $t \to \infty$; this is satisfied provided λk lies within the stability region S in the complex plane defined by

$$S = \{z : |L(z)| \leq 1\}. \tag{39}$$

Suppose now that a real square matrix A has a complete set of eigenvectors and can thus be diagonalised,

$$A = T \Lambda T^{-1}, \tag{40}$$

with Λ being the diagonal matrix of eigenvalues of A. The Runge-Kutta discretisation of the coupled system of o.d.e.'s,

$$\frac{dU}{dt} = AU, \tag{41}$$

can be written as

$$U^{(n+1)} = L(kA)\, U^{(n)} = T\, L(k\Lambda)\, T^{-1}\, U^{(n)}, \tag{42}$$

and hence

$$U^{(n)} = T\, (L(k\Lambda))^n\, T^{-1} U^{(0)}. \tag{43}$$

The necessary and sufficient condition for absolute stability as $n \to \infty$, requiring that there are no discrete solutions which grow exponentially with n, is therefore that $|L(k\lambda)| \leq 1$, or equivalently $k\lambda$ lies in S, for all eigenvalues λ of A. If this condition is satisfied, then using L_2 vector and matrix norms it follows that

$$\|U^{(n)}\| \leq \|T\| \|L(k\Lambda)\|^n \|T^{-1}\| \|U^{(0)}\| \leq \kappa(T) \|U^{(0)}\|, \qquad (44)$$

where $\kappa(T)$ is the condition number of the eigenvector matrix T.

If the matrix A is normal, meaning that it has an orthogonal set of eigenvectors then the eigenvectors can be normalised so that $\kappa(T) = 1$. In this case, $\|U^{(n)}\|$ is a non-increasing function of n and $\|U^{(n)}\|^2$ represents a non-increasing 'energy' which could be used in an energy stability analysis. If A is not normal, then the growth in $\|U^{(n)}\|$ is bounded by the condition number of the eigenvector matrix, $\kappa(T)$. Unfortunately, this can be very large indeed, allowing a very large transient growth in the solution even when for each eigenvalue $k\lambda$ lies strictly inside the stability region S and so $\|U^{(n)}\|$ must eventually decay exponentially. This problem can be particularly acute when the matrix A comes from the spatial discretisation of a p.d.e. in which case there is then a family of discretisations arising from a sequence of computational grids of decreasing mesh spacing h. It is possible in such circumstances for the sequence of condition numbers $\kappa(T)$ to grow exponentially, with an exponent inversely proportional to the mesh spacing (Reddy & Trefethen, 1992).

The stability of discretisations of systems of o.d.e.'s with non-normal matrices has been a major research topic in the numerical analysis community in recent years (Reddy & Trefethen, 1992; Kreiss & Wu, 1993; Kraaijevanger et al., 1987; Lenferink & Spijker, 1991; Reddy & Trefethen, 1990; Reddy, 1991; Lubich & Nevanlinna, 1991; van Dorsselaer & Kraaijevanger, 1993). Ideally, one would hope to prove *strong stability*,

$$\|U^{(n)}\| \leq \gamma \|U^{(0)}\|, \qquad (45)$$

with γ being a constant which is not only independent of n but is also a uniform bound applying to all matrices in the family of spatial discretisations for different mesh spacings h but with the timestep k being a function of h. However, at present, the conditions under which strong stability can be proved are too restrictive to be useful in practical computations. Instead, attention has focussed on weaker definitions of stability which are more easily achieved and are still useful for practical computations. One is *algebraic stability* (Reddy & Trefethen, 1992; Kraaijevanger et al., 1987; Lenferink & Spijker, 1991) which allows a linear growth in the transient solution of the form

$$\|U^{(n)}\| \leq \gamma n \|U^{(0)}\|, \qquad (46)$$

where γ is again a uniform constant. A sufficient condition for algebraic stability is that

$$\tau(kA) \subset S, \qquad (47)$$

where the numerical range $\tau(kA)$ is a subset of the complex domain defined by

$$\tau(kA) = \left\{ k \frac{W^* AW}{W^* W} : W \neq 0 \right\} \qquad (48)$$

in which W can be any non-zero complex vector of the required dimension and W^* is its Hermitian, the complex conjugate transpose. By considering W to be an eigenvector of A, it can be seen that $k\lambda \in \tau(kA)$ for each eigenvalue of A and so the requirement that $\tau(kA) \subset S$ is a tighter restriction on the maximum allowable timestep than asymptotic stability.

In the Navier-Stokes application, the main part of the analysis lies in bounding the range of the matrix $M^{-1/2}(C+D)M^{-1/2}$. The details are presented in Reference (Giles, 1997). The approach is to determine a timestep k such that $\tau(k M^{-1/2}(C+D)M^{-1/2}) \subset V \subset S$, with the subset V being either a rectangle or a half-circle. This leads to a sufficient condition for stability for time-accurate computations. With appropriate modifications to the matrix M, a sufficient stability limit for local timesteps for steady-state computations is also derived.

5.4. NUMERICAL EXPERIMENTS

Figure 7 shows two sets of numerical experiments used to verify the stability analysis and determine how close the predicted sufficient stability limit is to the actual stability limit. The numerical tests used a tetrahedral grid created from a $10 \times 10 \times 10$ Cartesian grid by cutting each hexahedron into six tetrahedra. Periodic boundary conditions were applied on all sides. In each case, a set of calculations was performed for a range of values for the CFL parameter r in increments of 0.25 starting from $r = 2.75$.

In the inviscid test case the Mach number was 0.5, and there was a grid stretching ratio of 10:1 in one direction. The algebraic stability theory predicts stability for $r < 2.828$. The numerical results shows stability up to $r \approx 3.4$ so the sufficient stability theory underpredicts the stability boundary by approximately 15%.

In the viscous test case, the grid stretching ratio was increased to 100:1, representative of a boundary layer grid. The cell Reynolds number was chosen to be 1.0, making the viscous and inviscid terms equally important. In this case the algebraic stability analysis predicts stability for $r < 2.616$. The actual stability boundary is at $r \approx 3.9$ so the theory underpredicts the maximum stable timestep by approximately 33%.

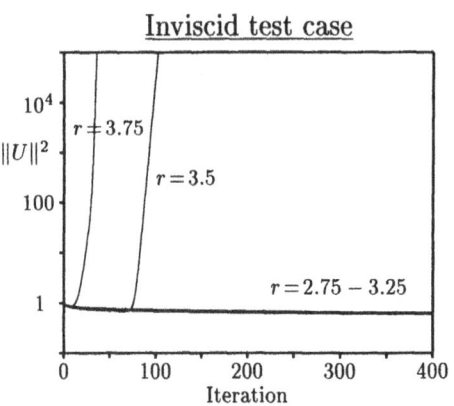

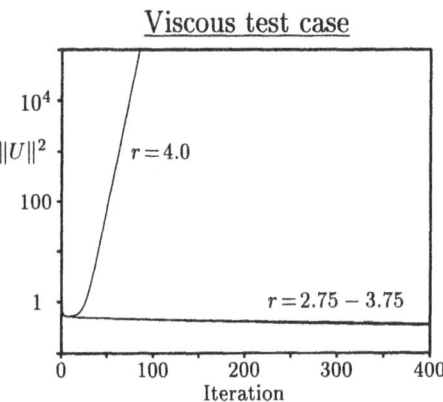

Figure 7.   Numerical energy growth in two test cases

5.5. CONCLUSIONS

The numerical experiments verify the usefulness of this algebraic stability analysis. The sufficient stability limits given by the theory do indeed lead to stable computations, and they are not very much smaller than the actual stability limits determined experimentally. The ability to analyse the stability of complex systems of equations such as the discrete Navier-Stokes equations is very useful. The same method of analysis could also be used to examine the stability of different forms of upwinding on unstructured grids, or the stability of discrete boundary conditions on either structured or unstructured grids.

6. Final comments

The four analyses in this paper have illustrated the application of both well-established and very recent numerical analysis theory to problems of immediate relevance to practical CFD computations. Referring back to the list of challenges posed in the introduction, these analyses have dealt with systems of equations, nonlinearity (1), boundary conditions and multidisciplinary applications (2, 3), and high Reynolds number viscous flow and unstructured grids (4).

As CFD researchers tackle increasingly difficult applications in the future, it is my belief that, more and more, algorithm development will have to be based on a firm foundation of numerical analysis of this kind, analysing model problems which retain much of the complexity of the real computations.

Acknowledgements

I wish to thank Rolls-Royce plc, EPSRC and DTI for funding the research reported in this paper.

References

Abarbanel, S. and Gottlieb, D., "Optimal time splitting for two- and three-dimensional Navier-Stokes equations with mixed derivatives," *Journal of Computational Physics*, Vol. 35, 1981, pp. 1–33.

Amano, R.S., Wang, K.D., and Pavelic, V., "A study of rotor cavities and heat transfer in a cooling process in a gas turbine," *J. Turbomachinery*, Vol. 116, 1994, pp. 333–338.

Bohn, D., Lang, G., Schönenborn, H., and Bonhoff, B., "Determination of thermal stress and strain based on a combined aerodynamic and thermal analysis for a turbine nozzle guide vane," *ASME Paper 95-CTP-89*, 1995.

Crumpton, P. and Giles, M.B., "Aircraft computations using multigrid and an unstructured parallel library," *AIAA Paper 95-0210*, 1995.

Chew, J., Taylor, I.J., and Bonsell, J.J., "CFD developments for turbine blade heat transfer," In *3rd International Conference on Reciprocating Engines and Gas Turbines, I. Mech E.*, London, 1994, pp. C499-035.

Dannenhoffer, J.F. III, "A comparison of two adaptive grid techniques," In *Numerical Grid Generation in Computational Fluid Mechanics*, Pineridge Press Ltd., 1988.

Giles, M.B., "Energy stability analysis of multi-step methods on unstructured meshes," Technical Report CFDL-87-1, MIT Dept. of Aero. and Astro., 1987.

Giles, M.B., "Analysis of the accuracy of shock-capturing in the steady quasi-1D Euler equations," *CFD Journal*, Vol. 5, No. 2, 1996, pp. 99–108.

Giles, M.B., "Stability analysis of a Galerkin/Runge-Kutta Navier-Stokes discretisations on unstructured tetrahedral grids," *Journal of Computational Physics*, Vol. 132, 1997, pp. 201-214.

Giles, M.B., "Stability analysis of numerical interface conditions in fluid structure thermal analysis," to appear in the *International Journal of Numerical Methods in Fluids*, 1997.

Giles, M.B., "Stability and accuracy of numerical boundary conditions in aeroelastic analysis," to appear in the *International Journal of Numerical Methods in Fluids*,

1997.

Godunov, S.K. and Ryabenkii, V.S., *The Theory of Difference Schemes–An Introduction*, North Holland, Amsterdam, 1964.

Gustafsson, B. and Sundström, A., "Incompletely parabolic problems in fluid dynamics," *SIAM J. Appl. Math.*, Vol. 35, No. 2, 1978, pp. 343–357.

Guruswamy, G.P., "ENSAERO — A multidisciplinary program for fluid/structural interaction studies of aerospace vehicles," *Comput. Systems Engrg.*, Vol. 1, Nos. 2-4, 1990, pp. 237–256.

Guruswamy, G.P., "Unsteady aerodynamic and aeroelastic calculations for wings using Euler equations," *AIAA J.*, Vol. 28, No. 3, 1990.

Heselhaus, A. and Vogel, D.T., "Numerical simulation of turbine blade cooling with respect to blade heat condution and inlet temperature profiles," *AIAA Paper 95-3041*, 1995.

Jameson, A., "Time dependent calculations using multigrid with applications to unsteady flows past airfoils, wings and rotors," *AIAA Paper 91-1596*, 1991.

Kraaijevanger, J.F.B.M., Jenferink, H.W.J., and Spijker, M.N., "Stepsize restrictions for stability in the numerical solution of ordinary and partial differential equations," *J. Comput. Appl. Math.*, Vol. 20, Nov. 1987, pp. 67–81.

Kreiss, H.O. and Wu, L., "On the stability definition of difference approximations for the initial boundary value problem," *Appl. Num. Math.*, Vol. 12, 1993, pp. 213–227.

Löhner, R., Morgan, K., and Zienkiewicz, O., *Adaptive grid refinement for the compressible Euler equations*, John Wiley & Sons Ltd., 1986, pp. 281–297.

Lubich, C. and Nevanlinna, O., "On resolvent conditions and stability estimates," *BIT*, Vol. 31, 1991, pp. 293–313.

Lenferink, H.W.J. and Spijker, M.N., "On the use of stability regions in the numerical analysis of initial value problems," *Math. Comp.*, Vol. 57, No. 195, 1991, pp. 221–237.

Marshall, J.G. and Imregun, M., "A 3D time-domain flutter prediction method for turbomachinery blades," In *Proc. Int. Forum on Aeroelasticity and Structural Dynamics*, Royal Aeronautical Society, Manchester, 1995.

Mavriplis, D. and Jameson, A., "Multigrid solution of the Euler equations on unstructured and adaptive meshes," In *Proceedings of the Third Copper Mountain Conference on Multigrid Methods: Lecture Notes in Pure and Applied Mathematics*, S. McCormick, ed., Marcel Dekker Inc., 1987.

Moore, J., Moore, J.G., Henry, G.S., and Chaudry, U., "Flow and heat transfer in turbine tip gaps," *Journal of Turbomachinery*, Vol. 111, July 1989, pp. 301–309.

Nomura, T. and Hughes, T.J.R., "An arbitrary Lagrangian-Eulerian finite element method for interaction of fluid and a rigid body," *Comput. Methods Appl. Mech. Engrg.*, Vol. 95, 1992, pp. 115–138.

Peraire, J., Peiró, J., and Morgan, K., "Finite element multigrid solution of Euler flows past installed aero-engines," *Comput. Mech.*, Vol. 11, 1993, pp. 433–451.

Peraire, J., Vahdati, M., Morgan, K., and Zienkiewicz, O.C., "Adaptive remeshing for compressible flow computations," *J. Comput. Phys.*, Vol. 72, 1987, pp. 449–466.

Rausch, R.D., Batina, J.T., and Yang, H.T.Y., "Three-dimensional time-marching aeroelastic analyses using an unstructured-grid Euler method," *AIAA J.*, Vol. 31, No. 9, 1993, pp. 1626–1633.

Reddy, S.C., *Pseudospectra of Operators and Discretization Matrices and an Application to Stability of the Method of Lines*, Ph.D. thesis, Massachusetts Institute of Technology, Cambridge, Massachusetts 02139, 1991. *Numerical Analysis Report 91-4*, 1991.

Richtmyer, R.D. and Morton, K.W., *Difference Methods for Initial-Value Problems*, Wiley-Interscience, 2nd edition, 1967. Reprint edn., Krieger Publishing Company, Malabar, 1994.

Reddy, S.C. and Trefethen, L.N., "Lax-stability of fully discrete spectral methods via stability regions and pseudo-eigenvalues," *Comput. Methods Appl. Mech. Engrg.*, Vol. 80, 1990, pp. 147–164.

Reddy, S.C. and Trefethen, L.N., "Stability of the method of lines," *Numer. Math.*, Vol.

62, 1992, pp. 235–267.

van Dorsselaer, J.L.M. and Kraaijevanger, J.F.B., "Linear stability analysis in the numerical solution of initial value problems," *Acta Numer.*, 1993, pp. 199–237.

Weatherill, N.P., Hassan, O., Marchant, M.J., and Marcum, D.L., "Adaptive inviscid flow solutions for aerospace geometries on efficiently generated unstructured tetrahedral meshes," *AIAA Paper 93-3390*, 1993.

EXAMPLES OF ERROR PROPAGATION FROM DISCONTINUITIES

BJÖRN ENGQUIST
Department of Mathematics
University of California at Los Angeles
Los Angeles, CA90024

AND

BJÖRN SJÖGREEN
NADA
KTH
100 44 Stockholm
Sweden

1. Introduction

We shall consider systems of hyperbolic partial differential equations,

$$\mathbf{u}_t + \mathbf{f}(\mathbf{u})_x = \mathbf{0}.$$

Let us assume that this problem is approximated by a numerical method of formal order of accuracy p. We then know that if:

- the solution is regular enough,
- the numerical method is regular enough,
- the numerical method is linearly stable,

then the numerical method is convergent and the actual convergence rate of the numerical approximation to to the exact solution is also of order p (Strang, 1964).

It is known that for some difference methods of ENO type, the actual convergence rate degenerates to become lower than the formal order of accuracy. In these methods ii) or iii) above or sometimes both are not satisfied. (Shu, 1990)

We shall here study problems where ii) and iii) hold, but where i) is not satisfied. We shall show examples where a discontinuity in the solution will lead to a degenerate convergence rate, even in domains outside a neighborhood of the discontinuity.

27

V. Venkatakrishnan et al (eds.), Barriers and Challenges in Computational Fluid Dynamics, 27-42.
© 1998 *Kluwer Academic Publishers.*

2. Errors on the downwind side of a shock

A typical example of how a discontinuity can cause degeneration of the convergence rate is the test problem

$$u_t + (u^2/4)_x = 0, \qquad -\infty < x < \infty, \; 0 < t$$
$$v_t + v_x + g(u) = 0,$$

with initial data for the first component,

$$u(x,0) = \begin{cases} 1 & x < 0 \\ -1 & x > 0 \end{cases}$$

Let $v(x,0)$ be an arbitrary smooth function. Let the source term $g(u)$ have the properties $g(1) = g(-1) = 0$, and $g > 0$ in $-1 < x < 1$. The solution of the u-equation is a steady discontinuity which, if computed by a shock capturing method, will have one or more intermediate grid values in the shock. These intermediate grid points represent order one errors, which are propagated into the post shock region by the v-characteristics, since these intersect the shock. The effect of these errors in v will be of order h. For more details, and numerical examples, see (Engquist & Sjögreen, 1995). We have here an example where a shock produces a first order error in a formally higher order method.

3. A combustion problem

We now consider the one reaction combustion model

$$\text{(3.1)} \quad \begin{pmatrix} \rho \\ m \\ n \\ e \\ \rho z \end{pmatrix}_t + \begin{pmatrix} m \\ m^2/\rho + p \\ mn/\rho \\ m(e+p)/\rho \\ mz \end{pmatrix}_x + \begin{pmatrix} n \\ mn/\rho \\ n^2/\rho + p \\ n(e+p)/\rho \\ nz \end{pmatrix}_y = \begin{pmatrix} 0 \\ 0 \\ 0 \\ 0 \\ -K\rho z e^{-T_i/T} \end{pmatrix}$$

describing the motion of a fluid in the x-y plane in which a one step irreversible chemical reaction is taking place. Initial data is given at $t = 0$. The dependent variables $\rho(x,y,t), m(x,y,t), n(x,y,t), e(x,y,t), z(x,y,t)$ are the density, x- and y-momentum, energy and the fraction of unreacted fluid respectively. The pressure is given by

$$p = (\gamma - 1)(e - \frac{1}{2}(m^2 + n^2)/\rho - q_0\rho z)$$

and the temperature is defined as $T = p/\rho$. We will use $u = m/\rho$ to denote the velocity. γ is the ratio between heat capacities and q_0 is the constant heat release due to the chemical reaction.

In this problem, the effect of one, or a few grid points in the shock can cause the solution to behave in a wrong way. It is possible to obtain numerical solutions of (3.1) where the wave structure is wrong due to grid points in the shock. The local $\mathcal{O}(1)$ errors at the discontinuity give rise to $\mathcal{O}(h)$ errors in large domains. This is more severe than in Section 2 above. However, this occurs only when the right hand side is very stiff.

In Figs. 3.1 and 3.2 we show the density and mass fraction unreacted fluid, using $K = 1 \times 10^5$ and $K = 3 \times 10^7$ respectively. The other parameter values were $T_i = 50, q_0 = 50, \gamma = 1.2$. We started with $p = 1, u = 0, \rho = 1$ to the right and obtained the initial profile by integration of the ODE which is obtained by putting the traveling wave ansatz $w(x,t) = w(x - st)$ into the system (3.1). Chapman-Joguet conditions were used to determine the wave speed. For the smaller K value, the solution profile is well resolved and a correct solution is obtained. For the larger K, the solution should be the same, except for a more narrow peak in the density. However, in the computed solution for $K = 3 \times 10^7$, we see a second wave emerging behind the combustion front.

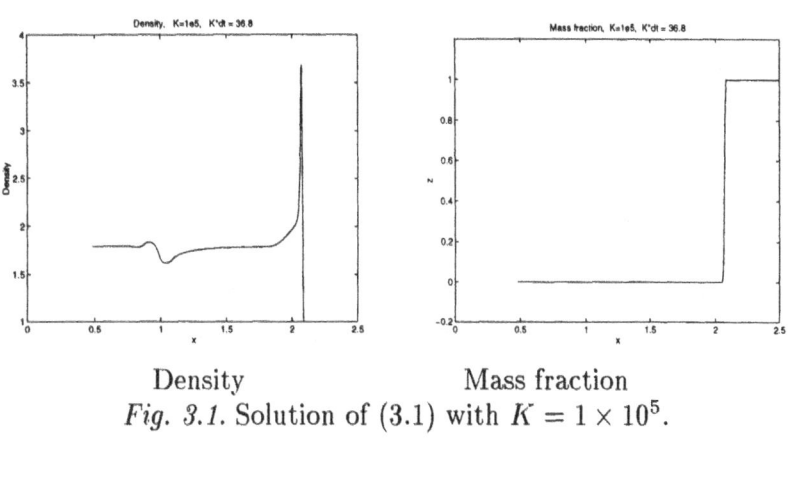

Density   Mass fraction
Fig. 3.1. Solution of (3.1) with $K = 1 \times 10^5$.

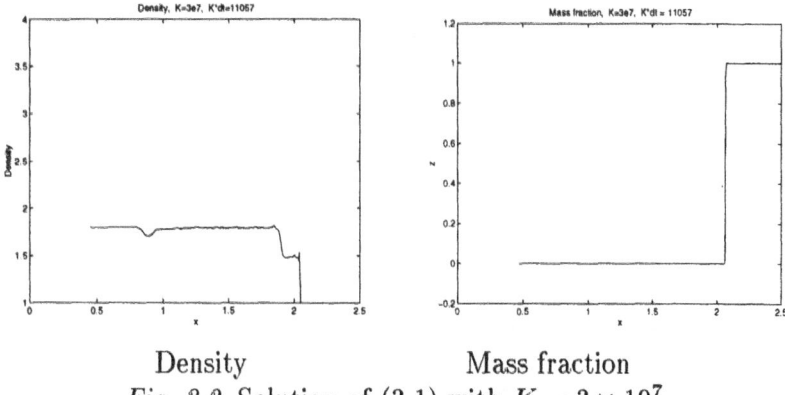

Density   Mass fraction
Fig. 3.2. Solution of (3.1) with $K = 3 \times 10^7$.

The solution is wrong, and large errors have appeared. See (Ben-Artzi, 1989)(Colella *et al.*, 1986) for similar results.

We shall describe the one dimensional setting for the numerical method used in these computations. Two space dimensions is implemented analogously. We introduce the grid points $x_j, j = \ldots, -1, 0, 1, \ldots$, with equal mesh spacing $\Delta x = x_{j+1} - x_j$. The time levels $t_0, t_1, \ldots$ are also uniformly spaced with space step $\Delta t = t_{n+1} - t_n$. We use u_j^n either to denote the approximate solution in the point (x_j, t_n), or sometimes to denote the approximate cell average of u in the cell $[x_{j-1/2}, x_{j+1/2}]$ at time t_n. The difference between cell averages and point values is significant only for order of accuracy three and higher.

The convective flux derivative

$$\begin{pmatrix} m \\ \rho u^2 + p \\ m(e+p)/\rho \\ mz \end{pmatrix}_x$$

is discretized in space, using an upwind method. The numerical results presented in the following sections, were computed using Roe's method.

The convective part of the equations deviates very little from the non reactive Euler equations, since the extra equation appears as an additional linear field in the characteristic variables. The formulas for eigenvectors and eigenvalues used to implement Roe's method are straightforward to derive, and differs very little from the non reactive case.

Higher order accuracy was obtained using piecewise polynomial interpolation of ENO type, (Harten *et al.*, 1986). Piecewise linear and piecewise parabolic reconstruction were implemented. In the problems solved here, it turned out that the interpolation in the physical variables, instead of using characteristic variables, was sufficient to produce a good solution. A TVD limiter function was used to the limit the slopes in the piecewise linear reconstruction.

In summary, the problem after flux discretization is on the form

$$\frac{du_j(t)}{dt} + \frac{h_{j+1/2} - h_{j-1/2}}{\Delta x} = g(u_j(t))$$

where

$$h_{j+1/2} = h(u_{j+1}^-(t), u_j^+(t))$$

with $h(u, v)$ the first order numerical flux function, and u_j^+, u_j^- are the values of u at the left and right ends of cell j obtained from the piecewise polynomial interpolation.

For the time discretization, a Runge-Kutta method is used. It is well known (Shu & Osher, 1988), that the second order method

$$
\begin{aligned}
u^{(1)} &= u^n - \lambda\Delta_+ h^n_{j-1/2} \\
u^{(2)} &= u^{(1)} - \lambda\Delta_+ h^{(1)}_{j-1/2} \\
u^{n+1} &= (u^n + u^{(2)})/2
\end{aligned}
$$

for the convective terms gives a good performance with respect to shocks. Here $h_{j-1/2}$ is the numerical flux function described above, and we use $\lambda = \Delta t/\Delta x$. The source term is added such that it does not impose any additional stability restrictions and such that the accuracy is the same as for the convective method (two). We therefore consider the following method

$$
\begin{aligned}
u^{(1)} &= u^n - \lambda\Delta_+ h^n_{j-1/2} + a\Delta t g(u^n) + b\Delta t g(u^{(1)}) \\
u^{n+1} &= (u^{(1)} + u^n)/2 - \tfrac{\lambda}{2}\Delta_+ h^{(1)}_{j-1/2} + c\Delta t g(u^{(1)}) + d\Delta t g(u^{n+1})
\end{aligned}
$$

where $g(u)$ is the source term. It is easy to derive the following one parameter family of second order accurate methods,

$$
a = 1, \quad b = 0, \quad c = 1/2 - d.
$$

For example the explicit method $a = 1, b = 0, c = 1/2, d = 0$ is included in this family. By performing standard ODE stability analysis on the problem with $h_{j-1/2} = \lambda_1 u, g(u) = \lambda_2 u$, it turns out that the only choice which contains the entire left hand plane of $\lambda_2\Delta t$ in the stability region (A stability) is

$$
a = 1, \quad b = 0, \quad c = 0, \quad d = 1/2.
$$

This is the method used for implicit treatment of the source term.

For the third order ENO discretization of the convective fluxes, we used a third order Runge-Kutta method for time discretization. This is another of the so called TVB-Runge-Kutta methods derived in (Shu & Osher, 1988). Similar to the second order case we add the source term to this method to obtain

$$
\begin{aligned}
u^{(1)} &= u^n - \lambda\Delta_+ h^n_{j-1/2} + a\Delta t g(u^n) + b\Delta t g(u^{(1)}) \\
u^{(2)} &= \tfrac{3}{4}u^n + \tfrac{1}{4}(u^{(1)} - \lambda\Delta_+ h^{(1)}_{j-1/2}) + c\Delta t g(u^{(1)}) + d\Delta t g(u^{(2)}) \\
u^{n+1} &= \tfrac{1}{3}u^n + \tfrac{2}{3}(u^{(2)} - \lambda\Delta_+ h^{(2)}_{j-1/2}) + e\Delta t g(u^{(2)}) + f\Delta t g(u^{n+1})
\end{aligned}
$$

A straightforward but lengthy analysis reveals that the following values

$$
a = 1, \quad b = 0, \quad c = -3/4, \quad d = 1, \quad e = 0, \quad f = 2/3
$$

give overall third order accuracy, and A-stability for the source term, when
the convective terms equal to zero. A further analysis yields that the method
is stable for the source term, when the convective term satisfies a CFL
condition.

It is necessary to treat the source term implicitly for problems with
very stiff chemistry, see, e.g., (Desideri *et al.*, 1990). It is an easy calculation to show that the nonzero eigenvalue of the linearized right hand side,
$\partial g(u)/\partial u$, is

$$\lambda_s = K e^{-T_i/T}((\gamma - 1)q_0 z \frac{T_i}{T^2} - 1).$$

The convective part of the equations has the maximum eigenvalue

$$\lambda_c = u + c.$$

The convective stability requirement is $\Delta t \leq \Delta x/(u+c)$, (assuming for
simplicity a CFL stability limit of one) and the ODE-stability limit due to
the source term is of the form $\Delta t \leq \lambda_s^{-1}$. Thus if

(3.2) $$\Delta t \lambda_c / \Delta x < \Delta t \lambda_s$$

the source term will impose an additional restriction on the time step, unless
approximated implicitly. Note that a large K will lead to a small Δt, but
when $K\Delta t \to 0$, and $\Delta t/\Delta x = $ constant, the CFL restriction from the
convective part will dominate. Condition (3.2) will have to be checked for
each particular problem. Some applications will require an implicit method
and some will not.

An implicit method for the source term means that a nonlinear equation
has to be solved in each grid point. The equation is (with X as unknown)

(3.3) $$X = (\rho z)_j^n - \lambda \Delta_+ h_{j-1/2}^n - K\Delta t X e^{-T_i/T_j^{n+1}}$$

where

$$T_j^{n+1} = (\gamma - 1)(e_j^{n+1}/\rho_j^{n+1} - \frac{1}{2}(u_j^{n+1})^2 - q_0 X/\rho_j^{n+1}).$$

The functions e, ρ, u are already know at t_{n+1} from the three equations
without source terms. Considering (3.3) as a function of X, it follows easily,
by elementary calculus, that there is always one root $X \in [0, \rho_j^{n+1}]$ if the
convective step guarantees that $(\rho z)_j^n - \lambda \Delta_+ h_{j-1/2}^n \in [0, \rho_j^n]$. We solve (3.3)
using Newton's method.

It is possible to avoid solving (3.3) numerically, if we neglect the dependence of the temperature on X. We can then solve (3.3) analytically.
The numerical experiments showed this approximation to lead to stiffness
problems and is thus *not* recommended.

This effect stems from the interaction of the points in the shock with the source term. It can be seen from the discretized equations that if $K\Delta t$ is large, then ρz will drop to zero immediately as T increases above T_i. All shock capturing methods will give a few grid points in the shock. This results in a local error of order 1 in the temperature. The error may trigger the chemistry at a wrong location. If, e.g., the difference method places one new point in the shock at each time step, the chemical reaction will propagate with one grid point per time step. Note that the time stepping method is made such that the source term is unconditionally stable in the sense of stability for ODE.

4. How to avoid stiffness problems

A detonation wave consists of a non reacting shock wave which first increases the temperature of the fuel mixture, so that ignition occurs behind the shock wave. No chemical reactions start before the shock wave has passed through. We shall emulate this behavior in the numerical method. Numerically, there are always a few grid points in the shock. We want to make sure that none of the points inside the numerical shock triggers unwanted chemical reactions. One simple way of ensuring this is to evaluate the right hand side a few grid points ahead of the shock. i.e., instead of solving (3.3) we solve

$$(4.1) \qquad X = (\rho z)_j^n - \lambda\Delta_+ h_{j-1/2}^n - K\Delta t X e^{-T_i/T_{j+d}^{n+1}}$$

where now

$$(4.2) \qquad T_{j+d}^{n+1} = (\gamma - 1)(e_{j+d}^{n+1}/\rho_{j+d}^{n+1} - \frac{1}{2}(u_{j+d}^{n+1})^2 - q_0 z_{j+d}^{n+1})$$

and d is the number of points in the shock, usually one or two. The accuracy will drop to first order, but higher order accuracy can be recovered by an extrapolation in the temperature, i.e., by using

$$(4.3) \qquad 2T_{j+1}^{n+1} - T_{j+2}^{n+1}$$

in place of T_{j+d}^{n+1} inside the exponentiation in (4.1). Here T_{j+1}^{n+1} and T_{j+2}^{n+1} depend on X through (4.2).

By computing in reverse order, $i = n, n-1, \ldots, 1$, we can solve the equation (4.1) analytically,

$$X = \frac{(\rho z)_j^n - \lambda\Delta_+ h_{j-1/2}^n}{1 + K\Delta t e^{-T_i/T_{j+d}^{n+1}}}.$$

In the formulas above, we have assumed that the detonation travels from the left to the right, so that the solution at $j+1$ have lower temperature than the solution at j. In a more general situation and in more than one space dimension, the direction of extrapolation is determined as the direction of decreasing temperature. E.g. in the two dimensional computations in Section 6, we use the minimization

$$\min_{|d|=1,|e|=1} T_{j+d,k+e}$$

to find the direction of extrapolation, where the temperature is evaluated after the convective part has been done.

$$T_{j,k} = T_{j,k}(\rho_{j,k}^{n+1}, u_{j,k}^{n+1}, v_{j,k}^{n+1}, e_{j,k}^{n+1}, z_{j,k}^{n})$$

once the direction is found, we can solve for the new X using (4.1).

It would have been possible to use a more complicated algorithm. We could, e.g., have used a sensor to indicate if a certain point is inside the shock layer, and modify the chemistry only at those points. The method described above is however simpler and works well, so we found no need to develop a more complicated method.

A theoretical justification for our type of technique is given in the thesis of Tegnér (Tegnér, 1992).

5. Numerical results

In Figs. 5.1 and 5.2 below we have made the same computations as in Figs. 3.1 and 3.2, but now with the temperature of the source term extrapolated by the linear extrapolation (4.3).

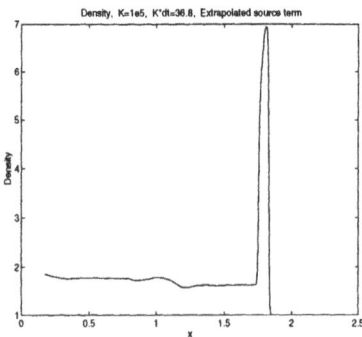

Fig. 5.1. Solution of (3.1) (density) with $K = 1 \times 10^5$.
Linear temperature extrapolation.

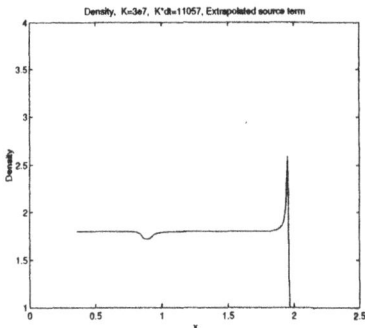

Fig. 5.2. Solution of (3.1) (density) with $K = 3 \times 10^7$.
Linear temperature extrapolation.

The unphysical waves for $K = 3 \times 10^7$ (Fig. 3.1) has disappeared and
the solution for $K = 1 \times 10^5$ is not seriously affected by the second order
error deriving from the extrapolation of the source term. Only a widening
of the density peak is seen.

The figures show that the correct velocity can be captured even with
coarse grids. The details of the profile are only approximate. For example,
the width of the pressure peak must always cover a few grid points and is
thus considerably wider than in the exact solution. The exact ZND profile
is displayed in Fig. 5.3, for the $K = 3 \times 10^7$ case. In Fig. 5.4 we display
the solution for $K = 3 \times 10^7$, but with the double number of grid points,
thereby confirming that the peak narrows with the number of grid points.

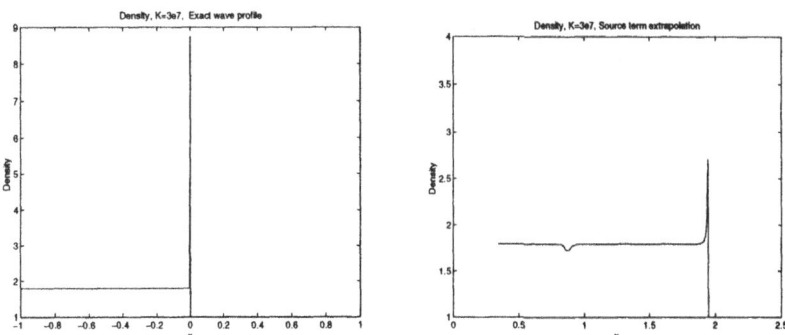

Fig. 5.3. Exact ZND profile.   *Fig. 5.4.* Twice the resolution of Fig. 5.2.

It is known that the traveling wave oscillates around this ZND profile
as it propagates, thus this profile can only be considered as an exact solution on the average. Next we shall investigate how well we can capture the
behavior of the peak pressure at the ZND spike. It is possible to increase
the pressure in the left hand state, and thereby give the detonation wave
a so called overdrive. For a detonation moving with speed s, the overdrive,
f, is defined as $f = (s/s_{CJ})^2$ where s_{CJ} is the speed of a Chapman-Joguet

detonation. Thus we always have $f \geq 1$, since weak detonations are not physically admissible. For certain values of f the pressure peak is oscillating in a periodic way, for other values it is stable, and for some values it is oscillating in a chaotic way. For an analysis of these phenomena, see (Abouseif & Toong, 1982)(Bourlioux *et al.*, 1991).

To capture this oscillatory behavior a large number of grid points is usually required. We next investigate how the resolution required is affected by the accuracy of the method used.

We here use the parameter values $T_i = 50, q_0 = 50, \gamma = 1.2, K = 10000$ and Arrhenius kinetics. The values are taken from (Abouseif & Toong, 1982) and (Bourlioux *et al.*, 1991) (except for K). An overdrive, $f = 1.6$, was given, a value corresponding to a case where the pressure peak oscillates periodically (Abouseif & Toong, 1982)(Bourlioux *et al.*, 1991). We compute on a domain of length 1, which is moved along with the wave. Initially the exact ZND profile is given by solving the traveling wave problem using a fourth order Runge-Kutta method. The boundary values are given as $p = 1, \rho = 1, u = 0$ to the right and extrapolated ($w_0 = w_1$) to the left, the wave travels to the right. All computations were run to time = 1.

Below we compare a few different numerical methods, one first order and two second order TVD schemes with different flux limiters and one third order ENO scheme. We did not use extrapolation of the source term in these computations. We used explicit time discretization of the source term, since condition (3.2) showed that the CFL condition for the convective part of the equations was sufficient to ensure stability for the overall method. For the second order scheme, we compare results from using the minmod flux limiter and van Leer's flux limiter (Sweby, 1984).

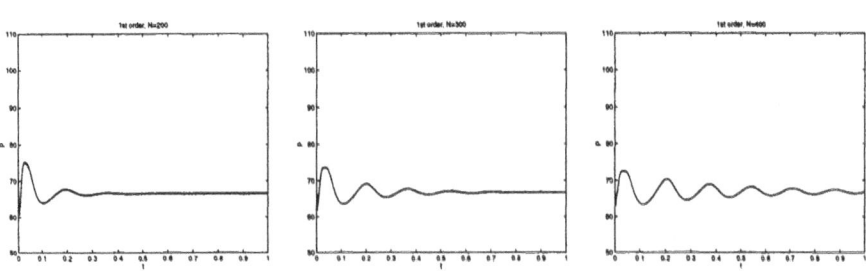

$f=1.6$, 200 points $f=1.6$, 300 points $f=1.6$, 400 points

Fig. 5.5. First order TVD method.

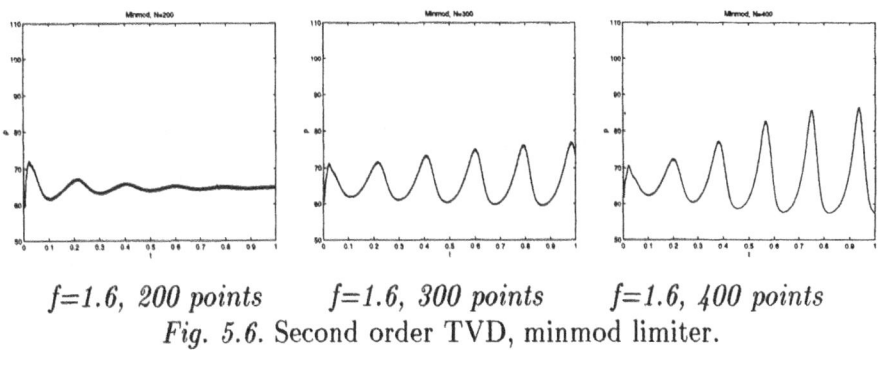

f=1.6, 200 points   *f=1.6, 300 points*   *f=1.6, 400 points*

Fig. 5.6. Second order TVD, minmod limiter.

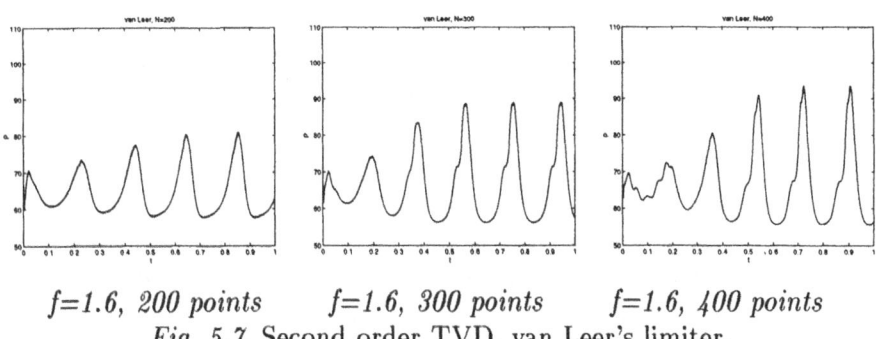

f=1.6, 200 points   *f=1.6, 300 points*   *f=1.6, 400 points*

Fig. 5.7. Second order TVD, van Leer's limiter.

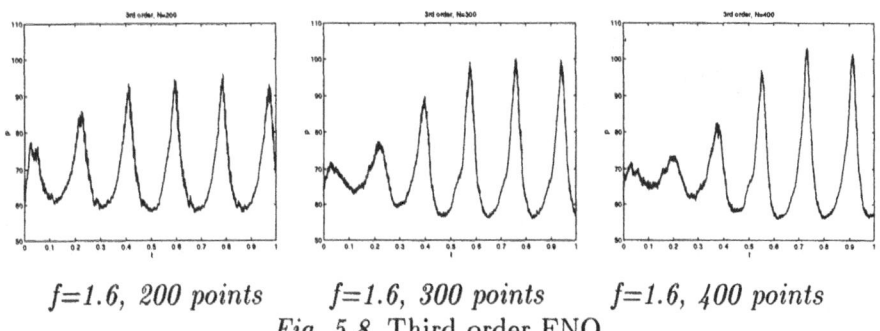

f=1.6, 200 points   *f=1.6, 300 points*   *f=1.6, 400 points*

Fig. 5.8. Third order ENO.

The first order scheme cannot capture the correct behavior, not even on the finest grid. This is hardly surprising, first order schemes are too dissipative. It is more interesting to note the difference deriving from different flux limiters in the second order methods. With van Leer's limiter, we can capture the right frequency, (but not amplitude) with as few as 200 grid points, while the minmod limiter requires at least 300 grid points for a similar resolution. The third order scheme performs best in the sense that fewer grid points are required to obtain a numerical result, similar to the results of lower order methods with more grid points. With the third order method, we get a solution with both frequency and amplitude reasonably well represented using 200 grid points.

In order to see an extremely well resolved solution, we show in Fig. 5.9 results obtained using 1000 grid points.

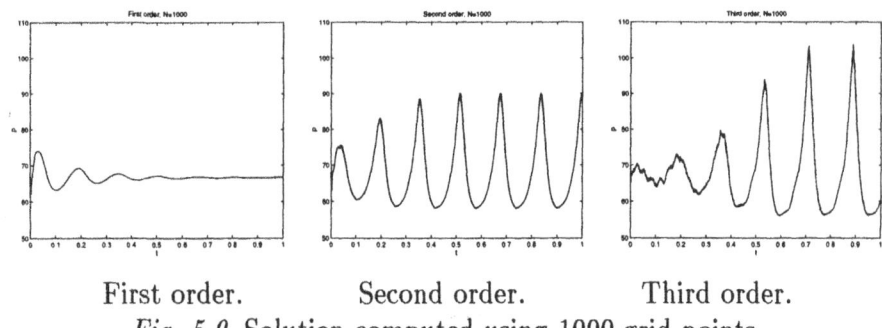

First order.   Second order.   Third order.

Fig. 5.9. Solution computed using 1000 grid points.

The frequency of the oscillations corresponds well with those we obtain on a coarser grid.

We did not use the half reaction time as time unit. Nevertheless, by numerical integration of the steady solution profile, we found that $t_{1/2} = 0.0212$ and the corresponding half reaction length $\Delta_{1/2} = 0.0193$. This means that 200 grid points corresponds to roughly 4 points per half reaction length, this seems to be the fewest possible number of points required to capture the oscillations using the best (3rd order) method. The period of the oscillations is 8 half reaction times, in accordance with previous results (Abouseif & Toong, 1982)(Bourlioux *et al.*, 1991).

In order to capture the dynamics correctly the time step had to be chosen so small that no instability of the type in Fig. 3.2 occurred. The extrapolation technique was thus not needed. Low order extrapolation introduces local truncation errors which may interfere with the dynamics and the advantage of the extrapolation method in this case when it is used for the details in the profile would be reduced.

6. Numerical results in two space dimensions

Computations of a two dimensional traveling detonation wave are presented. We used a second order TVD method in all computations in this section.

Consider a two dimensional channel of width $1/2$, the upper and lower boundaries are solid walls. We start a with a ZND profile in the x direction, but given a slightly curved shape through a sinusodial perturbation,

$$u_0(x, y) = w(x + \Delta x \sin 2\pi(2y + \frac{1}{4}))$$

where $w(x)$ is the one dimensional detonation wave. See Fig. 6.1 for the initial data.

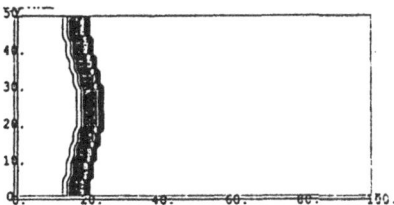

Fig. 6.1. Density contours.

The solution was computed on grid of 100×50 points which was moved along with the detonation wave. We solved equations (3.1) with Arrhenius kinetics, and parameter values $T_i = 50, q_0 = 50, \gamma = 1.2, K = 10000$. To the right, we give the free stream values $p = 3, \rho = 1, u = 0$ and the ZND profile $w(x)$ is obtained by numerical integration as in the one dimensional computations.

Fig. 6.2 below shows density contours at six different times.

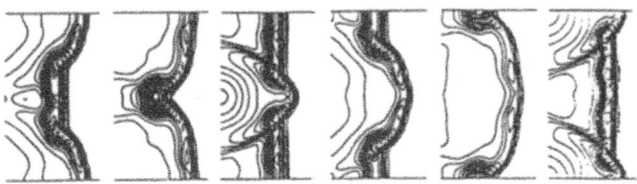

Fig. 6.2. Density contours

One important feature of this solution is the triple points, which travel in the transverse direction and bounces back and forth against the upper and lower walls.

An discussion of the mechanisms driving this solution is given in (Kailasanath *et al.*, 1985). The trace of the triple points can be seen in experiments with smoked screens as a cellular pattern (Bull *et al.*, 1982). By plotting contour lines in the y-t plane we can obtain a similar pattern for our numerical solution. Fig. 6.3 below shows 7 detonation cells.

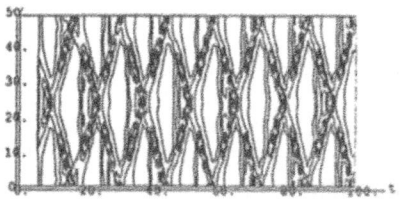

Fig. 6.3. Density contours in the y-t plane.

The CFL number in this computation was 0.4. We next increase the time step by taking a CFL number of 0.85. The solution at six different times are showed in Fig. 6.4

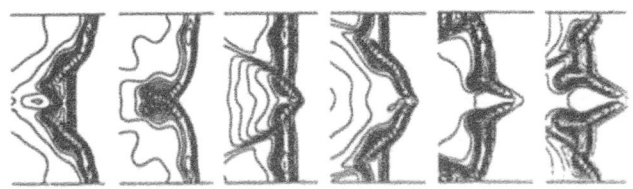

Fig. 6.4. Density contours.

After some time the triple points cease to move, and a triangular shape extends from the detonation front. In the y-t plane in Fig. 6.5 it is clearly seen that the triple point dynamics ceases.

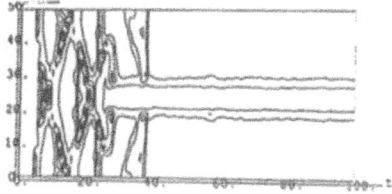

Fig. 6.5. Density contours in the y-t plane.

The solution is completely wrong, due to stiffness problems. The modified method will help to overcome this difficulty, as we show next. In Fig. 6.6 we display the same computation as in Fig. 6.4, but with the source term extrapolated, using $d = 1$. The direction of extrapolation is taken from the lowest temperature direction as described in Section 4. There is no artificial numerical detonation spike.

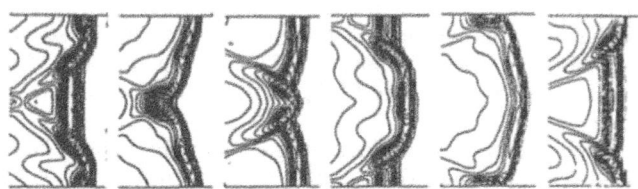

Fig. 6.6. Density contours.

The dynamics is recovered and the solution have the same structure as the one computed with smaller time step (Fig. 6.2). From the y-t plane plot, we conclude that the detonation cell sizes deviate very little from the solution in Fig. 6.3 even if due to the coarse grid in Figs. 6.6, 6.7 the details are different.

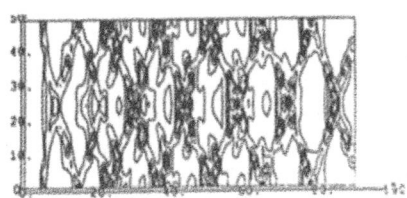

Fig. 6.7. Density contours in the y-t plane.

References

Abouseif, G. E. and Toong, T. Y., "Theory of Unstable One-Dimensional Detonations," *Combust. Flame*, Vol. 45, 1982, pp. 67–94.

Ben-Artzi, M., "The Generalized Riemann Problem for Reactive Flows," *J. Comp. Phys.*, Vol. 81, 1989, pp. 70–101.

Bourlioux, A., Majda, A., and Roytburd, V., "Theoretical and Numerical Structure for Unstable One-Dimensional Detonations," *SIAM J. Appl. Math.*, Vol. 51, 1991, pp. 303–343.

Bull, D. C., Elsworth, J. E., Shuff, P. J., and Metcalfe, E., "Detonation Cell Structures in Fuel/Air Mixtures," *Combust. Flame*, Vol. 45, 1982, pp. 7–22.

Colella, P, Majda, A., and Roytburd, V., "Theoretical and Numerical Structures for Reacting Shock Waves," *SIAM J. Sci. Stat. Comput.*, Vol. 7, 1986, pp. 1059–1080.

Desideri, J.-A., Glinsky, N., and Hettena, E., "Hypersonic Reactive Flow Computation," *Computers & Fluids*, Vol. 18, 1990, pp. 151–182.

Engquist, B. and Sjögreen, B., "High Order Shock Capturing Methods," in *Computational Fluid Dynamics Review*, M. Hafez and K. Oshima, eds., John Wiley & Sons Ltd., 1995, pp. 210–233.

Harten, A., Osher, S., Engquist, B., and Chakravarthy, S., "Some Results on Uniformly High-Order Accurate Essentially Nonoscillatory Schemes," *Applied Numerical Mathematics*, Vol. 2, 1986, pp. 347–377.

Kailasanath, K., Oran, E. S., Boris, J. P., and Young, T.R., "Determination of Detonation Cell Size and the Role of Transverse Waves in Two-Dimensional Detonations," *Combust. Flame*, Vol. 61, 1985, pp. 199–209.

Shu, C.-W., "Numerical experiments on the accuracy of eno and modified eno schemes," *Journal of Scientific Computing*, Vol. 5, 1990, pp. 127–149.

Shu, C.-W. and Osher, S., "Efficient Implementation of Essentially Non-oscillatory Shock-Capturing Schemes," *J. Comput. Phys.*, Vol. 77, 1988, pp. 439–471.

Strang, G., "Accurate Partial Difference Methods II," *Numerische Mathematik*, Vol. 6, 1964, pp. 37–46.

Sweby, P., "High Resolution Schemes Using Flux Limiters for Hyperbolic Conservation Laws," *SIAM J. Numer. Anal.*, Vol. 21, 1984, pp. 995–1010.

Tegnér, J. K., "Properties of Detonation Waves," Ph.D.Thesis, Royal Institute of Technology, Stockholm, Dec. 1992.

ACCURATE FINITE DIFFERENCE ALGORITHMS

JOHN W. GOODRICH
NASA Lewis Research Center
Cleveland, OH 44135

1. Introduction

Clear examples of the difficulties associated with applying CFD techniques to apparently simple problems are provided by computational aeroacoustics and computational electromagnetics. These applications require accurate wave propagation over long distances for a wide range of frequencies, placing a severe demand on numerical algorithms, and raising issues related to efficiency, accuracy, compatible space and time treatments, high frequency data, propagation along characteristic surfaces, isotropy, stable and accurate artificial boundary treatments, and nonrestrictive stability bounds. This paper briefly presents two methods for the development of finite difference algorithms which are intended to address these issues. High order single step explicit algorithms are possible, and examples with up to eleventh order accuracy will be shown. High resolution algorithms in the sense of (Lele, 1992) are also possible, with amplification factor and relative phase change per time step which are virtually 1 for normal mode frequencies in $[0, \pi]$ and CFL numbers in $[0, 1]$. If our most accurate algorithm is used to propagate an initial periodic sine wave, then after five periods with four grid points per wavelength, the maximum error is $O[10^{-6}]$, and after five hundred thousand periods with eight grid points per wavelength, the maximum error is $O[10^{-4}]$. High order algorithms are relatively more efficient (Kreiss & Oliger, 1972), and their relative efficiency tends to increase as the error bound decreases, as the simulation time increases, and as the spatial dimension increases. Our algorithms show several orders of magnitude difference in the number of multiplications required to meet an error bound of $O[10^{-4}]$ at five periods of propagation. Applied computations require boundary conditions. A new artificial boundary condition that has been developed by Hagstrom (Hagstrom, 1997) will be shown for the linearized Euler equations. This boundary condition is local in time and does

43

V. Venkatakrishnan et al (eds.), Barriers and Challenges in Computational Fluid Dynamics, 43-61.
© *1998 Kluwer Academic Publishers.*

not require information about the distance or direction of either an assumed source or an expanding wave front. The boundary and propagation
algorithms have a consistent derivation and similar properties (Goodrich
& Hagstrom, 1996). Additional issues are raised by shocks, but are not
addressed here.

2. Algorithms for the Linearized Euler Equations in 1D

As a first example we will consider the linearized Euler equations in one
space dimension. The equations for the isentropic case can be written in
nondimensionalized form as the system

$$
\begin{aligned}
\frac{\partial u}{\partial t} + M\frac{\partial u}{\partial x} + \frac{\partial p}{\partial x} &= 0, \\
\frac{\partial p}{\partial t} + M\frac{\partial p}{\partial x} + \frac{\partial u}{\partial x} &= 0,
\end{aligned}
\tag{1}
$$

where M is the constant mean convection Mach number, p is the disturbance pressure, and u is the disturbance velocity. System (1) can be diagonalized and solved by the Method of Characteristics to produce the general
solution

$$
\begin{aligned}
u(x,t) &= \tfrac{1}{2}(ui(x-(M+1)t) + pi(x-(M+1)t) \\
&\quad + ui(x-(M-1)t) - pi(x-(M-1)t)), \\[6pt]
p(x,t) &= \tfrac{1}{2}(pi(x-(M+1)t) + ui(x-(M+1)t) \\
&\quad + pi(x-(M-1)t) - ui(x-(M-1)t)),
\end{aligned}
\tag{2}
$$

with initial data $ui(x) = u(x,0)$, and $pi(x) = p(x,0)$.

The first step in producing algorithms for solution of (1) is to locally
interpolate u and p at t_n with order D polynomials in x:

$$
\begin{aligned}
u(x_i + x, t_n) &\approx ua(x) = \textstyle\sum_{\alpha=0}^{D} u_\alpha x^\alpha, \\
p(x_i + x, t_n) &\approx pa(x) = \textstyle\sum_{\alpha=0}^{D} p_\alpha x^\alpha.
\end{aligned}
\tag{3}
$$

The expansion coefficients are obtained by the Method of Undetermined
Coefficients using the known data on a given stencil. Note that there is no
specification of a particular mesh or data type, and that separate derivative
terms are not even considered. The use of single interpolation polynomials
of order D is equivalent to simultaneously approximating all derivatives up
to the Dth order, with

$$
\begin{aligned}
u_\alpha &= \frac{1}{\alpha!}\frac{\partial^\alpha ua}{\partial x^\alpha} \approx \frac{1}{\alpha!}\frac{\partial^\alpha u}{\partial x^\alpha}, \\
p_\alpha &= \frac{1}{\alpha!}\frac{\partial^\alpha pa}{\partial x^\alpha} \approx \frac{1}{\alpha!}\frac{\partial^\alpha p}{\partial x^\alpha}.
\end{aligned}
\tag{4}
$$

The local interpolants (3) for u and p at time t_n are used as initial data with the exact solution (2), producing an approximate local solution

$$
\begin{aligned}
u(x_i + x, t_n + t) &\approx \tfrac{1}{2}(ua(x - (M+1)t) + pa(x - (M+1)t) \\
&\quad + ua(x - (M-1)t) - pa(x - (M-1)t)), \\[2mm]
p(x_i + x, t_n + t) &\approx \tfrac{1}{2}(pa(x - (M+1)t) + ua(x - (M+1)t) \\
&\quad + pa(x - (M-1)t) - ua(x - (M-1)t)).
\end{aligned}
\tag{5}
$$

The approximate local solution (5) is a function of x and t, and is an exact solution to (1) with the local interpolants (3) as initial data, so that (5) is an exact local propagator for (1), correctly incorporating its dynamics. The fundamental viewpoint of this method is to approximate the solution of the system instead of particular terms in the governing equations. This fundamental idea is seen in the use of Riemann solvers for problems with shocks (Godunov, 1959), and in the use of separation of variables with local Fourier decompositions to develop algorithms for a wide variety of problems (Chen et al., 1980). The approximate local solution (5) is used to obtain a computational algorithm at the stencil center, with

$$
\begin{aligned}
u_i^{n+1} &= \tfrac{1}{2}(ua(-(M+1)k) + pa(-(M+1)k) \\
&\quad + ua(-(M-1)k) - pa(-(M-1)k)) \\
&\approx u(x_i, t_n + k), \\[2mm]
p_i^{n+1} &= \tfrac{1}{2}(pa(-(M+1)k) + ua(-(M+1)k) \\
&\quad + pa(-(M-1)k) - ua(-(M-1)k)) \\
&\approx p(x_i, t_n + k).
\end{aligned}
\tag{6}
$$

Algorithm (6) uses the exact local propagator (5) with approximate local data (3), so that the time evolution introduces no new error, but merely propagates what has been introduced by the interpolation. Note that finite difference forms are not specified, and that (6) represents a family of algorithms dependant upon the interpolants (3). If order D interpolants are used for a particular realization of (6), then the algorithm will have accuracy of order D in both space and time. There are several possible interpretations of the algorithm form (6). It can be viewed as an application of the Method of Characteristics, since this is how the general solution form for the exact local propagator for (1) is obtaind. If the interpolation coefficients are viewed as spatial derivatives (4), then the local solution in space and time (5) can be viewed as a Cauchy-Kowaleskaya expansion. The algorithm can also be reformulated as a truncated Taylor series expansion in the time step size k, with

$$
\begin{aligned}
u_i^{n+1} &= u_0 - k(p_1 + Mu_1) + k^2(2Mp_2 + (M^2+1)u_2) \\
&\quad -k^3((1+3M^2)p_3 + M(3+M^2)u_3) + \cdots,
\end{aligned}
$$

$$
\begin{aligned}
p_i^{n+1} &= p_0 - k(u_1 + Mp_1) + k^2(2Mu_2 + (M^2+1)p_2) \\
&\quad -k^3((1+3M^2)u_3 + M(3+M^2)p_3) + \cdots,
\end{aligned}
\tag{7}
$$

where the grid ratio $\lambda = \frac{k}{h}$ is implicit in (7), since the coefficients u_α and p_α include the factor $h^{-\alpha}$, where h is the space step size. A truncated Taylor expansion in time can be viewed as a generalized Lax-Wendroff method (Lax & Wendroff, 1960). Algorithm (6) can also be reformulated as a conventional finite difference method, since the underlying spatial interpolations use local polynomials. Further details are in (Goodrich, 1995).

We will introduce four relatively conventional realizations of algorithm (6), with central stencils for the interpolants (3), and with values for u and p at each grid point. These algorithms have three, five, seven, and nine point central stencils, they are second, fourth, sixth, and eighth order accurate in both space and time, and they are refered to as the "c3o0ex," "c5o0ex," "c7o0ex," and "c9o0ex" methods, respectively. These four algorithms are all single step explicit methods with dispersive truncation errors, and each is stable for $\frac{k}{h} \leq \frac{1}{1+|M|}$. Grid refinement data for these algorithms is presented and discussed below, with data from an additional class of high resolution Hermitian methods that are introduced below. The grid refinement data does confirm the order of accuracy of each of the four methods.

Stable high order boundary treatments can be derived if these algorithms are viewed as Cauchy-Kowaleskaya expansions. The interpolant from the data on a single stencil next to a boundary is used as initial data for the approximation of the evolution of the solution over the interval from the stencil center to the boundary point. Values for u and p are computed at each grid point that is not on the boundary in this interval, and outgoing Riemann variables are computed at the boundary point. The boundary treatment for the c9o0ex method must be modified for stability, and it uses a truncated eight point boundary stencil with additional data at the boundary point. Table 1 presents results for a simulation of the initial value problem with inital data $u(x,0) = 0$ and $p(x,0) = \sin(\pi x)$, on the numerical domain $-1 \leq x \leq 1$, with $\lambda = 0.8$ and $M = 0$. The boundary artificial treatments compute $u+p$ at the $x = 1$ boundary, and $u-p$ at the $x = -1$ boundary, with $u(-1,t)$ and $p(1,t)$ given as representative of typical CFD boundary conditions. The data in Table 1 shows that the propagation algorithms and boundary treatments are stable with from second to eighth order accuracy in both time and space. Further details are in (Goodrich, 1995).

TABLE 1. Maximum Error in u or p at $t = 10$.

$\frac{2}{h}$	n_{10}	c3d0ex	c5d0ex	c7d0ex	c9d0ex
8	50	1.87D-01	1.17D-02	1.01D-02	8.65D-05
16	100	4.57D-02	9.98D-04	5.26D-05	8.50D-07
32	200	1.32D-02	8.11D-05	8.79D-07	4.69D-09
64	400	3.50D-03	5.54D-06	1.29D-08	2.09D-11
128	800	8.94D-04	3.58D-07	1.89D-10	3.75D-13
256	1600	2.25D-04	2.27D-08	3.35D-12	4.91D-13
512	3200	5.66D-05	1.43D-09	1.89D-12	
1024	6400	1.42D-05	9.12D-11		

3. Hermitian Algorithms

We will introduce a second family of algorithms for (1), which use the exact local propagator (5), but which are distinguished from the relatively conventional algorithms introduced above by the use of Hermitian interpolants for (3). This particular family of algorithms uses and computes values for u and p, and for their spatial derivatives. Various orders of spatial derivative data are required at each grid point, depending upon the particular algorithm. If these algorithm forms are viewed as approximate local solutions for u and p, or as Cauchy-Kowaleskaya expansions that are locally defined as functions of x and t, then the local solutions for u and p can be differentiated in x to provide consistent local solutions for the spatial derivatives. The local derivative solutions obtained by differentiating (5) in x are used to obtain computational forms for the spatial derivatives that are analogous to and consistent with the computational forms (6) for u and p. These forms can be expressed as Taylor series time expansions in k, with

$$
\begin{aligned}
ux_i^{n+1} &= u_1 - 2k(p_2 + Mu_2) + 3k^2(2Mp_3 + (M^2+1)u_3) \\
&\quad -4k^3((1+3M^2)p_4 + M(3+M^2)u_4) + \ldots,
\end{aligned}
$$

$$
\begin{aligned}
px_i^{n+1} &= p_1 - 2k(u_2 + Mp_2) + 3k^2(2Mu_3 + (M^2+1)p_3) \\
&\quad -4k^3((1+3M^2)u_4 + M(3+M^2)p_4) + \ldots,
\end{aligned}
\tag{8}
$$

and with similar forms for second and higher derivatives. The interpolation step for the Hermitian algorithms produces interpolants (3) only for the functions u and p, using local polynomial approximations and the Method of Undetermined Coefficients with all of the data at each grid point on a given stencil. Note that there is no separate consideration given to interpolation for the derivative data, since the coefficients obtained from interpolating u and p are used in the time evolution algorithms for their spatial derivatives.

Algorithms using derivative data on an upwind stencil have been used previously (Takewaki *et al.*, 1985). Hermitian algorithms with central stencils have stability problems if the maximum order of accuracy is obtained. Stable algorithms with exact propagators can be obtained on alternating grids offset by a half mesh width. We will introduce four algorithms which use staggered two point grids, and three algorithms which use staggered four point grids. On staggered two point grids, if just u and p are used, then a first order exact propagator algorithm is obtained, which we will denote by "c3o0s2." If up to first, second, or third derivatives are used in addition to u and p, then stable algorithms are obtained which are third, fifth, or seventh order accurate in space and time, referred to as the "c3o1s2," "c3o2s2," and "c3o3s2" methods, respectively. On staggered four point grids, if just u and p are used, then a third order exact propagator algorithm is obtained, which we will denote by "c5o0s2." If up to first or second derivatives are used in addition to u and p, then stable algorithms are obtained that are seventh or eleventh order accurate in space and time, referred to as the "c5o1s2," and "c5o2s2" methods, respectively. These seven algorithms are all single step explicit methods on staggered grids with diffusive truncation errors, and each is stable for $\frac{k}{h} \le \frac{1}{1+|M|}$, where the staggered half step size is $\frac{k}{2}$. Note that two half time steps can be composed to produce single step algorithms with time step size k. The full time step algorithms resulting from staggered two point grids can be formulated as single step methods on a central three point stencil, and the algorithms from staggered four point grids as single step methods on a central seven point stencil. Other combinations are possible, such as composing the c3o3s2 and c5o1s2 methods, to produce a single step explicit seventh order Hermitian method on a single central five point stencil. It will be shown below that some of these Hermitian algorithms have extraordinary resolution and accuracy.

4. Numerical Experiments with the 1D Algorithms

Numerical experiments will be conducted with the eleven algorithms that have been introduced above, by computing solutions to (1) with inital data $u(x,0) = 0$ and $p(x,0) = \sin(\pi x)$ for $-1 \le x \le 1$, and with $\lambda = 0.8$ and $M = 0$. Periodic boundaries will be imposed, with $p(-1,t) = p(1,t)$ and $u(-1,t) = u(1,t)$ for $0 \le t$. Note that the initial data has wavelength and period 2. Figure 1 shows the $\log_{10}$ of the maximum absolute error at $t = 10$ in u or p versus the $\log_2$ of the number of grid points per wavelength. This data corroborates the order of accuracy of each of the methods. Note that the first order c3o0s2 method is incapable of producing accurate results with any reasonable resolution, and that the second order c3o0ex or Lax-Wendroff method requires 64 grid points per wavelength to produce $O[10^{-2}]$

accuracy after five periods. Note also that the c7o0ex and c9o0ex central methods show a greater sensitivity to roundoff errors than the high order Hermitian methods. An interesting feature of Figure 1 is the data at the coarsest resolution, with 4 grid points per wavelength. At this resolution, the simulations with the c3o2s2, c3o3s2, c5o1s2 and c5o2s2 algorithms have errors that range from $O[10^{-2}]$ to $O[10^{-6}]$, in contrast with errors that range from $O(\text{Chen} et al., 1980)$ to $O[10^{-1}]$ for the other algorithms. An $O[10^{-6}]$ error after five periods of propagation with four grid points per wavelength is exceptionally high resolution. Notice also the relative errors from methods which are of similar order. The c5o0s2 and c3o1s2 algorithms are both third order, but the c3o1s2 algorithm produces lower errors by about one order of magnitude at each grid resolution. The c5o1s2 and c3o3s2 algorithms are both seventh order, but the c3o3s2 algorithm produces lower errors by about two orders of magnitude at each grid resolution. The conventional sixth order c7o0ex and eighth order c9o0ex methods have larger errors at each grid level than the fifth order c3o2s2 and seventh order c5o1s2 Hermitian methods, respectively. In these comparisons of algorithms with similar orders of accuracy, the algorithm which produces lower errors has higher resolution from using more derivative information at each grid point. The accuracy of a numerical algorithm is determined by both its order of accuracy and its resolving power.

The periodic problem which produces Figure 1 is also used for propagation with 8 grid points per wavelength out to $t = 10$, $t = 1,000$, and $t = 100,000$. The data is presented in Table 2, where $O(\text{Chen} et al., 1980)$ errors are marked by asterisks, and where the algorithms are ranked by order. Note in Table 2 for $t = 10$, that the error data from the sixth order c7o0ex method is two orders of magnitude higher than from the fifth order c3o2s2 method, and that the error from the eighth order c9o0ex method is two orders of magnitude higher than from the seventh order c5o1s2 method, and four orders of magnitude higher than from the seventh order c3o3s2 method. A similar comparison is also seen in Table 2 for the data from $t = 1,000$ and $t = 100,000$. These comparisons once again show that both the order of accuracy and the resolution of a numerical algorithm determine its accuracy. Note that the only algorithms which produce errors that are not $O(\text{Chen} et al., 1980)$ at $t = 100,000$ are the high order and high resolution c5o1s2, c3o3s2 and c5o2s2 methods. These results show that far field can be redefined by several orders of magnitude, and that efficient propagation to a truly far field requires methods which are both high order and high resolution.

The superior accuracy of the high order and high resolution methods is obtained by using more complex algorithms, with more variables and equations, and with more operations per grid point per time step. It is

TABLE 2. Maximum Error in u or p at Various Times with $\frac{2}{h} = 8$.

Method	Order	$t = 10$	$t = 1,000$	$t = 100,000$
c3o0s2	1	****	****	****
c3o0ex	2	6.76D-01	****	****
c5o0s2	3	2.42D-01	****	****
c3o1s2	3	3.97D-02	****	****
c5o0ex	4	9.14D-02	6.74D-01	****
c3o2s2	5	2.27D-04	2.26D-02	****
c7o0ex	6	1.11D-02	1.04D-01	****
c5o1s2	7	4.52D-05	4.55D-03	3.75D-01
c3o3s2	7	6.74D-07	6.78D-05	6.75D-03
c9o0ex	8	1.39D-03	1.38D-02	****
c5o2s2	11	9.33D-10	9.37D-08	9.37D-06

natural to ask wether or not the simpler, more conventional algorithms are more or less efficient than the Hermitian algorithms. We will address this issue by replotting the data from Figure 1, this time with the $\log_{10}$ of the maximum error at $t = 10$ on the vertical axis, and the $\log_{10}$ of the total number of multiplications required for the entire simulation out to $t = 10$ on the horizontal axis. This data is presented in Figure 2. It has been shown in (Kreiss & Oliger, 1972) that relative computational efficiency increases with order of accuracy, and that the relative efficiency of the higher order methods increases as the error tolerance is lowered, and as the simulation time is increased. The data in Figure 2 clearly shows that for every level of error the efficiency in meeting that error tolerance increases with the order of accuracy of the algorithm, in spite of the fact that algorithms of radically different type are being used. In particular, the higher order Hermitian algorithms are more efficient even though they use more variables and solve more equations.

5. Normal Mode Analysis in 1D

The special properties of the full discretization Hermitian algorithms can be shown by a normal mode analysis. We will consider Hermitian algorithms applied to the linear first order wave equation

$$\frac{\partial u}{\partial t} + M\frac{\partial u}{\partial x} = 0. \tag{9}$$

The local exact propagator is defined by the Method of Characteristics. A normal mode for (9) can be written in local coordinates as

$$u(x,t) = \alpha \exp[i\theta \frac{(x - Mt)}{h}], \tag{10}$$

with amplitude α, frequency $\theta \in [0, \pi]$, and with space mesh size h. The symbol ρ for the algorithm can be written in terms of the normal mode (10), with

$$\rho(\lambda, \theta) = \frac{u^{n+1}}{u^n}, \tag{11}$$

where u^{n+1} is obtained from the algorithm with the normal mode as the known solution u^n at t_n, and where $\lambda = \frac{Mk}{h}$ is the CFL number. Spatial derivative data is needed by the Hermitian algorithms at each grid point, and since the normal mode (10) is perfectly known as a function of x at t_n, we will simply take and use its derivative values for this data. This heuristic procedure is intended to obtain qualitative insights about algorithms, and is not intended as a rigourous stability analysis, which would treat the total system of u and its spatial derivatives with independant error modes expected in each. From (11) we obtain the norm

$$\|\rho(\lambda, \theta)\| = (Re[\rho]^2 + Im[\rho]^2)^{\frac{1}{2}}, \tag{12}$$

and the phase change per time step

$$pc(\lambda, \theta) = \cos^{-1}[\frac{Re[\rho]}{\|\rho(\lambda, \theta)\|}]. \tag{13}$$

Note that we have varied from (Vichnevetsky & Bowles, 1982) by using a definition of phase change in terms of $\cos^{-1}$ rather than $\tan^{-1}$. For the phase speed properties of an algorithm, we will use the normalized relative phase change per time step

$$rpc(\lambda, \theta) = \frac{1}{\lambda \theta} \cos^{-1}[\frac{Re[\rho]}{\|\rho(\lambda, \theta)\|}], \tag{14}$$

rather than the phase change per time step (13).

All of the Hermitian methods with two point staggered grids were applied to the wave equation (9), and single step forms of the algorithms were obtained on three point central stencils. The norm of the amplification factor (12) and the relative phase change per time step (14) of these single step algorithm forms have been obtained numerically from an analytical expression. The most accurate of these four methods is the seventh order high resolution c3o3s2 algorithm, which uses u and its first three spatial derivative at every grid point. The norm of the amplification factor (12) for the c3o3s2 algorithm is plotted in Figure 3(a) as a function of the wave number $\theta \in [0, \pi]$ and the CFL number $\lambda \in [0, 1]$. Note that Figure 3(a) shows the norm of the amplification factor as less than or equal to 1 in the specified parameter range. The most dissipated behaviour is in the limit at $\theta = \pi$, and the norm of the amplification factor at $\theta = \pi$ is plotted as a function of λ in Figure 3(b). Figure 3(b) shows the norm of the amplification factor varying between approximately 0.9988 and 1 at $\theta = \pi$. The relative phase change per time step (14) for the c3o3s2 algorithm is plotted in Figure 4(a), also as a function of the wave number $\theta \in [0, \pi]$ and the CFL number $\lambda \in [0, 1]$. The most dispersed behaviour is in the limit at $\theta = \pi$, and the relative phase change at $\theta = \pi$ is plotted as a function of λ in Figure 4(b). Figure 4(b) shows the relative phase change per time step varying between approximately 0.9997 and 1.0002 at $\theta = \pi$. The amplification factor and relative phase change per time step plots in Figures 3 and 4 are extraordinary, and show truly "spectral like" qualities for the c3o3s2 algorithm. Recall from the numerical experiments reported above, that the c5o2s2 algorithm is of even higher resolution than the c3o3s2 algorithm.

6. Linearized Euler Equations in 2D

The algorithms that have been developed in one space dimension have all used the Method of Characteristics to obtain an exact solution form that propagates a local spatial inerpolant. The Method of Characteristics cannot be used in multiple space dimensions for nondiagonalizable hyperbolic systems. We will consider the linearized Euler equations in two space dimensions in order to indicate an approach for developing exact local propagators in multiple space dimensions. Consider the linearized Euler equations for the nondimensionalized isentropic case,

$$\frac{\partial u}{\partial t} + U\frac{\partial u}{\partial x} + V\frac{\partial u}{\partial y} + \frac{\partial p}{\partial x} = 0,$$

$$\frac{\partial v}{\partial t} + U\frac{\partial v}{\partial x} + V\frac{\partial v}{\partial y} + \frac{\partial p}{\partial y} = 0, \qquad (15)$$

$$\frac{\partial p}{\partial t} + U\frac{\partial p}{\partial x} + V\frac{\partial p}{\partial y} + \frac{\partial u}{\partial x} + \frac{\partial v}{\partial y} = 0,$$

where (U, V) is the constant mean convection velocity in Mach number, p is the pressure disturbance, and (u, v) is the velocity disturbance. This two dimensional system is nondiagonalizable, with wave propagation along characteristic surfaces.

We will briefly describe the development of a second order explicit algorithm for (15) on a symmetric 3×3 stencil. A second order local interpolant to u in x and y at t_n can be written as

$$
\begin{aligned}
u(x_i + x, y_j + y, t_n) \approx ua(x,y) &= u_{0,0} + u_{1,0}x + u_{2,0}x^2 \\
&+ (u_{0,1} + u_{1,1}x + u_{2,1}x^2)y \\
&+ (u_{0,2} + u_{1,2}x + u_{2,2}x^2)y^2,
\end{aligned}
\tag{16}
$$

with similar interpolants for v and p. The expansion coefficients are simultaneously obtained by the Method of Undetermined Coefficients, and can be interpreted as spatial derivatives, with

$$
u_{\alpha,\beta} = \frac{1}{\alpha!\beta!}\frac{\partial^{\alpha+\beta}ua}{\partial x^\alpha \partial y^\beta} \approx \frac{1}{\alpha!\beta!}\frac{\partial^{\alpha+\beta}u}{\partial x^\alpha \partial y^\beta}(x_i, y_j, t_n).
\tag{17}
$$

Notice that there are up to fourth order cross derivative terms in the interpolant (16). Exact polynomial solution forms for the linearized Euler equations (15) can be derived by substituting the expansion forms

$$
u(x_i + x, y_j + y, t_n + t) \approx ua(x,y,t) = \sum_{\alpha,\beta=0}^{2}\sum_{\gamma=0}^{4} u_{\alpha,\beta,\gamma}x^\alpha y^\beta t^\gamma,
$$

$$
v(x_i + x, y_j + y, t_n + t) \approx va(x,y,t) = \sum_{\alpha,\beta=0}^{2}\sum_{\gamma=0}^{4} v_{\alpha,\beta,\gamma}x^\alpha y^\beta t^\gamma,
$$

$$
p(x_i + x, y_j + y, t_n + t) \approx pa(x,y,t) = \sum_{\alpha,\beta=0}^{2}\sum_{\gamma=0}^{4} p_{\alpha,\beta,\gamma}x^\alpha y^\beta t^\gamma,
$$
$$\tag{18}$$

into system (15), and obtaining all the terms with $\gamma \neq 0$ by requiring the system (15) to be exactly satisfied for all x, y and t. Coefficients with $\gamma \neq 0$ are equivalent to time derivatives, and the resulting polynomial solutions are expressed entirely in terms of the spatial expansion coefficients. Note that there will be third and fourth order time terms present in the exact propagator solution forms. This procedure for obtaining an exact solution form is equivalent to the Cauchy-Kowaleskaya procedure (Harten *et al.*, 1987). The exact polynomial solution forms from (18) give exact propagator algorithms with correct multidimensional wave dynamics for the local spatial interpolants (16) as initial data. Since a biquadratic interpolant is used, the resulting algorithm will have $O[h^2 + k^2]$ truncation error, and it is dispersive. This procedure has been used with other symmetric stencils and interpolants to produce algorithms with fourth and sixth order accuracy in both space and time (Goodrich, 1995). Because a local exact propagator is used to develop these algorithms, the correct propagation of information along characteristic surfaces is automatically incorporated in the local approximation of the solution, so that this approach can be said to generalize the Method of Characteristics to nondiagonalizable systems in multiple space dimensions.

We have implemented new boundary conditions developed by Hagstrom (Hagstrom, 1997) as algorithms that are compatible with our propagation methods for the linearized Euler equations in two space dimensions.

Hagstrom actually provides a sequence of local approximations on the artifical boundary that are defined by a variable number of auxiliary functions. Each of the approximate problems are strongly well posed, and the approximate solutions converge exponentially to the exact solution on the open domain as the number of auxiliary functions increases (Hagstrom, 1997). As an example of these boundary conditions, consider the case of a subsonic mean flow in the positive x direction, with normal Riemann variables $r = u + p$ and $l = u - p$. The function r is outgoing at an artificial boundary on the right, and is viewed as being defined on the interior and the surface, while the function l is incoming and is viewed as being defined only on the surface. The boundary surface values of r and v are obtained essentially in the same way as the outgoing Riemann variables in the one dimensional case considered above, with the algorithm forms being interpreted as Cauchy-Kowaleskaya expansions in space and time, and all relevant solution values obtained over a common boundary stencil. The function l is obtained with auxiliary functions f_j and g_j, from the system

$$\frac{\partial l}{\partial t} + V\frac{\partial l}{\partial y} - U\frac{\partial v}{\partial y} + \sum_{j=1}^{m}(f_j + g_j) \;=\; 0,$$

$$\frac{\partial f_j}{\partial t} + (V - \alpha_j)\frac{\partial f_j}{\partial y} \;=\; \frac{\beta_j}{2}\frac{\partial^2 (l-r)}{\partial y^2}, \qquad (19)$$

$$\frac{\partial g_j}{\partial t} + (V + \alpha_j)\frac{\partial g_j}{\partial y} \;=\; \frac{\beta_j}{2}\frac{\partial^2 (l-r)}{\partial y^2},$$

where $\alpha_j = \sqrt{1 - U^2}\cos(\frac{j\pi}{2m+1})$, and $\beta_j = -\frac{1-U^2}{2}\sin^2(\frac{j\pi}{2m+1})$. Note that system (19) is forced by the interior solution propagating across the artifical boundary in the form of the $\frac{\partial v}{\partial y}$ term in the equation for l, and the $\frac{\partial^2 r}{\partial y^2}$ terms in the equations for the auxiliary functions. Note also that this boundary system does not require assumptions about solution form or source location. In practice, we have used this condition with disturbance data entirely contained within the computational domain, and with boundary data initialized as 0. Similar systems are defined on the left hand artifical boundary (Goodrich & Hagstrom, 1996), (Hagstrom, 1997). Boundary condition (19) and its left hand analog have been implemented in both second and fourth order algorithms, and numerical experiments with $m = 2$ auxiliary functions have shown no visible evidence of reflection (Goodrich & Hagstrom, 1996).

7. Conclusions

Single step methods with high order accuracy in both space and time are shown and compared. A particular class of methods using Hermitian interpolation on alternating grids has shown both high order accuracy and spectral like high resolution. The high order and high resolution methods are more efficient than the less accurate methods by orders of magnitude, even though the high order and high resolution methods are considerably more complex. Calculation to the "far field" can be redefined as propagation to more than $O[10^6]$ wavelengths or periods. Both high order accuracy and high resolution methods are required to compute to a true far field, with an example algorithm producing $O[10^{-4}]$ errors after 5×10^5 periods of propagation with eight grid points per wavelength. Algorithms for the linearized Euler equations in 2D are discussed, with propagation along characteristic surfaces, which generalizes the Method of Characteristics to nondiagonalizable Hyperbolic systems. Unobtrusive artificial boundary conditions are indicated.

References

Chen, C.-J., Naseri-Neshat, H., and Li, P., "The Finite Analytic Method," *Iowa Institute of Hydrolic Research Report No. 232*, Vol. 1-4, The University of Iowa, 1980.

Godunov, S. K., "Finite Difference Method for Numerical Computation of Discontinuous Solutions of the Equation of Fluid Dynamics," *Mat. Sbornik.*, Vol. 47, 1959, p. 271.

Goodrich, J. W., "An Approach to the Development of Numerical Algorithms for first Order Linear Hyperbolic Systems in Multiple Space Dimensions: The Constant Coefficient Case," NASA TM-106928, September 1995.

Goodrich, J. W. and Hagstrom, T., "Accurate Algorithms and Radiation Boundary Conditions for Linearized Euler Equations," AIAA Paper 96-1660, May 1996.

Hagstrom, T., "On High Order Radiation Boundary Conditions," *IMA Volume on Computational Wave Propagation*, B. Engquist and A. Majda, eds., Springer-Verlag, 1997.

Harten, A., Engquist, B., Osher, S., and Chakravarthy, S.R., "Uniformly High Order Accurate Essentially Non-oscillatory Schemes, III," *J. Comput. Phys.*, Vol. 71, 1987, p. 231.

Kreiss, H. O. and Oliger, J., "Comparison of accurate methods for the integration of hyperbolic equations," *Tellus*, Vol. 24, 1972, p. 199.

Lax, P. D. and Wendroff, B., "Systems of Conservation Laws," *Comm. Pure Appl. Math*, Vol. 13, 1960, p. 217.

Lele, S. K., "Compact Finite Difference Schemes with Spectral like Resolution," *J. Comput. Phys.*, Vol. 103, 1992, pp. 16-42.

Takewaki, H., Nishiguchi, A., and Yabe, T., "Cubic Interpolated Pseudo-particle Method (CIP) for solving Hyperbolic-Type Equations," *J. Comput. Phys.*, Vol. 61, 1985, pp. 261-268.

Vichnevetsky, R. and Bowles, J.B., *Fourier Analysis of Numerical Approximations of Hyperbolic Equations*, SIAM, Philadelphia, 1982.

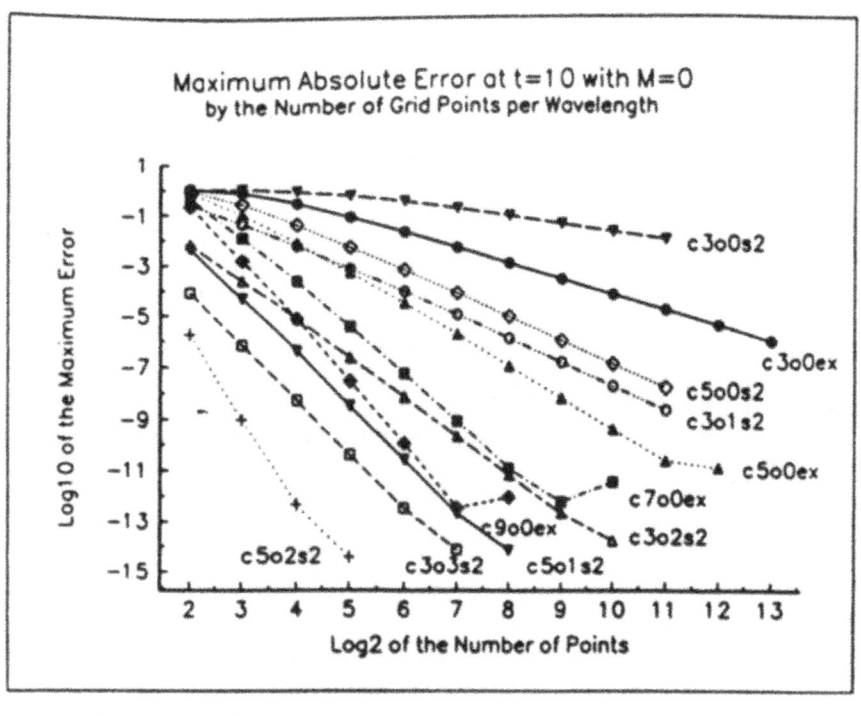

Figure 1. Maximum Error by the Number of Grid Points per Wavelength.

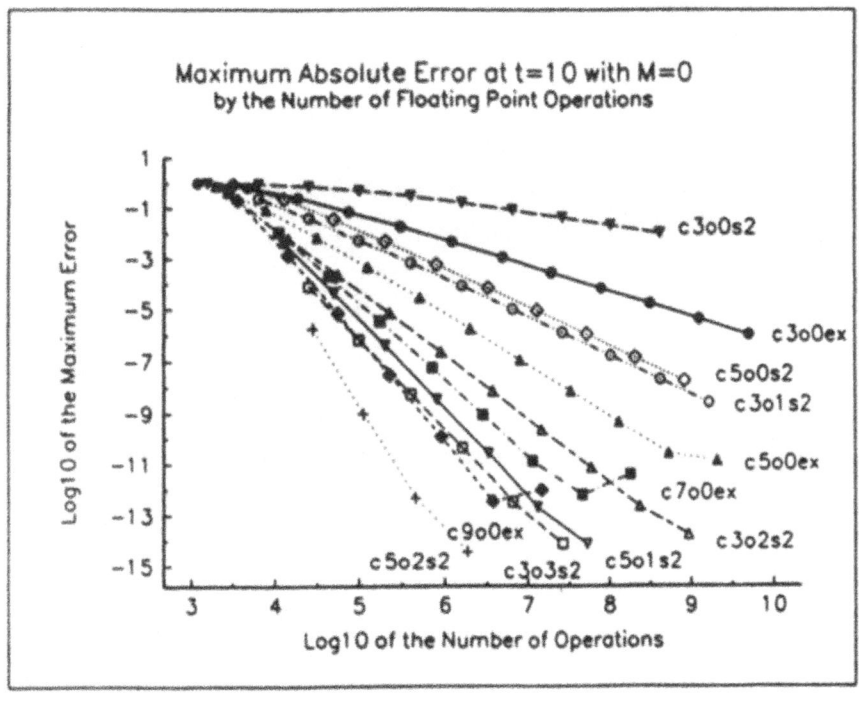

Figure 2. Maximum Error by the Total Number of Multiplications.

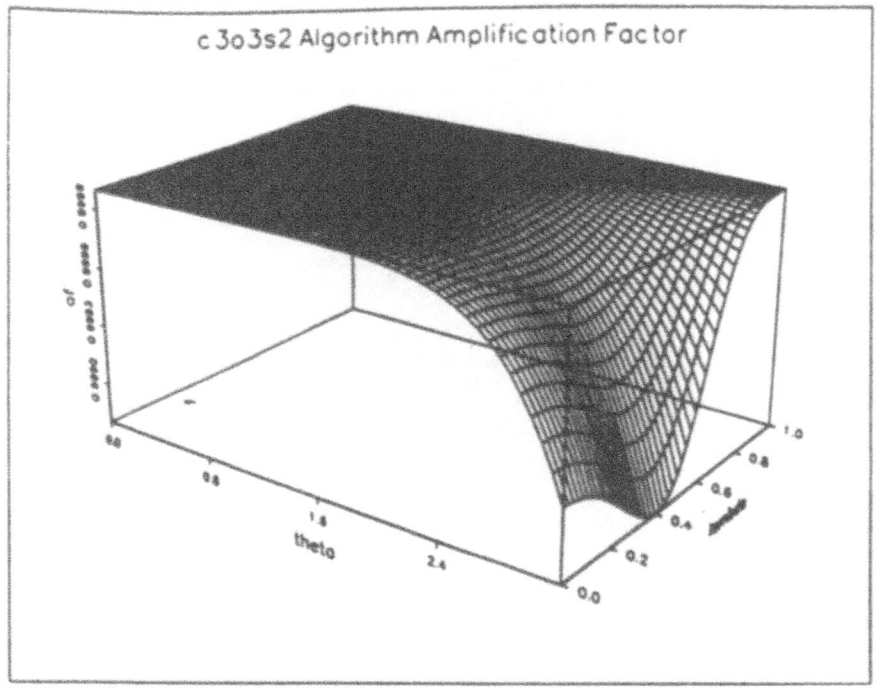

Figure 3a. Amplification Factor for the c3o3s2 Algorithm.

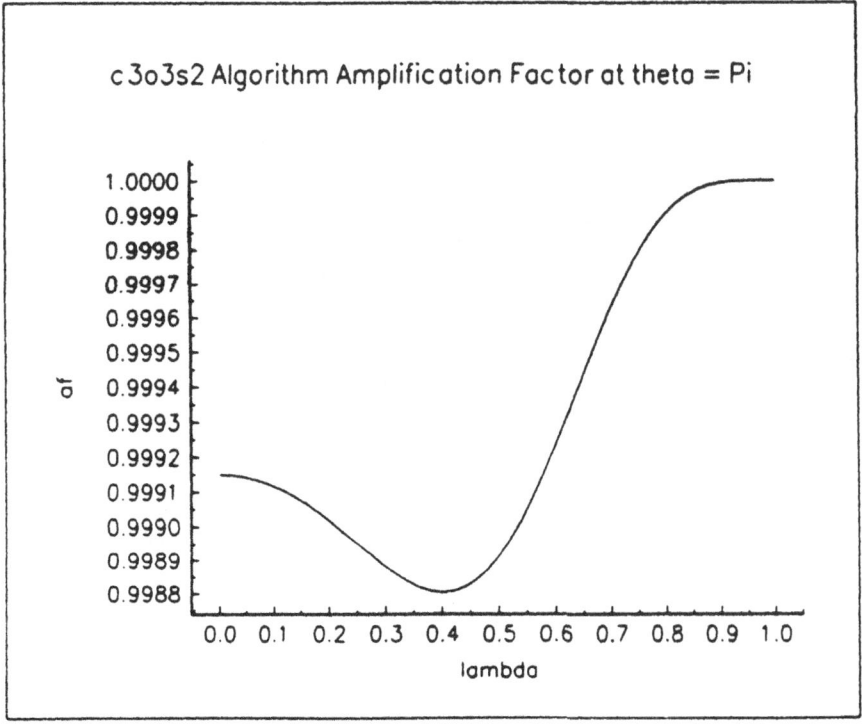

Figure 3b. Amplification Factor at $\theta = \pi$ for the c3o3s2 Algorithm.

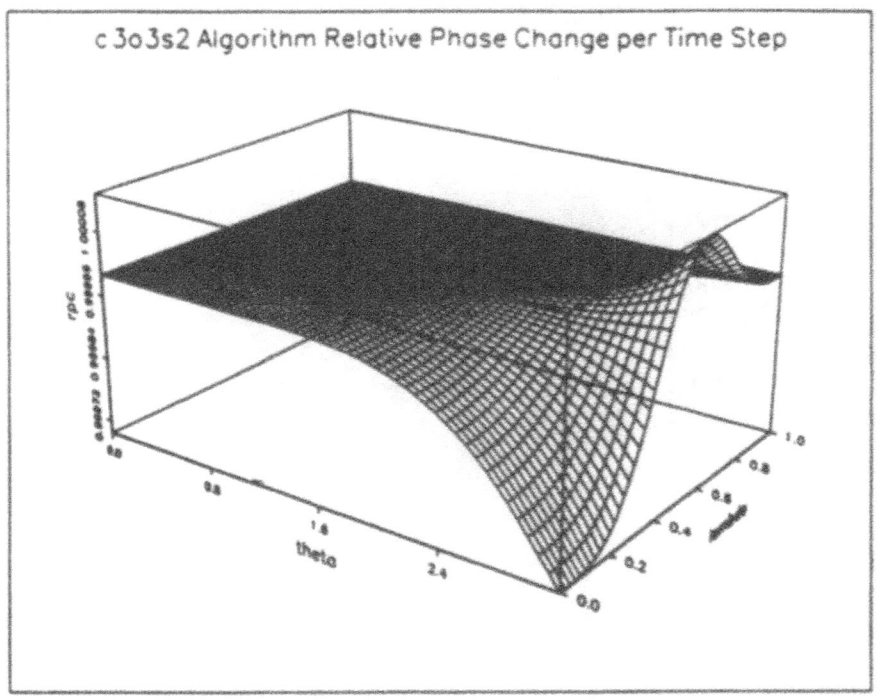

Figure 4a. Relative Phase Change for the c3o3s2 Algorithm.

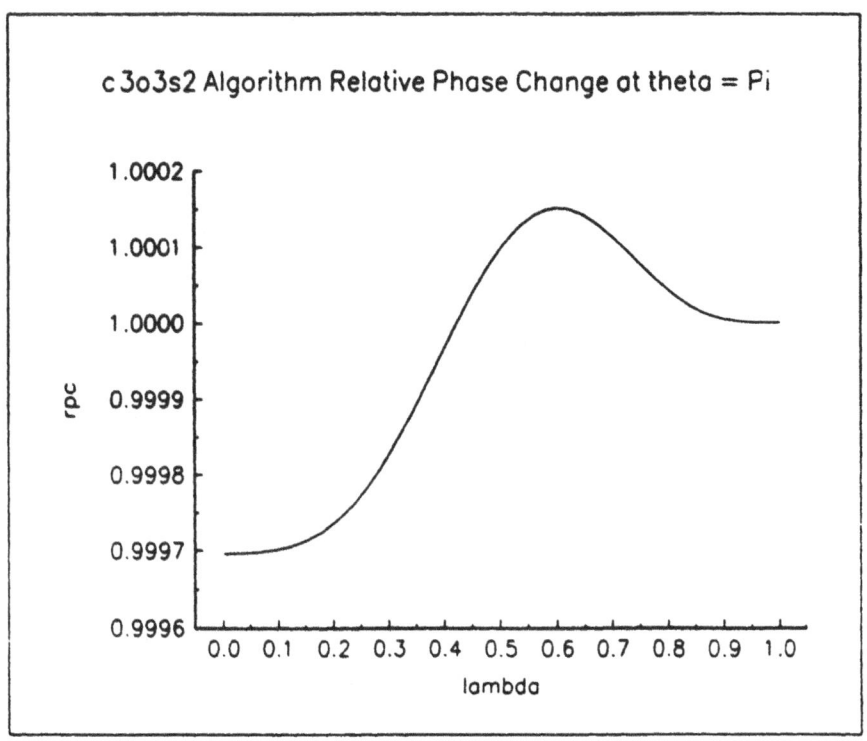

Figure 4b. Relative Phase Change at $\theta = \pi$ for the c3o3s2 Algorithm.

COMPUTATIONAL CONSIDERATIONS FOR THE SIMULATION OF DISCONTINUOUS FLOWS

MARK H. CARPENTER
NASA Langley Research Center
Hampton, Virginia

AND

JAY H. CASPER
Old Dominion University
Norfolk, Virginia

Abstract.

The numerical study of aeroacoustic problems places stringent demands on the choice of a computational algorithm, because it requires the ability to propagate disturbances of small amplitude and short wavelength. The demands are particularly high when shock waves are involved, because the chosen algorithm must also resolve discontinuities in the solution. In a previous work (Casper & Carpenter, 1998) the capabilities and deficiencies of shock-capturing methods for aeroacoustic problems were demonstrated using a high-order essentially nonoscillatory (ENO) numerical method. It was shown that first-order results are obtained when simulating time-dependent flows with discontinuities. The present study reaffirms this conclusion by comparing the ENO results with those obtained using a conventional linear scheme.

A sixth-order-accurate compact implicit finite difference scheme is used to investigate various discontinuous flows. The design order of accuracy is achieved in the smooth regions of a steady-state, quasi-one-dimensional Euler test case, as well as in the time-dependent Burgers' equation. However, in the unsteady Euler sound-shock interaction, first-order results are obtained downstream of the shock. A comparison is made between the linear and nonlinear results, noting the advantages of each method. A discontinuous linear model problem is then used to identify the cause of the first-order results. Here, the nature of the solution error is quantified as being predominantly a numerical phase shift, and a post-processing proce-

V. Venkatakrishnan et al (eds.), Barriers and Challenges in Computational Fluid Dynamics, 63-78.
© 1998 *Kluwer Academic Publishers.*

dure is demonstrated which increases the solution accuracy downstream of the discontinuity to second-order.

1. Introduction

This work is motivated by the desire to develop numerical methods that will be useful in the study of aeroacoustic phenomena that occur in flows with shocks. For example, shocks in jet flows, on wings, and in supersonic combustion inlets contribute significantly to sound generation. Problems such as these represent some of the more challenging aspects of ongoing research in the developing area of computational aeroacoustics (CAA). For a computational algorithm, obtaining acoustic information from a numerical solution that involves shock waves is a demanding proposition. In general, high-order accuracy is required for the propagation of high-frequency low-amplitude waves. In addition, the shock must be adequately captured.

In reference (Casper & Carpenter, 1998), a discussion was opened on the relative merits of the numerical methods which can simulate sound sources that are generated in flows with shocks. A fourth-order essentially nonoscillatory (ENO) numerical scheme was used to demonstrate the nature of solution error for steady and unsteady cases. It was shown that the quasi-one-dimensional Euler equations, in the presence of a shock, admit a steady-state solution which is of design order-of-accuracy. The time-dependent Euler equations revert to first-order accuracy downstream of the shock. The magnitude of the error, however, is much larger when using a uniformly first-order scheme.

Given the apparent inability of nonlinear schemes to recover design accuracy downstream of an unsteady shock, one must reassess the potential benefits of nonlinear schemes relative to linear schemes. In this work we begin to open the discussion on the relative merits of linear and nonlinear schemes for discontinuous flows. In addition, we identify a larger class of problems for which design accuracy can be obtained. Finally, we quantify the nature of the solution error downstream of a discontinuity when design accuracy is not obtained.

In the following sections, we introduce a fourth-order-accurate essentially nonoscillatory (ENO) method and a sixth-order compact method, and then use them to simulate a steady-state nozzle flow with a shock. In both cases design accuracy is obtained away from the discontinuity. It is then shown, using Burgers' equation, that design accuracy can be achieved in the time-dependent case, even for nonlinear equations. We then show that accuracy suffers if either numerical scheme is used on the time-dependent sound-shock interaction problem. The disappointing results in regard to the accuracy of this solution are explained through the study of a simpler linear model problem. The solution error is quantified as being phase error, to leading truncation order. We demonstrate in the linear case that post-processing the numerical solution, with the explicit knowledge of the phase error, increases the solution error to second order. Some of the difficult issues and ramifications for current methods that are raised by these results are discussed in the final section.

2. High-Order Accurate Shock-Capturing

Many methods are available in the literature that attempt to balance the properties of high-order accuracy and shock capturing. They can be classified into two categories : linear and nonlinear. We will not attempt to describe the details of either class of numerical methods, and refer those who are interested to a more complete discussion in reference (Casper & Carpenter, 1998). Rather, we present only a brief theoretical review of linear and the nonlinear strategies.

Within the linear class of numerical shock-capturing schemes, the interpolation set for the approximation of the solution or its derivatives is fixed as a function of grid location. Linear methods admit strong oscillations in regions in which physical gradients are inadequately resolved. Central-differencing operators and spectral methods are particularly prone to these numerical oscillations. For problems with discontinuities, limiters or filters are usually required to keep oscillations from growing without bound.

In the nonlinear class of schemes, the strategy with respect to discontinuities is to employ some sort of adaptive interpolation. The goal is to achieve formal high-order accuracy in smooth regions and high shock resolution without oscillations. The class of ENO schemes (Harten *et al.*, 1987) - (Shu & Osher, 1988) has been designed to have such properties. As originally presented, the local polynomial approximation operator adapts its interpolation set to the smoothest available part of the solution.

Although discontinuous solutions generated by a linear strategy are usually not as pictorially pleasing as solutions in which shock profiles are monotone, linear schemes are more computationally efficient than nonlinear schemes. The efficiency of a numerical algorithm is extremely important for aeroacoustic simulations because such problems are time dependent and require a fine computational mesh for the resolution of high-frequency disturbances. Because nonlinear methods are designed to avoid the production of spurious oscillations, the stability of a calculation of a flow with shocks is more readily obtained. However, their adaptive interpolation operator significantly hampers their efficiency relative to linear schemes.

3. Steady Shock in a Nozzle

A steady-state flow with a shock in a quasi-one-dimensional converging-diverging nozzle is numerically investigated. The governing equations are the quasi-one-dimensional Euler equations:

$$\frac{\partial}{\partial t}(AU) + \frac{\partial}{\partial x}(AF) = G \tag{1a}$$

where

$$U = \begin{bmatrix} \rho \\ \rho u \\ \rho E \end{bmatrix}, \ F = \begin{bmatrix} \rho u \\ \rho u^2 + P \\ (\rho E + P)u \end{bmatrix}, \ G = \begin{bmatrix} 0 \\ P\frac{dA}{dx} \\ 0 \end{bmatrix} \tag{1b}$$

The variables ρ, u, P, E, and A are the density, velocity, pressure, total specific energy, and nozzle area, respectively. The equation of state is

$$P = (\gamma - 1)\rho\left(E - \frac{1}{2}u^2\right)$$

where γ is the ratio of specific heats, which is assumed to have a constant value of 1.4.

The spatial domain of the nozzle is $0 \le x \le 1$. The nozzle shape is determined exactly through the requirement of a linear distribution of Mach number from $M = 0.8$ at the inlet to $M = 1.8$ at the exit, assuming the flow is isentropic and fully expanded. A schematic of the area distribution can be found elsewhere (Casper & Carpenter, 1998).

Given the prescribed area distribution, the Mach 0.8 inflow state is retained at $x = 0$, and the outflow condition at $x = 1$ is determined such that a shock forms at $x_s = 0.5$, which corresponds to a preshock Mach number of $M = 1.3$. Steady-state solutions are obtained by implementing a fourth-order accurate finite-volume ENO scheme and a sixth-order finite-difference scheme until residuals are driven to machine zero. In the ENO algorithm, the spatial accuracy is achieved by solving the equations in control-volume form as presented in Ref. (Harten et al., 1987). The equations are integrated in time via a third-order-accurate Runge-Kutta scheme (Shu & Osher, 1988). This numerical method will be referred to as "ENO-4-3." As has been established in previous research, (Rogerson & Meiberg, 1990) - (Casper et al., 1994) the adaptive stencils employed in the spatial operator are biased in smooth regions toward those that are linearly stable.

The finite-difference algorithm uses a compact implicit spatial operator, shown elsewhere to be time-stable and spatially a sixth-order accurate discretization (Carpenter et al., 1993). A small amount of background tenth-order numerical dissipation was added to stabilize the calculation. No explicit filtering is needed in these calculations. The time advancement scheme is a five stage fourth-order low-storage RK scheme (Carpenter & Kennedy, 1994). This scheme which is formally sixth-order in space and fourth-order in time will be referred to as the "LIN-6-4" scheme.

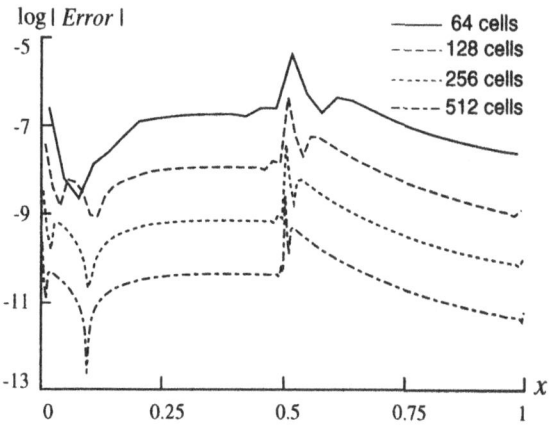

Figure 1. Quasi-1D Nozzle: Entropy error, ENO-4-3.

One of the simpler methods of determining the error of this solution relies on the fact that the value of the entropy-like quantity $S \equiv P/\rho^\gamma$ is piecewise constant. The exact magnitude of the jump is known analytically. The pointwise entropy error for this solution on four successively refined meshes is illustrated in Fig. 1. It is evident from Fig. 1 that the accuracy is fourth order on either side of the shock (error decay by a factor of 16 with each grid doubling).

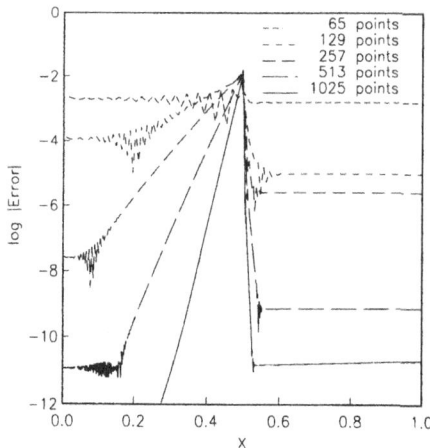

Figure 2. Quasi-1D Nozzle: Entropy error, LIN-6-4.

Fig. 2 shows the steady-state Entropy error as determined by the LIN-6-4 scheme. Note that an order one error always exists at the location of the discontinuity. This error which is not present in the ENO-4-3 results, is inherent in the linear scheme as a consequence of the Gibbs' oscillations. Away from the discontinuity the scheme converges at the design accuracy of sixth-order. The ENO-4-3 scheme is considerably more accurate than the LIN-6-4 scheme on coarse grids, despite its lower formal accuracy. Similar results were obtained for test cases in which the shock location did not coincide with a grid point.

The time independence of the solution makes this a convenient example for the demonstration of high-order accuracy in the presence of a shock. Both the linear and nonlinear schemes were able to obtain higher-order accuracy away from the discontinuity, and the nonlinear schemes' behavior in the neighborhood of the shock was exceptionally good. We assert that these results are representative of linear and nonlinear schemes, and are general for the steady-state Euler equations in one spatial dimension. The generalization to multiple spatial dimensions is not straightforward and is addressed elsewhere (Carpenter & Casper, 1997).

4. Time Dependent Interactions

A moving shock presents a greater challenge in regard to high-order-accurate shock capturing. As will be seen, many of the desirable attributes of the nonlinear schemes are lost in the time-dependent case. We begin our discussion with a simple test case in which the equation is nonlinear and supports a time-dependent discontinuity.

4.1. BURGERS' EQUATION

Studying Burgers' equation helps to clarify an important issue about solution accuracy in the presence of discontinuities. It is common to use Burgers' equation as a model problem to mimic the nonlinear nature of the Euler equations. The quadratic nonlinearity of the spatial flux in Burgers' equation is similar to the convective terms in the Euler equations. An assertion frequently implied in the literature is that solution error will be qualitatively similar for Burgers' equation and the Euler equations. We show by counter-example that the nature of the solution error for discontinuous flows is fundamentally different. Thus, Burgers' equations is of only limited value in studying time-dependent discontinuous flows.

In the absence of viscosity, Burgers' equation can be expressed as $\frac{\partial U}{\partial t} + \frac{1}{2}\frac{\partial U^2}{\partial x} = 0$. We employ initial conditions $U(x,0) = 1; 0 \leq x < \frac{1}{2}, U(x,0) = -1; \frac{1}{2} < x \leq 1$ and boundary conditions $U(0,t) = 1 + \epsilon sin(\omega_1 t); t \geq 0$,

$U(1,t) = -1 + \epsilon sin(\omega_2 t)$; $t \geq 0$. Here $\epsilon = 0.01$, $\omega_1 = 12\pi$, and ω_2 is related to ω_1 by a phase shift. A simulation using the LIN-6-4 scheme is run to a physical time $T = 10$. An "exact" solution is obtained using a two-domain Chebyshev discontinuity fitting code. Sufficient resolution in space and time is used to ensure that further refinement does not change the interpolated solution on the uniform grid.

Large oscillations near the discontinuity can be observed in the numerical solution obtained with the LIN-6-4 scheme. Fig. 3 shows the pointwise error on four successive grids, plotted on a Logarithmic scale. Qualitatively, the solution error is similar to that obtained in the steady-state nozzle using the LIN-6-4 scheme. An order one error exists at the discontinuity, but the solution converges at the design rate of sixth-order on both sides of the discontinuity. We note here, and clarify later in this section, that this high-order behavior is not achieved in more general hyperbolic equations.

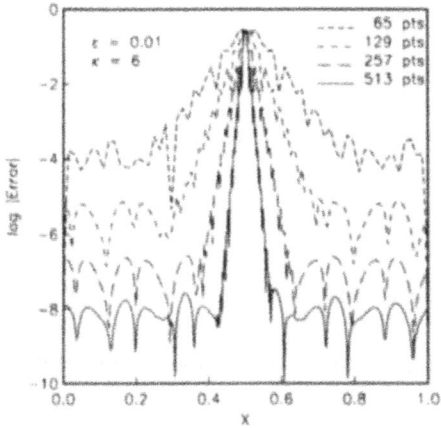

Figure 3. Burgers' equation: solution error, LIN-6-4.

4.2. SOUND-SHOCK INTERACTION

The effects of shocks on sound waves, (and vice versa), are important to the acoustics and performance of aircraft design. Therefore, the ability to obtain an accurate solution to such a model initial-boundary-value problem (IBVP) is important in the development of shock-capturing methods for CAA research.

The governing equations are the one-dimensional Euler equations:

$$\frac{\partial}{\partial t} U + \frac{\partial}{\partial x} F(U) = 0 \qquad (2a)$$

where the components of U and $F(U)$ are identical to those given in Eq. 1b. The equation of state is also the same as in the previous example.

The spatial domain is $0 \leq x \leq 1$. The piecewise constant initial conditions, U_L and U_R, are those of a steady shock located at $x_s = 0.5$. The flow is from left to right, and the state U_L is a Mach 2 flow upstream of the shock. The flow variables are normalized with respect to this upstream flow. At $t = 0$, an acoustic disturbance is introduced at $x = 0$:

$$
\begin{aligned}
P(0,t) &= P_L \left(1 + \epsilon \sin \omega t \right) \\
\rho(0,t) &= \rho_L \left[\frac{P(0,t)}{P_L} \right]^{1/\gamma} \\
u(0,t) &= u_L + \frac{2}{\gamma - 1} [\, c(0,t) - c_L \,]
\end{aligned} \tag{2b}
$$

where ω is the circular frequency, ϵ is the amplitude, and $c = \sqrt{\gamma P / \rho}$ is the local sound speed.

The numerical solution of this problem is obtained through the implementation of both the ENO-4-3 and the LIN-6-4 algorithms. Both algorithms should achieve their design spatial accuracy for a suitably small temporal step size. The exact solution is obtained by a two-domain Chebyshev spectral technique (Carpenter et al., 1994). Shock fitting is used to divide the domain into two computational regions. A Chebyshev collocation method is used in each region for the spatial discretization. A fourth-order Runge-Kutta scheme is used to discretize time. Sufficient spatial and temporal resolution are used to guarantee machine precision of the solution.

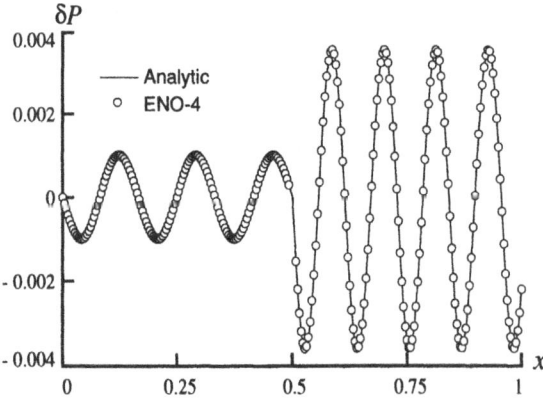

Figure 4. Solution of the sound-shock interaction problem at $t = 30\,T_\lambda$, ENO-4-3.

Fig. 4 depicts the pressure perturbation $\delta P(t) = P(x,t) - P(x,0)$ at $t = 30\,T_\lambda$, where $T_\lambda = 2\pi/\omega$ is one period of the incoming acoustic wave. The

acoustic wave amplitude is $\epsilon = 0.001$, and $\omega = 2\pi k(u_L + c_L)$ is determined
by requiring a wave number $k = 6$ with respect to unit length and a mean
wave speed $u_L + c_L$. The calculation, represented by circles, was performed
on a uniform mesh of 256 cells and a Courant number of 0.5, with the ENO-
4-3 code. The exact solution is represented by a continuous line. In this
pictorial measure, the numerical algorithm performs well with respect to
its prediction of the amplified sound wave at higher frequency downstream
of the shock. The missing circle values near the shock are off the plot and are
due to the use of the stencil-biasing parameters near a moving discontinuity.

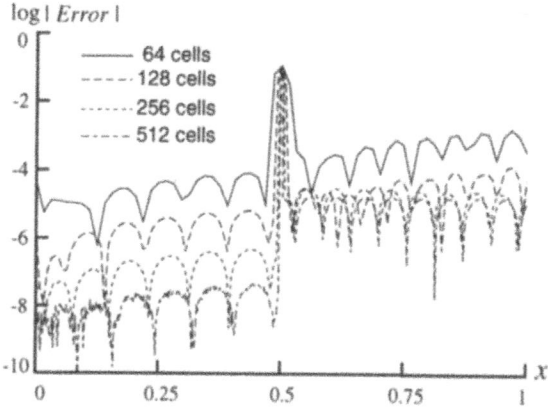

Figure 5. Pointwise error for the sound-shock interaction problem, ENO-4-3.

Even more instructive, however, is the pointwise error made by this
calculation with respect to the mesh width. Fig. 5 illustrates this error
on four successively refined meshes. The solution is clearly fourth-order
accurate upstream of the shock, but only first-order downstream of the
shock, as shown by the L_∞ error data in Table 1. The errors are computed
on two spatial subdomains: $0 \le x \le 0.45$ and $0.55 \le x \le 1$. In this manner,
the first-order error that is generated in the vicinity of the shock is avoided.

TABLE 1. L_∞ Pressure Errors : ENO-4-3

N_c	$x \le 0.45$	r_c	$x \ge 0.55$	r_c
64	8.358 E-05		1.677 E-03	
128	6.540 E-06	3.68	1.392 E-04	3.59
256	4.758 E-07	3.78	3.087 E-05	2.17
512	4.511 E-08	3.40	1.689 E-05	0.87

Fig. 6 depicts the solution error for the sound-shock interaction problem, using the LIN-6-4 scheme. (Here, in addition to the tenth-order damping, the solution was filtered with a tenth-order filter every 5 time-steps). The qualitative features of the solution and its error are nearly identical to those obtained with the ENO-4-3 scheme. The order one error near the discontinuity is a bit more localized in the ENO-4-3 case, but the magnitude of solution error downstream of the discontinuity is nearly the same. The design accuracy is obtained upstream of the discontinuity, but the solution accuracy is first-order downstream of the shock.

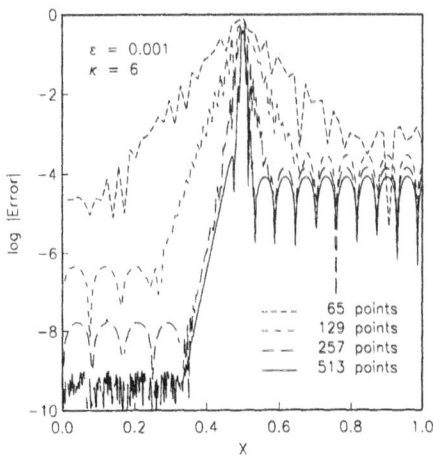

Figure 6.   Pointwise error for the sound-shock interaction problem, LIN-6-4.

Comparing the error differences in the Euler and Burgers' equations can lend insight into when a first-order solution will occur at a time-dependent discontinuity. We begin by noting how error propagates in each respective equation. In Burgers' equations, all information propagates towards the discontinuity. Thus, error which is generated at the discontinuity is localized at the discontinuity. In the Euler equations, error that is generated at the discontinuity is swept downstream on the down running characteristics. (It can be shown that all information traveling towards the discontinuity is design-order accurate on both sides of the discontinuity, until nonlinear corruption occurs near the shock). Information traveling away from the discontinuity is first-order. All flow variables being combinations of the two, are no better than first-order accurate. Burgers' equation is not a good model problem for determining the error characteristics of a time-dependent numerical scheme.

The general observation that both linear and nonlinear schemes are first-order downstream of the shock raises considerable concern in regard to the use of high-order-accurate methods in the study of unsteady flows with shocks. If high-order methods only yield first-order results, why use them? Before this question can be fully answered, it must first be determined whether the first-order error from a high-order method is significantly smaller than that of a lower order method. Fig. 7 depicts the error of the sound-shock interaction problem with a first-order upwind method. Upon comparison with the Fig. 5 or Fig. 6, it is clear that the solution downstream of the shock is more accurate when using either higher-order method. This result is by no means conclusive, as there are many other considerations. For instance, what about second- or third-order methods? What about the cost of using a given method with respect to its accuracy? These and a more general comparison of linear and nonlinear scheme are topics for future research.

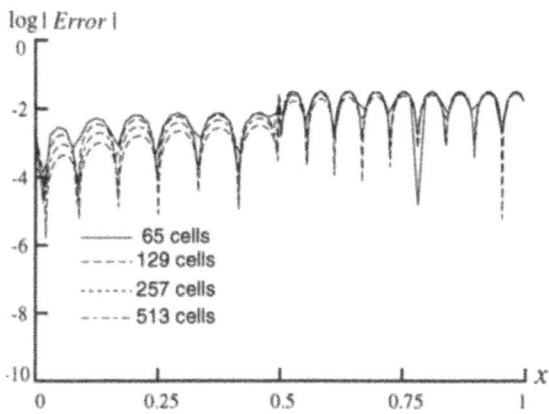

Figure 7. Pointwise error for the sound-shock interaction problem, using a first-order upwind method.

The first-order nature of the solution downstream of the discontinuity is difficult to study using the full Euler equations. In the next section, we construct a simple model problem which contains the underlying mechanism for the degradation to first-order accuracy.

5. A Linear Model Problem

The lower-order accurate results of the previous section can be analyzed through the study of a discontinuous linear scalar model problem, wherein we isolate the important phenomenon of propagation of information through a discontinuity. This trait will be common to almost any aeroacoustic problem that involves shock waves. Consider the scalar equation: $\frac{\partial u}{\partial t} + \frac{\partial [a(x)u]}{\partial x} = 0$, where the piecewise constant wave speed $a(x)$ is $a(x) = a_L$, $x < \frac{1}{2}$ and

$a(x) = a_R$, $x > \frac{1}{2}$ on the interval $0 \leq x \leq 1$. (For the finite difference code, the jump value at $x = \frac{1}{2}$ is assigned to be the average of the left and right states.) The initial condition is given by $u(x,0) = \sin(\omega x)$ on $0 \leq x \leq 1$, and the boundary condition is $u(0,t) = \sin(\omega(-a_L t))$ at $x = 0$, $t \geq 0$. The constant ω is determined by requiring a wave number $k = 4$ with respect to unit length and the downstream wave speed a_R. The exact solution is given at long times (times after the initial solution has been swept out of the domain) by $u(x,t) = \sin[\omega(x - a_L t)]$ on the interval $0 \leq x \leq \frac{1}{2}$ and $u(x,t) = \frac{a_L}{a_R} \sin[\omega(\frac{a_L}{a_R}(x - \frac{1}{2}) - a_L t + \frac{1}{2}]$ on the interval $\frac{1}{2} \leq x \leq 1$. The exact solution is obtained by requiring conservation of fluxes at the interface.

The loss in accuracy in numerical solutions of linear problems with discontinuous initial data has been the subject of previous research by other authors (Majda & Osher, 1977) - (Donat & Osher, 1990). All of these previous studies involved solutions of coupled linear systems. It is, therefore, instructive to note that the solution of this discontinuous linear problem is analogous to that of a coupled system in the following way. The flux conservation condition couples at the discontinuity location the equations in the left and right domains. The coupling occurs at a discrete point (like a boundary) and the information transfer is from one equation to the other. As in the present study, the work of Majda and Osher (Majda & Osher, 1977) was concerned with the inherent degradation in accuracy in a region where information is numerically propagated across a discontinuity.

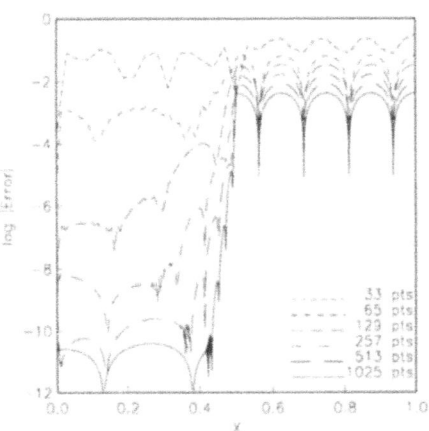

Figure 8.   Pointwise error for discontinuous linear model problem run with LIN-6-4

Fig. 8 shows the pointwise error for the discontinuous linear model problem with a wave-speed ratio of $\frac{a_L}{a_R} = 2$. The simulation is run to time $T = 4$

on a series of six grids with the LIN-6-4 scheme. The solution is filtered every five time steps. As with the Euler equations, the solution accuracy is sixth-order upsteam, and first-order downstream of the discontinuity.

Inspection of the numerical solution downstream of the discontinuity reveals that the dominant component of the error is related to phase. Specifically, the amplitude of the waves downstream of the discontinuity is nearly correct, but the waves are displaced by some fraction of the grid spacing. To quantify this behavior, the Fourier transform of the solution is formed on the interval $\frac{3}{4} \leq x \leq 1$. The exact transform of the downstream sine wave on the interval (made periodic by construction) is the $k = 1$ mode of unit amplitude. Fig. 9 plots the amplitude of the Fourier modes obtained from the numerical solution on this interval. The energy in the spurious higher modes decays very rapidly, indicating that the amplitude portion of the solution error is higher-order.

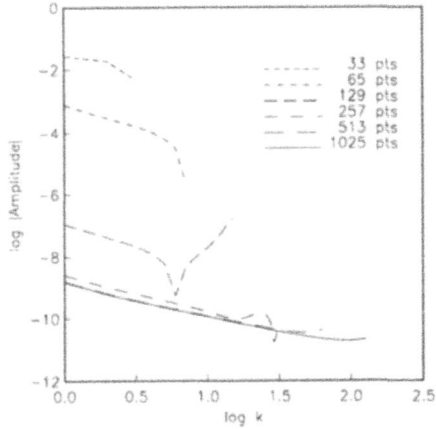

Figure 9. Solution error as a function of Fourier wave number k

The phase shift of the numerical solution as a percentage of grid spacing (as determined by the phase angle of the Fourier transform), asymptotes to a nonzero constant, however. For sufficient resolution, the numerical solution is shifted a finite percentage of the grid spacing, independent of the resolution. If the numerical solution is adjusted to account for this phase shift, then the solution error can be reduced. Fig. 10 plots the data shown in Fig. 8 where the exact solution downstream of the discontinuity is shifted by a constant multiple of the grid spacing. It is apparent that the solution accuracy is increased by this procedure. Unfortunately, the convergence rate downstream of the discontinuity is still only second-order and not the design order of six as would be expected with the LIN-6-4 scheme.

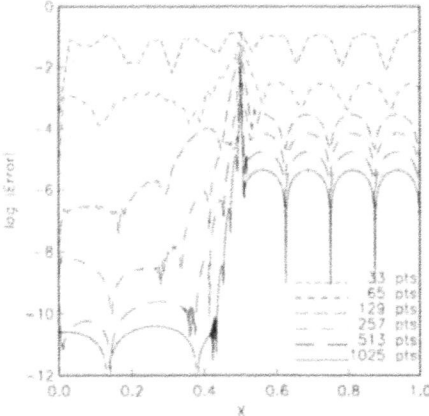

Figure 10. Pointwise solution error after correcting for the numerical phase shift, with LIN-6-4

A more troubling observation about the numerical phase shift, is that the shift is amplitude dependent. Table 2 shows the dependence of the phase shift on the wave-speed ratio $\frac{u_L}{u_R}$. Multiple waves interact at a shock in the Euler equations, each having a different wave-speed ratio. To correct the multiple wave situation would require component by component shifting and would be more difficult to implement.

TABLE 2. Phase shift verses $\frac{U_L}{U_R}$.

$\frac{U_L}{U_R}$	$\frac{\delta}{\Delta x}$
4/1	0.2351
2/1	0.08412
1/2	0.1700
1/4	0.9674

6. Discussion

In this work, and elsewhere (Casper & Carpenter, 1998), we provide a systematic study of the effects of discontinuities on solution accuracy. We focus on the ENO-4-3 and LIN-6-4 formulations as being representative of high-order nonlinear and linear schemes in use today. We show for higher-order formulations, that the solution accuracy is strongly dependent on the mathematical character of the governing equations, and only weakly dependent on the numerical method.

Specific conclusions include: 1) The quasi-one-dimensional Euler equations admit design accuracy away from a steady discontinuity. The ENO-4-3 results are significantly better than the LIN-6-4 results for the steady case. 2) The unsteady Burgers' equation admits design accuracy away from a discontinuity, in contrast to the unsteady Euler equations. Equations that focus information at the discontinuity maintain design accuracy away from that discontinuity. 3) The unsteady Euler equations are first-order accurate downstream of a discontinuity. The ENO-4-3 and LIN-6-4 schemes achieve similar solution accuracy downstream of the shock for the sound-shock interaction problem.

We use a linear model problem with a discontinuous wave speed, to study the nature of discontinuity related solution error. We recover first order solution error downstream of a discontinuity in the linear case, and quantify the error as being phase related to leading truncation order. Second-order results, are obtained by post-processing the solution with the known phase shift. It is doubtful whether this post-processing step is general for systems of equations in multiple dimensions.

Research continues in several directions. It is questionable whether design accuracy can be obtained for the discontinuous steady Euler equations in multiple spatial dimensions. Further investigation continues into the relative merits of high-order-accurate shock-capturing schemes, and in particular, the relative merits of linear and nonlinear schemes. Attention focuses on shock resolution, design accuracy, and computational efficiency. Work continues into developing general techniques which recover design accuracy in the presence of discontinuities. Linear schemes suffer from a first order phase shift which is wave-speed jump dependent. It is possible, therefore, to correct the time-dependent solution for the full Euler equations making use of phase shift information.

We conclude by noting that the inability to recover design accuracy downstream of a discontinuity is not a proof of inherent first-order accuracy; but rather only a demonstration that design accuracy has not been found!

References

Atkins, H., "High-Order ENO Methods for the Unsteady Navier-Stokes Equations," AIAA 91-1557, June 1991.

Carpenter, M. H., Atkins, H. L. and Singh, D. J., "Characteristic and Finite-Wave Shock-Fitting Boundary Conditions for Chebyshev Methods," *Transition, Turbulence, and Combustion*, Vol. II, M. Y. Hussaini, T. B. Gatski, and T. L. Jackson, eds., Kluwer Academic Publishers, 1994.

Carpenter, M.H. and Casper, J., "The Accuracy of Shock Capturing in Two Spatial Dimensions," *13th Computational Fluid Dynamics Conference*, AIAA-97-2107, 1997.

Carpenter, M. H., D. Gottlieb, and S. Abarbanel, "The Stability of Numerical Boundary Treatments for Compact High-Order Finite-Difference Schemes," *Journal of Computation Physics*, Vol. 108, No. 2, October, 1993.

Carpenter, M. H. and Kennedy, C. A., "Fourth-Order 2N-Storage Runge-Kutta Schemes," NASA-TM-109111, April 1994,

Casper, J., and Carpenter, M. H., "Computational Considerations for the Simulation of Shock-Induced Sound," NASA TM 110222, Dec., 1995, to appear *SIAM J. of Sci. Comput.*, Vol. 19, No. 1, January 1998.

Casper, J., Shu, C. W., and Atkins, H., "A Comparison of Two Formulations for High-Order Accurate Essentially Non-Oscillatory Schemes," *AIAA Journal*, Vol. 32, No. 10, October 1994, pp. 1970-1977.

Donat, R. and Osher, S., "Propagation of Error in Regions of Smoothness for Nonlinear Approximations to Hyperbolic Equations," *Computer Methods in Applied Mechanics and Engineering*, Vol. 80, 1990, pp. 59-64.

Harten, A., Engquist, B., Osher, S., and Chakravarthy, S., "Uniformly High Order Accurate Essentially Non-Oscillatory Schemes III," *Journal of Computational Physics*, Vol. 71, No. 2, 1987, pp. 231-323.

Majda, A. and Osher, S., "Propagation of Error in Regions of Smoothness for Accurate Difference Approximations to Hyperbolic Equations," *Comm. Pure and Applied Math*, Vol. 30, 1977, pp. 671-705.

Mock, M. S. and Lax, P. D., "The Computation of Discontinuous Solutions of Linear Hyperbolic Equations," *Comm. Pure and Applied Math*, Vol. 31, 1978, pp. 423-430.

Rogerson, A. and Meiberg, E. "A Numerical Study of the Convergence Properties of ENO Schemes," *Journal of Scientific Computing*, Vol. 5, No. 2, 1990, pp. 151-167.

Shu, C., "Numerical Experiments on the Accuracy of ENO and Modified ENO Schemes," *Journal of Scientific Computing*, Vol. 5, No. 2, 1990, pp. 127-150.

Shu, C. W., "Numerical Solutions of Conservation Laws," Ph. D. Dissertation, UCLA, 1986.

Shu, C. and Osher, S., "Efficient Implementation of Essentially Non-Oscillatory Shock-Capturing Schemes," *Journal of Computational Physics*, Vol. 77, No. 2, 1988, pp. 439-471.

SPACE-TIME METHODS FOR HYPERBOLIC CONSERVATION LAWS

ROBERT B. LOWRIE
Los Alamos National Laboratory
Scientific Computing Group

AND

PHILIP L. ROE AND BRAM VAN LEER
University of Michigan
Aerospace Engineering Deparment

1. Introduction

Two major challenges for computational fluid dynamics are problems that involve wave propagation over long times and problems with a wide range of amplitude scales. An example with both of these characteristics is the propagation *and* generation of acoustic waves, where the mean-flow amplitude scales are typically orders-of-magnitude larger than those of the generated acoustics. Other examples include vortex evolution and the direct simulation of turbulence. All of these problems require greater than second-order accuracy, whereas for nonlinear equations, most current methods are at best second-order accurate. Of the higher-order (greater than second-order) methods that do exist, most are tailored to high-spatial resolution, coupled with time integrators that are only second or third-order accurate. But for wave phenomena, time accuracy is as important as spatial accuracy.

One property of successful second-order methods is that they attempt to be faithful to the physics of hyperbolic problems. To develop higher-order methods, particularly for unsteady problems, it is tempting to violate this philosophy. Typically, higher accuracy is obtained by increasing the size of the update stencil. Instead, our aim is to develop time-accurate methods that minimize the size of the update stencil.

The approach in this study is strongly motivated by the physics of hyperbolic conservation laws. Specifically, we insist that a numerical method's discrete zone of dependence should only be slightly larger (for stability) than the physical zone of dependence. Consider a two-time-level method

V. Venkatakrishnan et al (eds.), Barriers and Challenges in Computational Fluid Dynamics, 79-98.
© 1998 *Kluwer Academic Publishers.*

with a Courant number less than 1. In one dimension, only three cells should contribute to a cell's update; the cell itself and its immediate neighbors. We will refer to such methods as *compact*.

Compact methods have the following potential benefits:

- The methods are explicit.
- Parallel implementations using domain decomposition only need a single buffer cell at domain boundaries, which is updated only once per time step.
- Boundary procedures are straightforward.
- The effects of discontinuities are localized.

Additional benefits will also be realized, but to discuss them will first require some analysis.

Nearly all of the popular higher-order methods are not compact. In particular, ENO-based methods, and methods based on multi-stage time integration, are not compact. For a higher-order method to be compact, either more time levels must be included in the update, or more data must be carried in each cell. The choice in this study is to carry more data in each cell, with a two-time-level approach.

Methods that carry more data in each cell, than simply a cell-average or point-value, are not a new idea. In fact, the approach outlined here is a generalization of Van Leer's Scheme III (van Leer, 1977), which carries the solution average and first moment (or derivative) in each cell. Scheme III is third-order accurate in space and time.[1] This scheme can also be derived in a finite-element context (Johnson & Pitkaranta, 1986), and is referred to as the Discontinuous Galerkin (DG) method. It is the possibility of deriving schemes with improved order-of-accuracy that makes this approach potentially superior to using the extra storage for mesh refinement.

For a bibliography of other work using DG, see (Lowrie, 1996). A notable multi-stage implementation is given by (Atkins & Shu, 1996), where efficiency is gained by eliminating the need for quadrature. The multi-stage DG method may be ideal for steady-state problems; however, no multi-stage scheme can be compact. Nevertheless, since each *stage* of the method is compact, it has many of the advantages of compact schemes that are outlined above.

The DG method in this study follows the 'space-time' approach. Control volumes (elements) are defined in space and time, and then a polynomial representation (which includes the time variable) of the solution is found in each element. The elements are arranged so that when solving a linear equation, each element can be solved for explicitly in a marching procedure. For nonlinear equations, the method is *point-implicit*, in that the solution in

---

[1] In a norm that measures the error in the least-damped mode of the solution.

each element requires the solution of a small system of equations, and only a small number of elements may be coupled. Moreover, the implicitness is weak, arising solely from the nonlinearity, so that rapid iteration is possible.

The approach taken in this study is by no means simple, and at this time the computational cost is high.[2] However, the method has a strong theoretical foundation and many of its properties are highly desirable. Future work should make the present method, or related methods, practical for a broader range of problems.

2. Conservation Laws

Consider a conservation law of m-equations in d-space dimensions, written as

$$\partial_t \mathbf{u} + \underline{\nabla} \cdot \underline{\mathbf{f}}(\mathbf{u}) = \mathbf{0}, \tag{1}$$

where $\mathbf{u}$ is the vector of conservation variables, $\underline{\nabla} \equiv (\partial_{x_1}, \ldots, \partial_{x_d})$, and $\underline{\mathbf{f}} \equiv (\mathbf{f}_1, \ldots, \mathbf{f}_d)$. Another form that will be used in this study is

$$\vec{\nabla} \cdot \vec{\mathbf{f}} = \mathbf{0}, \tag{2}$$

where $\vec{\nabla} \equiv (\underline{\nabla}, \partial_t)$ and $\vec{\mathbf{f}} \equiv (\underline{\mathbf{f}}, \mathbf{u})$. The notation is that $\underline{v}$ is a vector in space, while $\vec{v}$ is a space-time vector.

Define the matrix

$$\mathbf{A}_\ell \equiv \underline{\ell} \cdot \underline{\mathbf{A}}, \tag{3}$$

with $\underline{\ell}$ a unit vector, and each component of $\underline{\mathbf{A}}$ is the $m \times m$ Jacobian matrix corresponding to $\underline{\mathbf{f}}$. We assume (1) is hyperbolic; that is, for every $\underline{\ell}$, $\mathbf{A}_\ell$ has real eigenvalues, $\lambda_{\ell,k}$, $k = 1, 2, \ldots, m$, and distinct eigenvectors. The Euler equations of gas dynamics are an example of a hyperbolic system.

3. Discontinuous Galerkin

3.1. FORMULATION

Let the solution domain Ω be divided into a set of N_e non-overlapping control volumes (elements), $\{\Omega_e\}$. Each Ω_e is allowed to be any type of polygon, with boundary $\partial\Omega_e$. A sample space-time mesh for $d = 1$, between two time levels, is shown in Figure 1.

The solution in each element is written in terms of the parameter vector, $\mathbf{w}(\mathbf{u})$ (Roe, 1981). For many conservation laws, the quantities $\mathbf{u}$ and $\mathbf{f}_i$ can

---

[2]Note that the "quadrature-free" idea of (Atkins & Shu, 1996) has not yet been implemented, which may increase performance substantially.

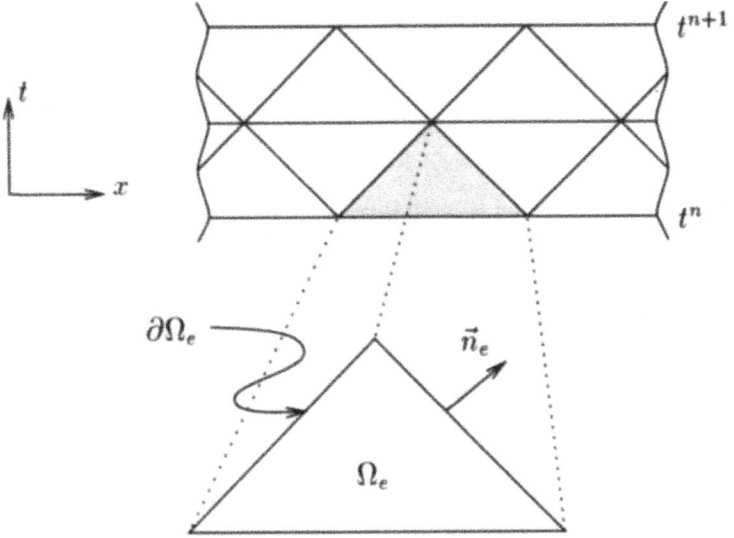

Figure 1. Sample Space-Time Mesh and Element Definition.

be written as quadratic functions of the components of $\mathbf{w}$. This property is used to allow exact evaluation of the flux integrals.

The numerical solution sought will be continuous within each Ω_e, but possibly discontinuous across element boundaries. The solution within each element-e will be referred to as $\mathbf{w}_e(\vec{x})$, $\forall \vec{x} \in \Omega_e$, and on the element boundary by $\mathbf{w}_b(\vec{x})$, $\forall \vec{x} \in \partial\Omega_e$. Note that in general

$$\mathbf{w}_e(\vec{x}) \neq \mathbf{w}_b(\vec{x}), \quad \forall \vec{x} \in \partial\Omega_e.$$

Indeed, the boundary value may be equal to one of the two neighboring element values, or some combination thereof. The precise definition of the boundary value will be given in the next section.

In weak form, (2) may be written for each element as

$$\oint_{\partial\Omega_e} \Phi \, \vec{\mathbf{f}}_b \cdot \vec{n}_e \, dS - \int_{\Omega_e} \vec{\mathbf{f}}_e \cdot \vec{\nabla}\Phi \, dV = \mathbf{0}, \tag{4}$$

where $\vec{n}_e$ is the outward boundary unit normal, and $\Phi = \Phi(\vec{x})$ is a suitable test function, to be defined later. Note that Equation (4) can be related to the finite-volume approach by taking $\Phi(\vec{x}) = 1$.

3.2. FACE DEFINITIONS

In this section, the element-boundary values, $(\cdot)_b$, are defined. Let ∂w_e be an element face that separates two elements Ω_e and Ω_{e^*}. Borrowing from the finite-volume approach, the interface value is written as a function of the values in the adjacent elements;

$$\mathbf{w}_b(\vec{x}) = \mathcal{F}(\mathbf{w}_e(\vec{x}), \mathbf{w}_{e^*}(\vec{x})), \quad \forall \vec{x} \in \partial w_e.$$

The function $\mathcal{F}$ will depend on the face "type." Our present code permits two types of faces (refer also to Figure 2):

1. *Riemann Face:* A face that is aligned with the t-axis ($n_{e,t} = 0$). On such faces, a (approximate) Riemann solver is used.
2. *Explicit Face:* The face orientation is such that all of the characteristic paths cross the face in the same direction; that is, the quantity

$$\mu = (\ell, \lambda_{\ell,k}) \cdot \vec{n}_e,$$

is either strictly positive (also referred to as an "Outflow Face"), or strictly negative ("Inflow Face"), for all k and ℓ, and all $\vec{x} \in \partial w_e$. The vector $(\ell, \lambda_{\ell,k})$ is defined via (3). In practice, the calculation of $\lambda_{\ell,k}$ for all ℓ and $\vec{x} \in \partial w_e$ is not needed, as long as a reasonable local value is used, along with a safety factor. The boundary value is then set as

$$\mathbf{w}_b(\vec{x}) = \begin{cases} \mathbf{w}_{e^*}(\vec{x}) & \text{if } \mu < 0 : \text{ "Inflow"} \\ \mathbf{w}_e(\vec{x}) & \text{if } \mu > 0 : \text{ "Outflow"} \end{cases} \tag{5}$$

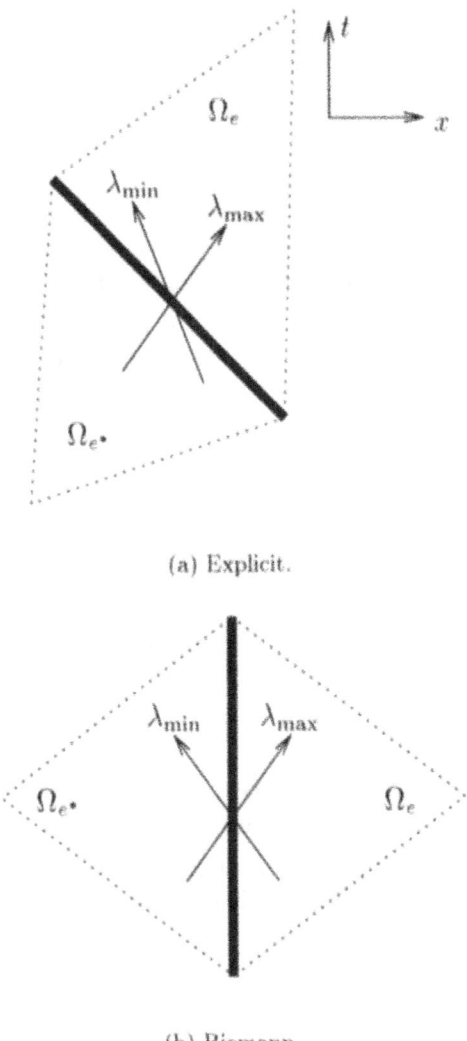

(a) Explicit.

(b) Riemann.

Figure 2. Face Definitions for $d = 1$. $\lambda_{\min}$ and $\lambda_{\max}$ correspond to the local minimum and maximum eigenvalues, respectively.

for all $\vec{x} \in \partial\omega_e$. Note that $\mu \equiv 0$ is only permissible for Riemann Faces.

The above definitions are what give the scheme an "upwind" character. In fact, an Explicit Face can be thought of as a Riemann Face on which the Riemann solution is known *a priori*. Also, in a region where all of the eigenvalues have the same sign, a Riemann Face satisfies the definition of an Explicit Face.

4. Implementation

4.1. SPACE-TIME MESHES

This section defines the space-time meshes on which the discrete form of the conservation law will be solved. How the mesh is defined will greatly influence the cost of the DG method. The underlying principle will be to form each space-time mesh in such a way that the numerical method is at worst point implicit.

4.1.1. *1-D Meshes*

In 1-D, two space-time meshes are used. The Riemann Mesh, shown in Figure 3a, contains Riemann Faces which couple pairs of elements implicitly. To avoid Riemann Faces altogether, the Staggered Mesh in Figure 3b will also be used. For both meshes, the diagonal Explicit Faces satisfy Equation (5) as long as the Courant number is less than 1. This condition is in fact the stability constraint.

Figure 3 also indicates the order in which the elements are solved. The solution procedure for the Riemann Mesh is as follows:

1. Solve in each of the elements of family $\boxed{1}$. The flux on the bottom face is either from the initial condition, or the solution in element $\boxed{3}$ from the previous time level.
2. Solve in the elements of family $\boxed{2}$. In general, these elements are coupled in pairs by the Riemann Faces. The two elements are not coupled implicitly if all of the eigenvalues are of the same sign.
3. Solve in each of the elements of family $\boxed{3}$.
4. Proceed to the next time level.

The Staggered Mesh is solved in a similar fashion, the only difference being that no elements are coupled implicitly. For the Euler equations, the computational cost of using the Riemann Mesh is approximately 70% more than the Staggered Mesh.

A mesh can also be defined that is made up of segments of the Riemann and Staggered Meshes. In particular, to enforce boundary conditions (other than periodic) the Staggered Mesh uses a Riemann Mesh segment at the boundary.

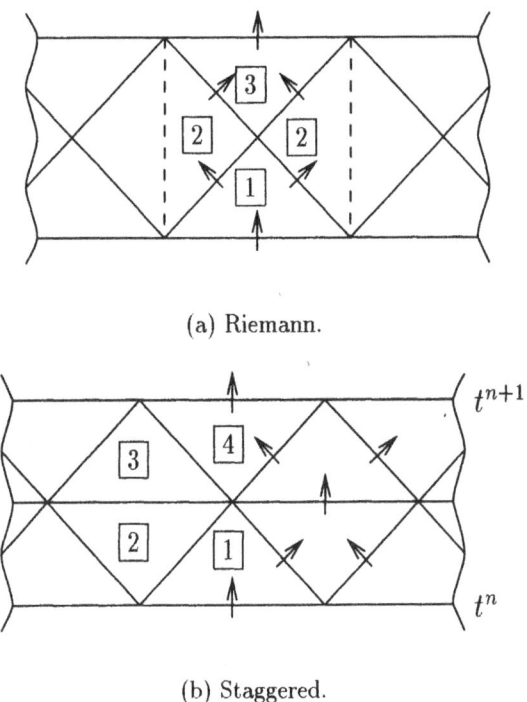

(a) Riemann.

(b) Staggered.

Figure 3. Various Space-Time Mesh Segments. Solid lines denote Explicit Faces, with arrows indicating the flow of information. Dashed lines are Riemann Faces. The element types are numbered according to the order in which they are solved.

4.1.2. *2-D Meshes*

The 2-D space-time mesh is an extension of the 1-D Staggered Mesh of Figure 3b. The underlying 2-D spatial mesh is a quadrilateral mesh. To visualize the space-time mesh, Figure 4 shows the order in which elements are solved, over $\frac{1}{2}\Delta t$. In the next half-time step, the mesh is staggered; the Step (1) pyramid base, in the $\frac{1}{2}\Delta t < t \leq \Delta t$ interval, is coincident with the Step (3) pyramid base of the $0 < t \leq \frac{1}{2}\Delta t$ interval. Note that at each step, none of the element solutions are coupled, just as in 1-D Staggered Mesh. For Step (2), the same final solution results if the y-axis "valley" elements are solved before the x-axis "valley" elements. The method does not use "operator splitting." As a consequence, the results show that this mesh does not exhibit a loss of accuracy when the advection direction is skewed with respect to the spatial mesh.

A 2-D analogy to the 1-D Riemann Mesh also exists, along with meshes based on a 2-D triangular spatial mesh and a 3-D hexahedral mesh; see (Lowrie, 1996). Only the Staggered Mesh has been used for 2-D problems, with Riemann Faces used on the domain boundaries to enforce boundary conditions.

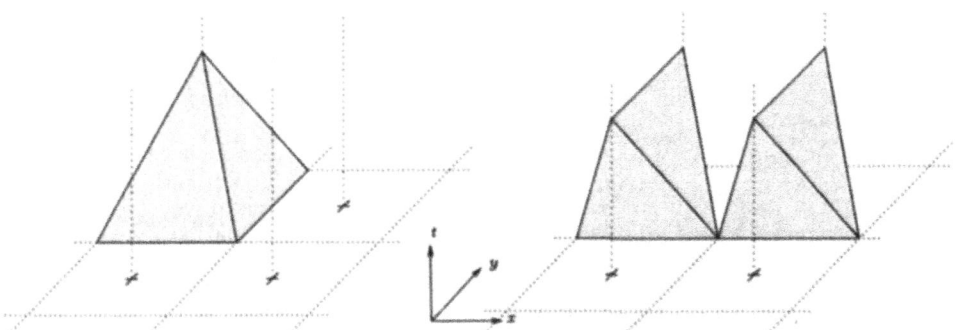

Step (1): Solve in the pyramid elements covering each x, y mesh cell.

Step (2): Solve in the tetrahedral elements that fill the x-axis valleys of the pyramid elements in Step [1].

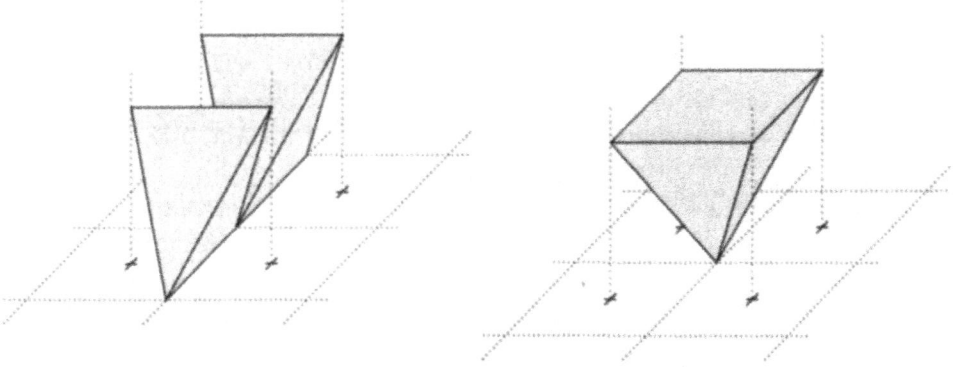

Step (2)-continued: Solve in the tetrahedral elements that fill the y-axis valleys of the pyramid elements in Step [1].

Step (3): Solve in the inverted-pyramid elements.

Figure 4. Element Solution Order for the 2-D Staggered Mesh. Only $\frac{1}{2}\Delta t$ of mesh is shown.

4.2. ALGEBRAIC SYSTEM

To solve Equation (4) numerically, in each element the solution is approximated as

$$\mathbf{w}_e(\vec{x}) = \sum_{j=1}^{N} \phi_{e,j}(\vec{x}) \mathbf{c}_{e,j}, \quad \forall \vec{x} \in \Omega_e,$$

where the $\{\phi_{e,j}\}$, $j = 1, 2, \ldots, N$ are polynomial basis functions over the element Ω_e. DG(k) will refer to the DG method with each $\phi_{e,j} \in \mathcal{P}_k(\Omega_e)$, where $\mathcal{P}_k(\Omega_e)$ is the space of polynomials of maximum order k defined on Ω_e. The variable $\mathbf{w} = \mathbf{w}(\mathbf{u})$ is the parameter vector. There are N-unknown m-vectors in each element, namely the expansion coefficient vectors, $\mathbf{c}_{e,j}$.

To generate the necessary N vector equations, the Galerkin approach is to choose $\Phi = \phi_{e,i}$, $i = 1, 2, \ldots, N$, in Equation (4). This choice results in the minimization property described in (Lowrie, 1996). For a nonlinear conservation law, the resulting system of equations is nonlinear in the $\mathbf{c}_{e,j}$. To solve this system, a Newton-Kantorovich approach is taken.

Consider the general case where the solution of N_c elements is coupled. Linearizing (4) gives a system of equations of the form

$$\mathbf{M}(\delta \mathbf{c}) = -\mathbf{r}, \tag{6}$$

where $\delta \mathbf{c}$ is the update vector to the expansion coefficients, and $\mathbf{M}$ is a $N_c \times N_c$ block matrix. Each submatrix block is an $(mN) \times (mN)$ matrix. From the boundary definitions, the coupling of elements can only occur across Riemann Faces. Therefore, the maximum $N_c = 2$ for the 1-D Riemann Mesh, and $N_c = 1$ for the Staggered Meshes.

The integrals are computed numerically using Gaussian quadrature. Note that if Equation (1) is a linear system, then $\mathbf{M}$ is a constant matrix, which can be inverted once and stored for each class of coupled elements. For nonlinear systems, the cost can be reduced without a significant decrease in accuracy by performing only a single iteration of (6), and by computing $\mathbf{M}$ using the DG(0) solution (which can often be found explicitly). This eliminates the need for quadrature when computing $\mathbf{M}$.

Boundary conditions are applied by defining a "ghost" Gauss point at each physical Gauss point that is on the boundary. A Riemann problem is then solved to determine the boundary flux. The boundary procedure is the same for any order-of-accuracy, unlike methods that require special difference stencils.

5. Accuracy

The DG method has the following properties:

1. Conservation.
2. Stability, and satisfies an entropy condition, for Courant numbers less than 1 and any order-of-accuracy.
3. A minimization property.
4. High accuracy.

Each of these properties are discussed in (Lowrie, 1996). A brief overview of the accuracy will be given in this section.

For a method using an order-k interpolant, the expectation is that at best the error will converge as $\mathcal{O}(h^{k+1})$, where h is some measure of the mesh size. However, a Fourier analysis shows that DG(k) converges as $\mathcal{O}(h^{2k+1})$ in a certain norm, indicating that DG(k) has a *superconvergence* property. This norm, denoted by L_p^{ev}, measures only the error in the evolution of the initial condition projected onto the accurate mode of the update operator.[3] The L_p^{ev}-norm will be related to more standard norms in the remainder of this section.

For a calculation to any fixed time $t > 0$, in the limit $h \to 0$, all of the spurious modes will be damped out. Therefore, any other norm will include a term proportional to the evolution error, plus the particular norm's measure of the initial condition projected onto the accurate mode (a one-time contribution).

A norm typically used to measure the solution accuracy at a given time-level is

$$
L_p(v) = \left\{ \frac{1}{|\Omega_d|} \int_{\Omega_d} |v(\underline{x}) - v_{\text{exact}}(\underline{x})|^p \, d\underline{x} \right\}^{1/p},
$$

where Ω_d is the spatial domain, and v, v_{exact} are the numerical and exact solutions of a representative variable of the conservation law. A finite-element analysis shows that DG(k) converges in L_2 as $\mathcal{O}(h^{k+1/2})$ (Jaffre *et al.*, 1995). However, for many smooth solutions on 'regular' meshes, practitioners often realize $\mathcal{O}(h^{k+1})$ accuracy (Jaffre *et al.*, 1995; Richter, 1988). By including the initial-projection error, the Fourier analysis gives a convergence rate of $\mathcal{O}(h^{k+1})$ in L_p.

Another norm studied here is the error in the cell averages, denoted by $\bar{L}_p(v)$. The Fourier analysis for this norm gives $\mathcal{O}(h^{k+2})$ for $k > 0$, and $\mathcal{O}(h)$ for $k = 0$. Although L_p and $\bar{L}_p$ have slower convergence than L_p^{ev}, the consequences of the superconvergence property will be apparent in the results of Section 6.1.

Note that the semi-discrete version of DG, with $(k+1)$-multi-stage, is $\mathcal{O}(h^{k+1})$ accurate (Cockburn & Shu, 1989), and therefore does not have

---

[3]The accurate mode is the least-damped mode. There are k-spurious modes.

the superconvergence property. However, (Atkins & Shu, 1996) later observed $\mathcal{O}(h^3)$ convergence in the cell-averages for $k = 1$, using a 3-stage time integration. Since the 3-stage method is a third-order time integrator, obviously there is some mode that is advected with third-order accuracy. What Atkins and Shu have apparently shown is that the projection of the initial condition onto this mode is also third-order.

6. Results

6.1. SCALAR ADVECTION

Results are now presented for the linear equation

$$\partial_t u + \partial_x u = 0,$$

on $0 \le x \le 1$, with the initial condition $u(x, 0) = \sin(2\pi x)$, and periodic boundary conditions. Figure 5 shows the order-of-accuracy history for the Staggered Mesh, $\nu = 0.8$, using the L_1 and $\bar{L}_1$-norms. Similar results are obtained for the Riemann Mesh. At least for early times, DG(k) follows the Fourier analysis' prediction of $\mathcal{O}(h^{k+1})$ in L_p. An interesting phenomenon can occur, however, as shown by the results for DG(1). Given enough time, the evolution error will accumulate, and overcome the initial-projection error, denoted as L_1^0. Past this time, for a given stage in mesh refinement, the accuracy convergence is dictated by the evolution error, L_1^{ev}, which converges as $\mathcal{O}(h^{2k+1})$.

Such behavior is also evident in the $\bar{L}_1$-norm for $k = 3$. Since the initial error $\bar{L}_1^0 \ll L_1^0$, the order asymptotes to apparent $\mathcal{O}(h^{2k+1})$-accuracy much more quickly in $\bar{L}_1$ than L_1. The oscillations in the order are the result of the spurious modes. Note that the time at which the evolution error overcomes the initial-projection error increases with mesh size, and therefore strictly speaking the convergence in $\bar{L}_1$ is $\mathcal{O}(h^{k+2})$. However, the finest mesh used in each case here gives results that are well-resolved. At the final time shown in each plot, DG(1) has an error given by $\log_{10}(\bar{L}_1) = -3.56$, while DG(3)'s error is -7.51. For many practical problems where long-time integration is required, there is the possibility of realizing $\mathcal{O}(h^{2k+1})$-accuracy, particularly in $\bar{L}_p$.

For smooth solutions, nonlinearity does not seem to destroy the superconvergence property, but the time-histories are more complicated. See (Lowrie, 1996) for results when solving the Euler equations.

6.2. SHU-OSHER PROBLEM

The problem of (Shu & Osher, 1989) corresponds to a Mach 3 shock propagating into a sinusoidal density wave. This problem requires a limiter; see

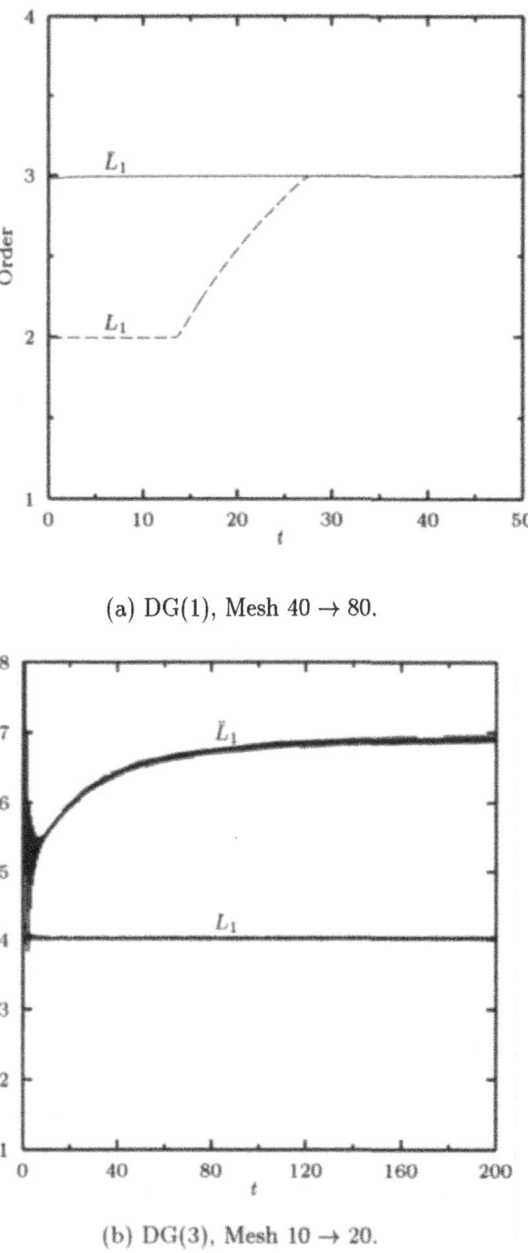

(a) DG(1), Mesh 40 → 80.

(b) DG(3), Mesh 10 → 20.

Figure 5. Order-of-Accuracy History, Sine Wave, Staggered Mesh, $\nu = 0.8$, using L_1 and $\bar{L}_1$-norms. The two mesh sizes used for the order calculation are indicated.

(Lowrie, 1996) for specifics on the limiter used. Figure 6 shows the DG(1) cell-average densities, using 400 cells on the Staggered Mesh. The "exact" solution here is the 1600-cell solution. For this problem, the Riemann Mesh solution is very similar. Also, the 800-cell solution is nearly indistinguishable from the exact solution. The results here compare favorably with the TVD results of (Huynh, 1995; Cockburn *et al.*, 1989).

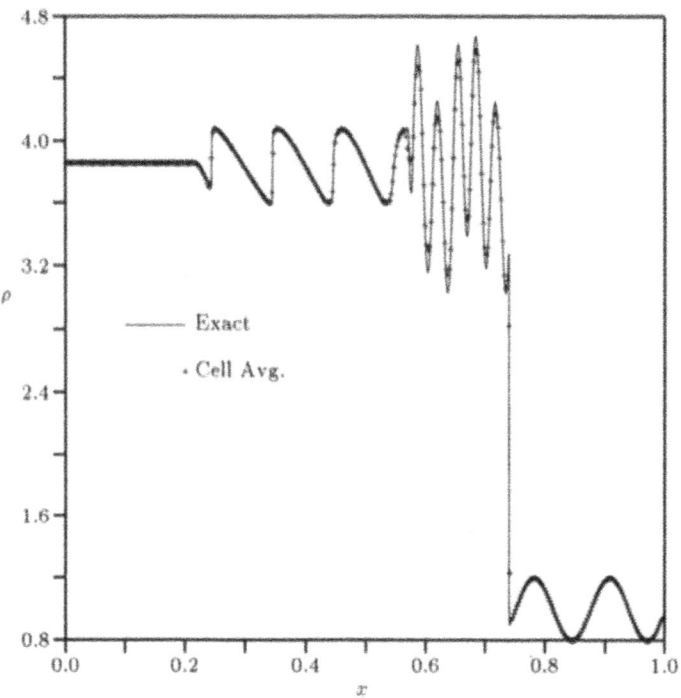

Figure 6. Shu-Osher Problem, DG(1), Staggered Mesh, $\nu = 0.8$, 400 cells, $t = 0.18$.

6.3. PRESSURE PULSE IN FREESTREAM OVER WALL

This problem has an initial condition of a Gaussian-pressure pulse above a plane wall, immersed in a steady flow. This is the "Category IV" problem described in the (Hardin *et al.*, 1995) workshop proceedings. The Workshop participants used a 200×200 mesh for this problem. Figure 7 shows a solution of the full Euler equations using DG(3) on a 25×25 mesh.[4] On such a coarse mesh, the initial Gaussian is spread over approximately three cells.

---

[4]In 2-D, DG(3) uses $\sqrt{10}$ more degrees-of-freedom per mesh direction than a conventional difference method. By this measure, DG(3) on a 25×25 mesh is equivalent to a conventional method on a 79×79 mesh.

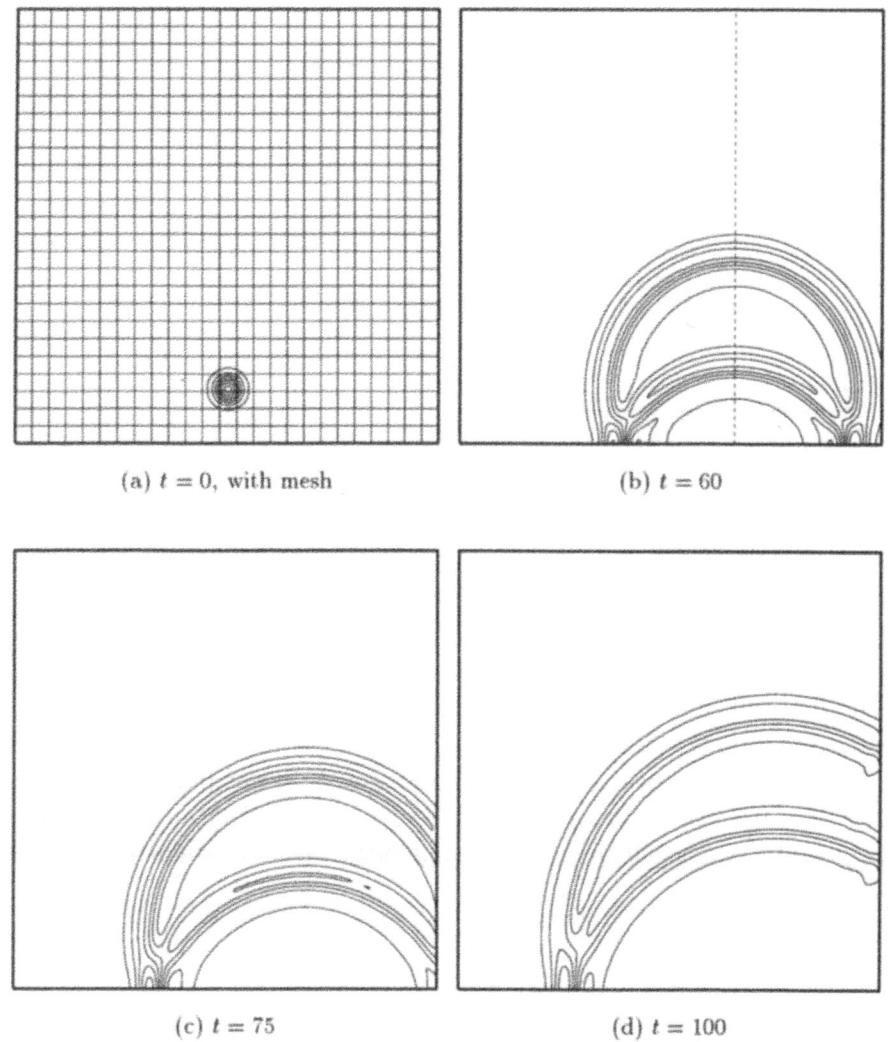

(a) $t = 0$, with mesh

(b) $t = 60$

(c) $t = 75$

(d) $t = 100$

Figure 7. Gaussian-Pressure Pulse over Wall. DG(3) results, 25×25 mesh, $\nu = 0.8$. On the $t = 60$ plot, the dashed line indicates a cut along which comparisons will be made.

The method does a reasonable job of maintaining symmetry, but wave reflections at the downstream boundary are noticeable at $t = 100$. These reflections are a result of the characteristic-normal boundary condition, and do not disappear with mesh refinement. A more sophisticated farfield boundary condition is needed to suppress the reflections of oblique waves.

Since the waves in this problem are weak, a full Euler solution should compare well with linear theory. Figure 8 compares the pressures at $t = 60$

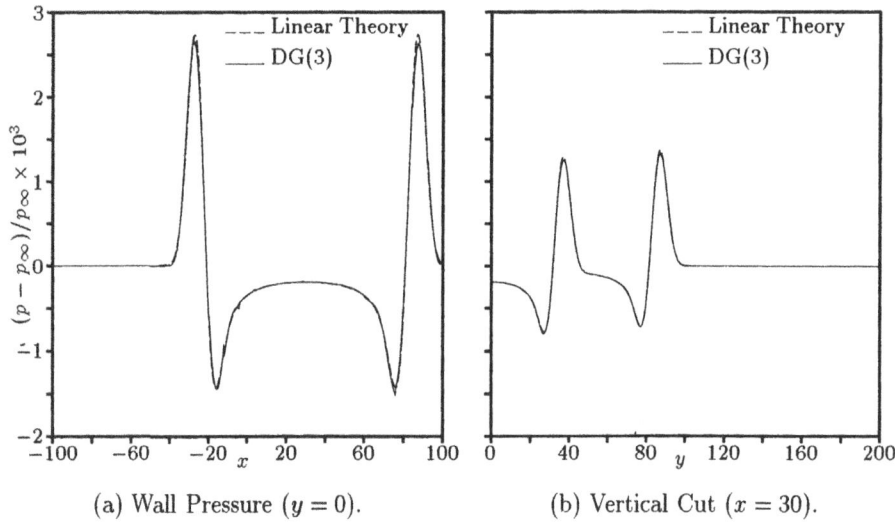

(a) Wall Pressure $(y = 0)$.   (b) Vertical Cut $(x = 30)$.

Figure 8. Gaussian-Pressure Pulse over Wall, Comparison of DG(3) with Linear Theory, Mesh 25 × 25. See Figure 7b for location of vertical cut.

along the wall, and through a vertical cut at $x = 30$. The agreement is good, except at the peaks, where DG(3) under-predicts the pressure. Note that the glitches in the predicted wall pressure, most apparent in the region $-20 \leq x \leq 0$, show the mismatch at the cell boundaries of the discontinuous interpolant. On a 50×50 mesh, DG(3) and linear theory are indistinguishable (Lowrie, 1996).

6.4. PRESSURE PULSE AND VORTEX INTERACTION

This problem is the interaction of a pressure pulse and a vortex, immersed in a steady flow. The "Category III" problem in (Hardin *et al.*, 1995) is very similar to the problem in this section. The solution domain is $-100 \leq x, y \leq 100$, with farfield conditions specified at all four boundaries. The initial conditions are

$$p = 1 + 0.01\gamma \mathcal{E}_p(x, y),$$
$$\rho = \gamma + 0.01\gamma \left[\mathcal{E}_p(x, y) + 0.1\mathcal{E}_v(x, y)\right],$$
$$u = 0.5\cos(\alpha) + 0.0004(y - y_v),$$
$$v = 0.5\sin(\alpha) - 0.0004(x - x_v),$$

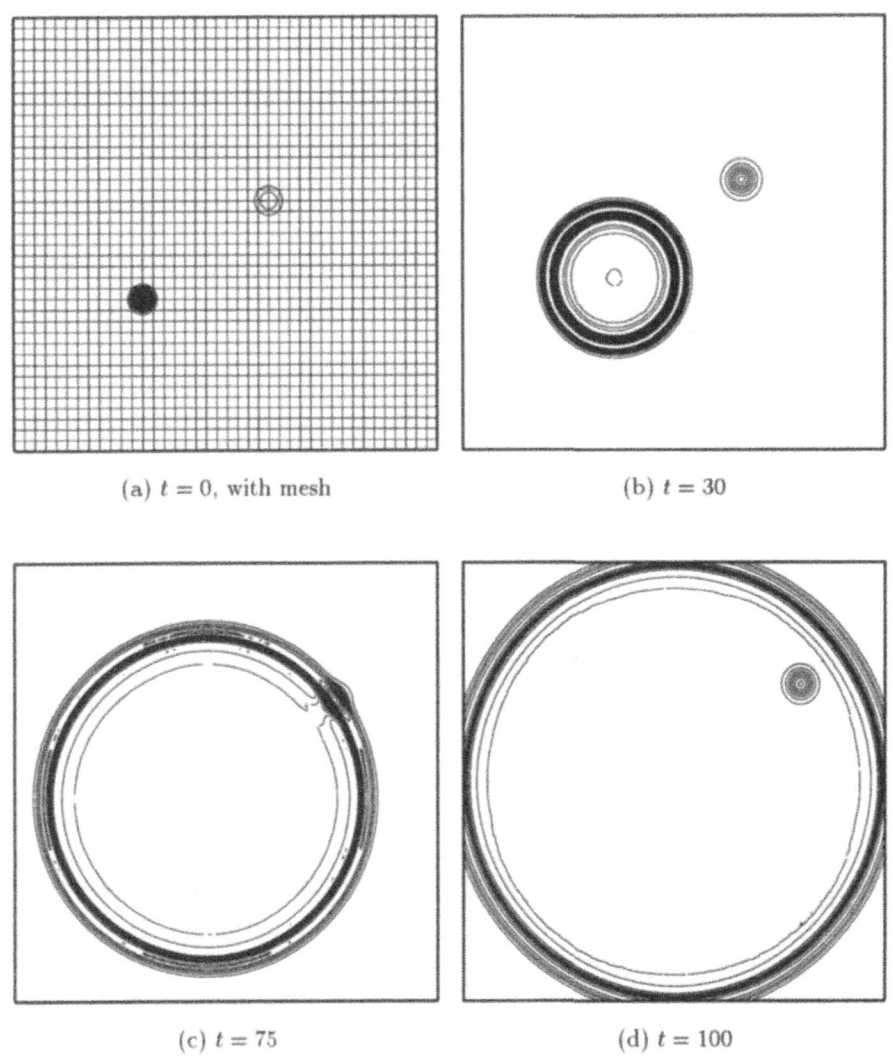

(a) $t = 0$, with mesh   (b) $t = 30$

(c) $t = 75$ (d) $t = 100$

Figure 9. Pressure-Pulse / Vortex Interaction, $\alpha = \arctan(3/4)$. DG(3) Density Contours, 40×40 Mesh, $\nu = 0.8$.

where α is the freestream-flow angle, and

$$\mathcal{E}_p(x, y) = \exp\left[-\ln(2)\frac{(x - x_p)^2 + (y - y_p)^2}{9}\right],$$

$$\mathcal{E}_v(x, y) = \exp\left[-\ln(2)\frac{(x - x_v)^2 + (y - y_v)^2}{25}\right],$$

$$(x_p, y_p) = -50(\cos(\alpha), \sin(\alpha)),$$

$$(x_v, y_v) = 25(\cos(\alpha), \sin(\alpha)).$$

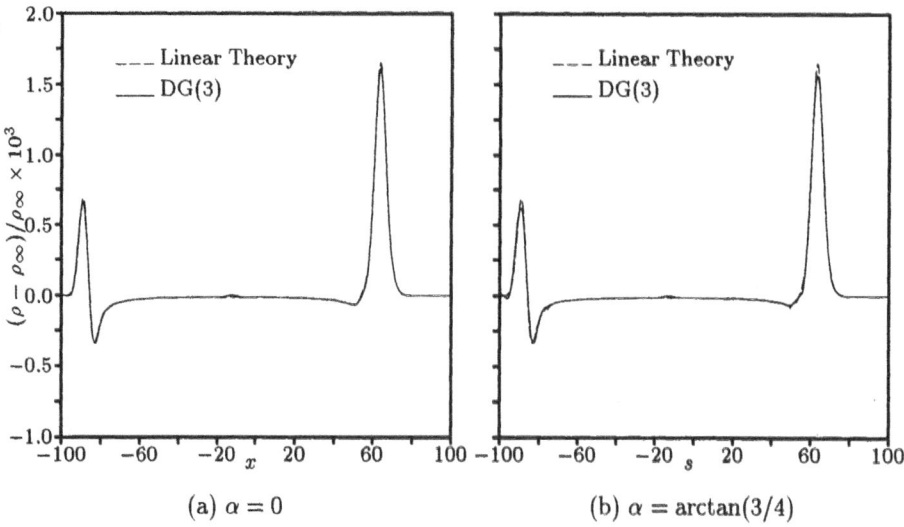

Figure 10. Pressure-Pulse / Vortex Interaction, Comparison with Linear Theory, 40×40 Mesh, $\nu = 0.8$. Cut is along line connecting pressure pulse and vortex.

Here (x_p, y_p) and (x_v, y_v) represent the initial centers of the pressure pulse and vortex, respectively. The above initial conditions differ from those in (Hardin *et al.*, 1995) in that the distance between the pressure pulse and vortex is independent of α.

DG(3) density contours are shown in Figure 9 for $\alpha = \arctan(3/4)$. Comparing the density contours at $t = 30$ and $t = 100$, the vortex is not distorted by the acoustic wave, as predicted by linear theory. A comparison with linear theory is made for both $\alpha = 0$ and $\alpha = \arctan(3/4)$ in Figure 10. Both solutions compare reasonably well, with those for $\alpha = \arctan(3/4)$ slightly worse. The density glitch at $x = -12.5$ corresponds to the center of the pressure pulse, and is present in the DG(k) results because the initial condition is not isentropic. Results for DG(3) on a 60×60 mesh are within 2% of linear theory (Lowrie, 1996).

7. Summary

A time-accurate method has been developed that is based on the Discontinuous Galerkin method. In deriving the method, the idea of *compactness* has been strictly followed. That is, that the discrete domain of dependence should contain a minimum amount of data outside of the physical domain of dependence. For any order-of-accuracy, the method is stable for Courant numbers less than 1, satisfies an entropy condition, and a minimization property.

We've given an argument as to why the superconvergence property is important for problems that require long-time integration. For the long-time integration of linear problems, the error is driven by a norm that converges as $\mathcal{O}(h^{2k+1})$. Although not shown here, numerical experiments indicate that this property extends to some nonlinear cases.

The use of the Staggered Mesh allows for the elimination of the Riemann problem. The disadvantage of *any* time-staggered approach is the presence of diffusion at low Courant numbers. Although seemingly avoidable in 1-D problems, in 2-D diffusion effects are unavoidable, resulting in 'cross-diffusion'; see (Lowrie, 1996) for more discussion.

A comparison of DG(k) with the multi-stage version (Cockburn *et al.*, 1989; Atkins & Shu, 1996) shows two main advantages in keeping the method compact. The first advantage is that the Courant restriction for the multi-stage version is inversely proportional to k, at least for $k = 1, 2$. Secondly, DG(k) has a superconvergence property, although there is some evidence that increasing the order of the multi-stage integration may give the same benefit, along with a modest increase in stability. In terms of simplicity, the multi-stage version clearly shows an advantage.

The accuracy of DG(k) is impressive. The typical arguments against moving beyond second-order accuracy are that higher-order methods are less robust, as a result of stability restrictions, limiting, and boundary procedures. Although much work is still needed in developing a limiter, the method described here shows that by strictly following the physics, many desirable properties can be obtained including any order-of-accuracy. The DG(k) method actually improves with increasing k, in terms of the cost for a specified error tolerance. This work should open the door to the development of more practical methods that are ideal for problems that demand high accuracy.

References

Atkins, H. and Shu, C.-W., "A quadrature-free implementation of the discontinuous Galerkin method for hyperbolic problems," *AIAA Paper 96-1683*, 1996.

Cockburn, B. and Shu, C., "TVB Runge-Kutta local projection discontinuous Galerkin finite element method for conservation laws II: General framework," *Mathematics of Computation*, Vol. 52, 1989, pp. 411-435.

Cockburn, B., Lin, S., and Shu, C., "TVB Runge-Kutta local projection discontinuous Galerkin finite element method for conservation laws III: One-dimensional systems," *Journal of Computational Physics*, Vol. 84, 1989, pp. 90-113.

Hardin, J.C., Ristorcelli, J.R., and Tam, C.K.W., eds., "ICASE/LaRC Workshop on Benchmark Problems in Computational Aeroacoustics (CAA)," NASA Conference Publication 3300, May 1995.

Huynh, H.T., "Accurate upwind methods for the Euler equations," *AIAA Paper 95-1737*, 1995.

Jaffre, J., Johnson, C., and Szepessy, A., "Convergence of the discontinuous Galerkin finite element method for hyperbolic conservation laws," *Mathematical Models and*

Methods in Applied Sciences, Vol. 5, 1995, pp. 367-386.

Johnson, C. and Pitkaranta, J., "An analysis of the discontinuous Galerkin method for a scalar hyperbolic equation," *Mathematics of Computation*, Vol. 46, 1986, pp. 1-26.

Lowrie, R.B., "Compact Higher-Order Numerical Methods for Hyperbolic Conservation Laws," Ph.D. thesis, University of Michigan, 1996. Available at http://www.engin.umich.edu/research/cfd/research/-publications/.

Richter, G.R., "An optimal-order error estimate for the discontinuous Galerkin method," *Mathematics of Computation*, Vol. 50, 1988, pp. 75-88.

Roe, P.L., "Approximate Riemann solvers, parameter vectors and difference schemes," *Journal of Computational Physics*, Vol. 43, 1981.

Shu, C.-W. and Osher, S., "Efficient implementation of essentially non-oscillatory shock capturing schemes, II," *Journal of Computational Physics*, Vol. 83, 1989, pp. 32-78.

van Leer, B., "Towards the ultimate conservative difference scheme. IV. A new approach to numerical convection," *Journal of Computational Physics*, Vol. 23, 1977.

ANISOTROPIC MESH ADAPTATION:
A STEP TOWARDS A MESH-INDEPENDENT
AND USER-INDEPENDENT CFD

W. G. HABASHI

Computational Fluid Dynamics Laboratory (CFD Lab),
Concordia University, 1455 de Maisonneuve Blvd W., ER 301,
Montreal, (Qc), H3G 1M8, Canada.

M. FORTIN

Groupe Interdisciplinaire de Recherche en Éléments Finis (GIREF),
Université Laval, Ste-Foy, (Qc), G1K 7P4, Canada.

J. DOMPIERRE

CFD Lab and Centre de Recherches Mathématiques (CRM),
Université de Montréal, C. P. 6128, succ. Centre Ville,
Montréal, (Qc), H3C 3J7, Canada.

M.-G. VALLET

Institut des Matériaux Industriels, CNRC,
75, boul. de Mortagne, Boucherville, (Qc), J4B 6Y4, Canada.

AND

Y. BOURGAULT

CFD Laboratory

Abstract.

This paper presents an anisotropic remeshing strategy to obtain accurate numerical solutions of problems showing distinct directional features. The adaptation criterion is measured on the mesh edges and is independent of the problem or the discretization scheme used. The strategy is shown to converge and is demonstrated for two-dimensional inviscid and viscous laminar flows, at a variety of speeds.

99

V. Venkatakrishnan et al (eds.), Barriers and Challenges in Computational Fluid Dynamics, 99-117.
© 1998 *Kluwer Academic Publishers.*

1. INTRODUCTION

Mesh adaptation is one of the current promising technologies for the improvement of numerical predictions. It provides the ability to position the degrees of freedom in an optimal way. This is especially useful in Computational Fluid Dynamics where features are localized, and directional, like boundary layers, slip-lines, wakes and shock waves.

This paper presents an anisotropic adaptation strategy to yield accurate solutions of such flows. The adaptive technique is general and can be applied to a wide range of Finite Element, Finite Volume or Finite Difference problems.

In the following sections, the anisotropic adaptation criterion is first presented, then the adaptive strategy and its implementation for triangular and quadrilateral meshes is shown. Some remarks are made about the coupling of the adaptive library to a FEM solver.

The anisotropic strategy is extensively validated through the monitoring of the remeshing steps and convergence of the overall process. In the last section, the efficiency of the strategy is demonstrated on two examples of external flow computations.

2. THEORY

The adaptive strategy is briefly exposed in this section. We refer to previous papers (Vallet, 1992; Ait-Ali-Yahia et al., 1996a; Vallet et al., 1996) for more details.

2.1. THE ERROR ESTIMATOR

Consider a 1-D problem in which the solution variable u is approximated by u_h with linear interpolation. A local error, e_E is defined over the element E to be:

$$e_E = u - u_h. \tag{1}$$

By expanding with Taylor series the solution u at one end of the element E and provided that the nodal error is zero, the error e_E for a linear element may be cast into the form:

$$e_E = \frac{1}{2}x(h-x)\frac{d^2u(\xi)}{dx^2} \tag{2}$$

where h represents the element length, x is measured from one end of the element E and ξ is a point in the element E. As the solution u is unknown, its second derivative is also unknown and it is approached by the second derivatives of u_h. Once the second derivative has been recovered, the error field e_E is known and the root-mean-square value e_E^{RMS} over an element E is a good local error estimator (Peraire *et al.*, 1987) and is defined as:

$$e_E^{RMS} = \left(\frac{1}{h}\int_E e_E^2(x)\,dx\right)^{1/2} = \frac{1}{\sqrt{120}}h^2\left|\frac{d^2u_h}{dx^2}\right|_E. \tag{3}$$

Thus, the interpolation error for this 1-D problem is proportional to the product of the second derivative and the square of the characteristic length of an element. Mesh adaptation may then be carried out by equidistributing the error over the elements, so that the adaptation process is guided by the requirement that the local spacing has to verify $h^2|u''| = $ constant.

Extending these ideas to the 2-D case, the second derivatives can now be replaced by the symmetric Hessian matrix

$$H_{ij} = \partial^2 u_h \,/\, \partial x_i \partial x_j. \tag{4}$$

Second derivatives of a piecewise linear function can be recovered either by local reconstruction, projection or variational recovery (see work of Zienkiewicz & Zhu, for a classical one). In the present work, a weak formulation, combined with mass lumping, is applied to recover an estimate of the second derivatives. This yields the following expression

$$H_{ij}(x_I) = \int_{\Omega_I}\frac{\partial^2 u_h}{\partial x_i \partial x_j}(x)\varphi_I(x)\,d\Omega \,\bigg/ \int_{\Omega_I}\varphi_I(x)\,d\Omega \tag{5}$$

where x_I is a node of the mesh, Ω_I represents the elements sharing node x_I and φ_I is the linear basis function at this node. After integration by parts of the numerator of Eq. (5), the nodal values of the Hessian reduce to

$$H_{ij}(x_I) = \frac{\displaystyle\int_{\Gamma_I}\frac{\partial u_h}{\partial x_i}\varphi_I\,n_j\,d\Gamma - \int_{\Omega_I}\frac{\partial u_h}{\partial x_i}\frac{\varphi_I}{\partial x_j}\,d\Omega}{\displaystyle\int_{\Omega_I}\varphi_I(x)\,d\Omega}. \tag{6}$$

The Hessian matrix, given by the Eq. (4) may be diagonalized as

$$H = R(\alpha)\,\Lambda\,R^T(\alpha) \tag{7}$$

where Λ is the diagonal matrix of eigenvalues of H and R is the matrix of the eigenvectors. The transformation Λ is a scaling in the axes directions and R is a rotation with angle α, that the eigenvector corresponding to the smallest eigenvalue, λ_1, makes with the x_1-axis.

In order to obtain a symmetric positive-definite matrix, the Hessian is modified by taking the absolute values of its eigenvalues (Vallet, 1992). This results in

$$M = R(\alpha)\,|\Lambda|\,R^T(\alpha) = S(\alpha)\,S^T(\alpha) \tag{8}$$

where $S(\alpha) = R(\alpha)\sqrt{|\Lambda|}$. The transformation S of a unit circle would be an ellipse, rotated through an angle α, whose semi-major and minor axes are inversely proportional to the square roots of the eigenvalues $|\lambda_1|$ and $|\lambda_2|$, respectively. Therefore, one may obtain a directionally stretched mesh by mapping a uniform mesh using the transformation S.

The length $e(\Gamma)$ of a curve Γ in this transformed plane is given by

$$e(\Gamma) = \int_0^1 \sqrt{\Gamma'(s)^T M(s)\, \Gamma'(s)}\, ds \tag{9}$$

where $\Gamma(s)$, $s \in [0, s]$ is a parametric representation of the curve Γ. In fact, Eq. (9) define a Riemannian metric and the length of a curve Γ is the evaluation of the interpolation error over that curve. Making a mesh with the same length in the metric for each edge is making a mesh that equidistributes the interpolation error over the edges. The metric M is computed and stored on a background mesh, and thus the value of M at any position of the domain can be interpolated during the adaptation on this mesh. The edge-based error estimate can then be numerically evaluated from Eq. (9) for each edge of the mesh.

2.2. LOCAL IMPROVEMENTS

To adapt a mesh with all edges having approximately the same value of the error estimator, the following approach is used: from an existing mesh and a description of the domain boundaries, the mesh optimization library *modifies* iteratively this mesh. All the modifications to the mesh are local. These local operations are: adding a node, removing a node, moving a node and swapping an edge. Modifying an existing mesh may be cheaper than generating a new one, especially when there are few differences between the initial mesh and the adapted one. In the following, the operations involved are discussed.

First, the mesh is refined by sweeping over all edges and by carrying out local modifications where needed. If the edge error estimate $e(\Gamma)$ is greater than e_{max}, then this edge is cut by introducing a new node and two new edges. See Fig. 1.

The mesh is then coarsened by sweeping over all edges and by carrying out local modifications where needed. If the edge error estimate $e(\Gamma)$ is lower than e_{min}, then one of the two nodes of the edge is removed, creating a "hole" in the mesh. This hole is then remeshed by some triangulation process. See Fig. 2.

Following this, edge-swapping is used. It is a classical procedure for building the Delaunay triangulation which maximizes the minimum of the angles. An edge between two triangles is a diagonal of a quadrilateral (see

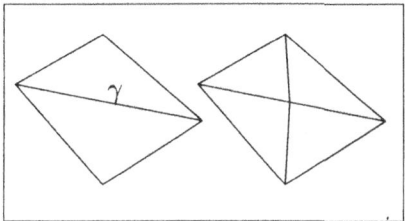

Figure 1. Edge refinement.

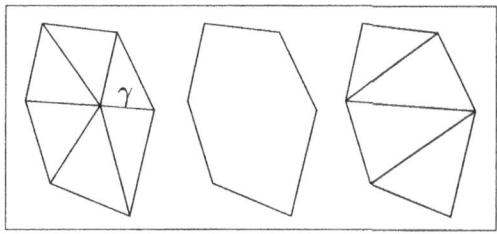

Figure 2. Edge coarsening.

Fig. 3). One has to choose between its two diagonals to get the "best" triangles. Maximizing the minimum of

$$\text{shape}(\triangle) = 27 \left(\frac{r}{p}\right)^2 = 27 \frac{(p-a)(p-b)(p-c)}{p^3} \tag{10}$$

over the two possible configurations is a way to build an equilateral mesh. Here, p, r, a, b and c stand for the half-perimeter, the radius of the inscribed circle and the lengths of the edges of the triangle, respectively. While this does not lead exactly to the so-called Delaunay triangulation (Vallet, 1992), this criterion has the advantage of depending only on the edge lengths. In the present optimization process these lengths are taken to be the edge-based error estimates so that a mesh is built having all edge errors approximatively the same.

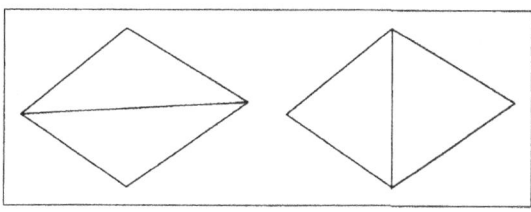

Figure 3. Node reconnection.

Finally, node displacement is used in order to get a smoother mesh. It is a standard part of many mesh generation codes. The simplest version of node movement, for which the nodes are relocalized at the center of their neighbors, would destroy the previous adaptation efforts. To avoid that, one has to include the error estimates in the procedure. Another way of moving nodes is to make an analogy with a network of springs (Gnoffo, 1982; Gnoffo, 1983; Nakahashi & Deiwert, 1987; Baruzzi, 1995). For moving a node x, all other nodes are supposed to be fixed and all the edges around this node are assumed to be springs. See Fig. 4. The node x is then displaced to minimize the energy of this spring network. The coordinates of x are then the solution of the following minimization problem

$$\min_{x \in \mathbf{R}^2} E(x) = \min_{x \in \mathbf{R}^2} \sum_{I=1}^{n} (x_I - x)^2 k_I(x), \tag{11}$$
$$\text{with } k_I(x) = e(x_I - x)/\|x_I - x\|,$$

where $k_I(x)$ is the spring constant between x and x_I, $e(x_I - x)$ is the error estimate of the edge $\overline{x\,x_I}$ and $\|x_I - x\|$ is its Euclidean length. This problem is equivalent to finding a point x that is the root of

$$\sum_{I=1}^{n} F_I(x) = \sum_{I=1}^{n} (x_I - x) k_I(x) = 0 \tag{12}$$

where by Hooke's law, $(x_I - x)k_I(x)$ is the force $F_I(x)$ related to the edge $\overline{x\,x_I}$. This method works well by itself and needs only some small modifications to work in combination with other optimization techniques. It can be noticed that even if some authors (Gnoffo, 1983; Nakahashi & Deiwert, 1985; Palmerio, 1994) prefer to control the grid orthogonality while using the spring analogy, the only limitation on node movement in the present implementation, as in Ait-Ali-Yahia *et al.*, is the convexity of the elements.

Node displacement is the most powerful tool in the current optimization strategy. It can be used solely when the mesh topology has to be kept unchanged, such as with structured grids or unstructured grids with a fixed number of nodes, and it greatly improves the mesh quality when combined with other techniques. In fact, all other processes are discontinuous or sporadic: one chooses to do something or not depending on whether a criterion is above a threshold value. Displacement, however, is a continuous adaptive process and it can perform surgical improvements after a discontinuous process.

To conclude this section, there is an analogy between a CFD solver and a mesher. From a mesh (nodes and connectivity between the nodes), a CFD solver finds a solution (ρ, u, v, T, for example) which minimizes the residual of the equations to be solved. From a solution, the current

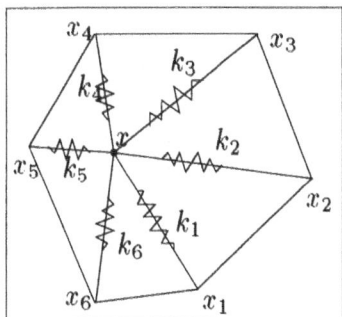

Figure 4. Node displacement

mesher finds a mesh which minimizes the edge error estimate. Adaptation
has been traditionally used to improve the mesh, but no one tried to reach
the convergence of the mesh. In fact, by letting the remesher converge, a
better solution is obtained because the more adapted the mesh is, the more
accurate the solution.

2.3. ADAPTIVE STRATEGIES

Particular attention has to be paid to the different criteria driving the adaptation process. For example, the refinement process converges in few sweeps
over all edges, and the same is true for mesh coarsening. But, alternating refinement and coarsening can loop indefinitely when the threshold value for
cutting an edge into two, and those for removing an edge to create greater
ones, are too close to each other. So all the criteria governing the local
operations must be set in such a way that the overall process converges.

After many tests, the following algorithm has been retained:

1. Smooth the mesh after estimating the error by alternatively

    (a) moving all the nodes iteratively

    (b) swapping all the edges until convergence.

2. Adapt the mesh by iterating the following loop

    (a) refine all edges above a threshold error estimate, then move the
        nodes,

    (b) remove all nodes whose edges have an error estimate below a
        threshold value, then move the nodes,

    (c) swap the edges until convergence, then move the nodes.

3. Finally, smooth the mesh by repeating loop 1 before solving the equations again, starting from an interpolated solution.

2.4. COUPLING WITH A SOLVER

The goal of the approach is not only to have a few improved meshes and solutions and get better results, but to converge the adaptive mesher and solver to an optimal solution, on an optimal mesh. This is done by coupling the solver with the mesher by looping in the following way:

Given (M_n, S_n), a mesh and a solution on this mesh at step n, the mesher produces a new mesh M_{n+1} and a solution $S_{n+1/2}$, the reinterpolation of S_n on M_{n+1}. A solution S_{n+1} on M_{n+1} is then obtained with the solver starting with $S_{n+1/2}$ as an initial guess. The iterations go over until convergence is reached.

A close coupling provides a maximum of flexibility to the mesh and permits to follow the evolution of the solution during the iterative resolution. This is done by frequent adaptation. In fact, it is useless to exactly solve the equations on an inadequate mesh. On the contrary, better results are obtained when the mesh and the solution converge in a coupled manner. Typically, about one hundred adaptive steps are used to reach a converged steady flow, starting from a uniform initial solution.

In this mesher/solver loop, only a mesh file and a solution file are exchanged between the mesher and the solver. Hence, the adaptive process is usable to improve the quality of solutions of a wide variety of problems using different discretization techniques.

3. VALIDATION

3.1. CONVERGENCE OF THE ADAPTIVE STEP

Algorithms to adapt a given mesh to a fixed error field (§ 2.3) may not be guaranteed to converge. All local modifications tend to equidistribute the edge error, but rigorously speaking not with the same significance. To get confidence in this method the convergence of the adaptation algorithm is itself checked, i.e. the last iterations must have a negligible effect on the mesh.

A test case is presented for which a mesh is adapted to a viscous laminar flow around a NACA 0012 profile with a freestream Mach number of 2.0 and a Reynolds number of $10,000$ (see Fig. 5). The adapted mesh has nothing in common with the initial one (Fig. 6) despite the fact that it is deduced from it using successive local alterations. The number of alterations is represented in Table 1 at each iteration of the overall loop and indicates that changes becomes negligible after 5 iterations.

Now that some convergence is demonstrated, the question is to characterize the converged state. A statistical graphic (Fig. 7) indicates that the edge error estimate is more equally distributed as adaptation proceeds. In

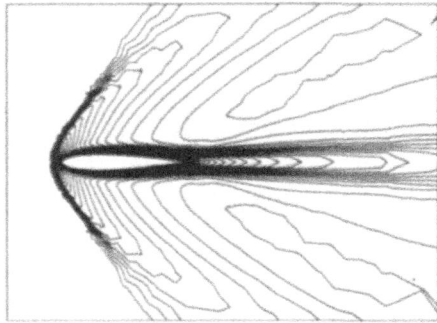

Figure 5. The error estimator was derived from second derivatives of the Mach number field shown here.

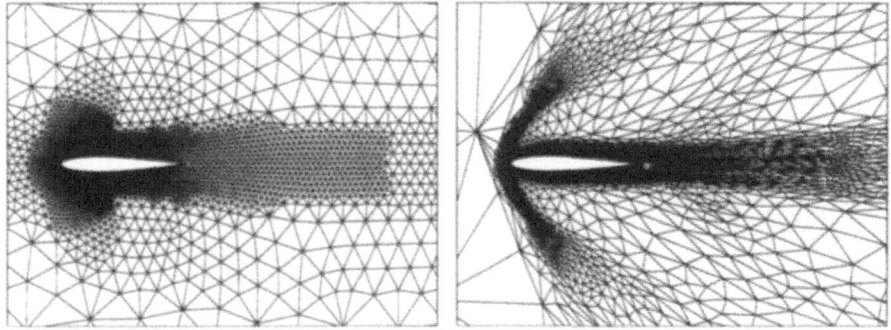

Figure 6. Initial mesh (also used as the background mesh to support the error estimator) and optimal adapted mesh.

particular, the ratio between the maximum and minimum values decreases from 5 000 to 3 (Fig. 8 left) and the standard deviation is reduced by an order of magnitude (Fig. 8 right).

Even if the convergence indicators are not monotonic (refinement results in a decrease of the minimum error, and coarsening in an increase of the maximum) the algorithm of § 2.3 converges towards a mesh that can be considered optimal for the flow conditions at hand.

3.2. CONVERGENCE OF THE OVERALL LOOP

Validating the coupling strategy is not an easy task. The topology of the meshes changes in a discrete way and no common metric can be used to quantify the convergence of the meshes. We will consider here the convergence of the number of nodes on the successive adapted meshes as an indication of the convergence of the coupled problem. In fact, the number of nodes of an adapted mesh is controlled by a target error. Since this target

TABLE 1. Convergence of local improvements during one
step of mesh adaptation.

refinement		coarsening		swapping	
# edges	%	# nodes	%	# edges	%
1671	15.32	2225	41.54	2697	24.73
721	7.77	290	7.53	102	1.10
62	0.59	69	1.90	45	0.43
25	0.24	25	0.70	17	0.16
19	0.18	14	0.39	14	0.13
9	0.09	13	0.36	11	0.10
10	0.10	10	0.28	8	0.08
8	0.08	8	0.22	10	0.10
12	0.11	10	0.28	2	0.02
7	0.07	5	0.14	4	0.04

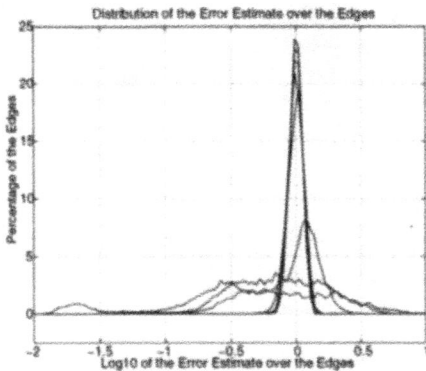

Figure 7. Distribution of the edge error over the mesh at different iterations of the mesh
adaptation process.

is kept fixed, the number of nodes is directly related to the global error of
the solution.

The same example as above is used to illustrate this point. Figure 9
shows the number of nodes of the meshes as the overall loop goes on. An
indication of convergence is the stagnation of the number of nodes after
enough remeshing steps. The meshes at steps A, B, C, and D marked on
Fig. 9 are presented in Fig. 10.

One surprising fact is the initial increase of nodes followed by a grad-
ual decrease to the asymptotic value. The few first meshes being not so
well adapted, the solutions on these meshes are polluted with spurious oscillations. An over-refinement of the meshes results at the beginning, just

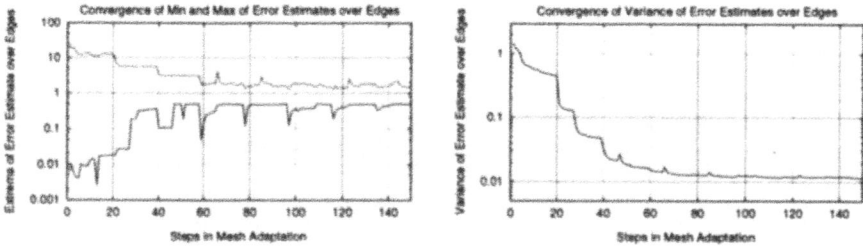

Figure 8. Evolution of the extremum values and the variance of the edge error during one step of mesh adaptation.

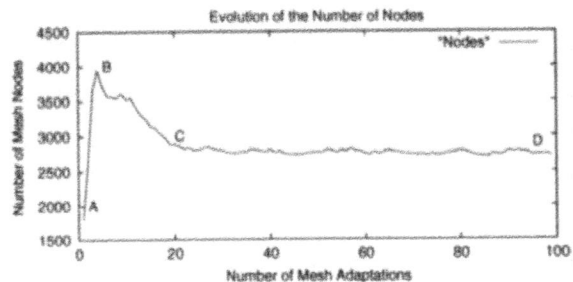

Figure 9. Total number of nodes versus adaptation steps for the flow over a NACA 0012 at $Ma = 2.0$ and $Re = 10\ 000$.

to correct the solutions and detect the salient features of the flow. When the solution improves, the mesher reduces gradually the number of nodes, bringing to bear all the techniques presented above.

4. APPLICATIONS TO EXTERNAL FLOWS

4.1. INVISCID FLOW (STRUCTURED GRIDS)

To validate the mesher/solver coupling loop on structured meshes, the AGARD03 test case of the AGARD Working Group 07, a subpanel of the AGARD Fluid Dynamics Panel, is computed. This inviscid test case consists of a NACA 0012 airfoil at a free stream Mach number of 0.95 and an angle of attack of 0°. This flow is characterized by a "fish tail" shock structure emanating from the trailing edge as shown on the right in Fig. 11. The downstream normal shock is relatively weak with a Mach number smaller than 1.1 ahead of the shock. A brief analysis indicates that its location is very sensitive to the grid.

It is clearly impossible to obtain an accurate solution with sharp and oblique shocks on the coarse initial grid (see Fig. 11 left). Figure 11 (middle) shows the same grid, same boundaries, same number of nodes and elements

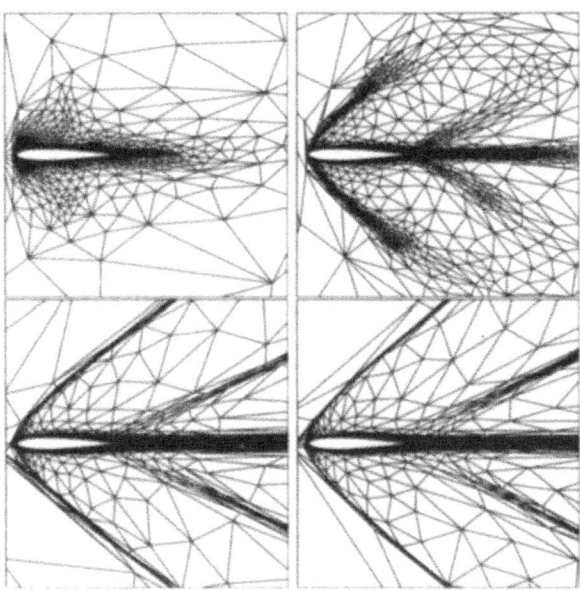

Figure 10.   Adapted meshes after one, 4, 20 and 100 iterations (left-right, top-bottom) of the coupling mesher/solver.

and the same connectivity, but adapted to the flow. The only method used to adapt the mesh in that case is node movement. Figure 11 (right) shows the Mach contours computed on the adapted grid.

Pulliam & Barton, carried out this test case with a 561×65 C-grid (4.5 times more grid nodes than present) on a 48 chord-length domain. Our result, drag = 0.1092, is in good accordance with their result, drag = 0.1103 (approximately 1% difference).

The pressure coefficient (Fig. 12) is also compared on the airfoil and behind the trailing edge on a horizontal cut. Here again, there is a good agreement between the Cp of Pulliam and Barton (cross) and the Cp computed on the adapted mesh (solid line). However, the weak normal shocks differ. As with inviscid flows, shocks should be perfect with infinite slope, the shock computed on the adapted grid has a more realistic shape than the one of Pulliam and Barton. The position of the weak normal shock is approximately 3.1 chord-lengths behind the trailing edge for the Pulliam and Barton solution and 3.37 chord-lengths on the adapted grid. These two compuations may be too coarse to decide which is the good one. With a 2049×765 O-grid (approximately 200 times more than our grid) in a 100 chord-length domain, Warren *et al.*, get a normal weak shock at 3.35 chord-lengths behind the trailing edge. This confirms the cost-effectiveness of mesh adaptation.

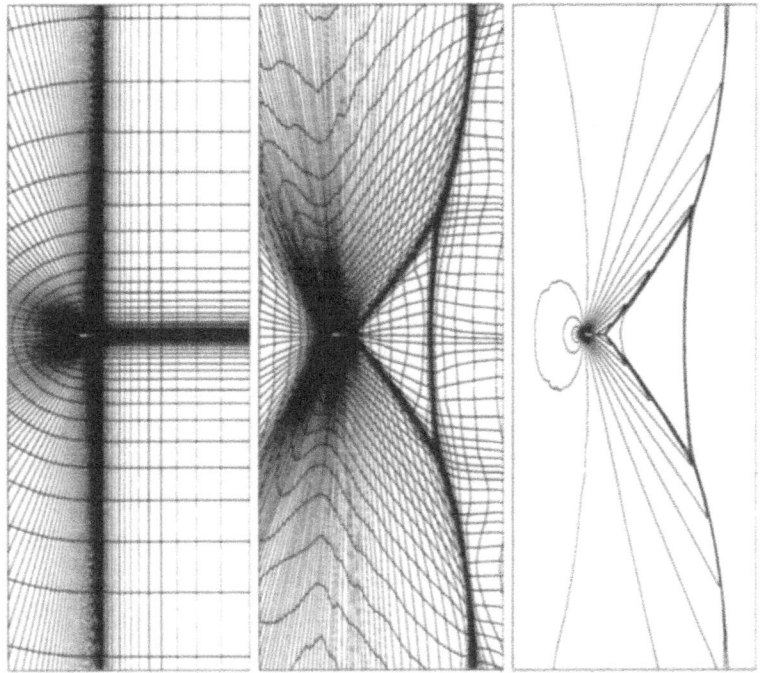

Figure 11. The same 200×40 C-grid in a 50 chord-lengths domain, before (left) and after (middle) mesh adaptation (30 mesher/solver iterations). Mach contours of the solution on the final grid (right).

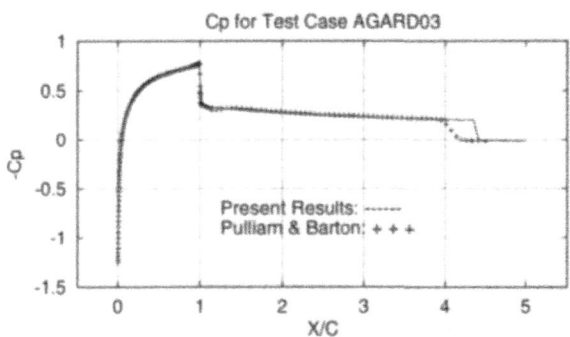

Figure 12. Pressure coefficient on the airfoil and behind the trailing edge. Comparison with the result of Pulliam & Barton.

4.2. VISCOUS FLOWS (UNSTRUCTURED MESHES)

One of the problems all CFD developers deal with is that of spurious oscillations. Generally speaking, oscillations may appear during the solution of flow equations for two different reasons.

First, the approximation scheme may be unstable. For example, while solving the Stokes or Navier-Stokes equations, an equal-order approximation of the velocity and pressure gives rise to spurious oscillations (see for example Brezzi & Fortin, Girault & Raviart and Pironneau). As long as the discretization scheme does not satisfy the inf-sup condition of stability, some terms must be added to the equations to eliminate oscillations and ensure convergence.

Oscillations can also appear in convection dominated flow computation, even with a stable approximation. They are mainly caused by the mesh which is too coarse with regard to the magnitude of convective terms. The simplest cure is to add a diffusive term into the equations. Artificial viscosity methods are often combined with oscillation detectors to add extra diffusion in the vicinity of oscillations and prevent excessive smearing of the solution. Upwinding methods, like SUPG and GLS, are more elegant because they dissipate only in the flow direction and in a consistent manner.

Mesh adaptation proves to be a good way to reduce the amount of artificial viscosity needed. More than that, we postulate that any artificial viscosity is bad viscosity and that none should be needed provided,

1. the discretization scheme is stable,
2. and the mesh can be adapted enough to capture physical features.

Of course, adaptation has its own limitations and cannot cure all spurious oscillations, but it has a stabilizing effect. This concept is nicely illustrated by the following example: a compressible viscous solution is computed on a NACA 0012 at a freestream Mach number of 2 and Reynolds numbers of 125, 250, 500, ..., 32 000.

The code uses primitive variables discretized on triangular meshes by the Bercovier-Pironneau element (Pironneau, 1988) i.e. temperature and density are element-wise linear while velocity components are linear on subelements obtained by splitting each triangle into four, from the mid-edges. It is well established that this element satisfies the stability condition, at least for the Stokes equations.

The mesh adaptation algorithm makes it possible to compute external viscous shocked flows with *absolutely no artificial viscosity*, at least upto a Reynolds number of 32 000. We did not try to increase any more the Reynolds number because the flow computation becomes more and more costly, due to unsteadiness of the wake at Re above 2 000. So, the limitation of stabilization by using only mesh adaptation was not reached in this test-case.

To show that the computed solutions contain the amount of viscosity prescribed by the physics, without any extra artificial viscosity, cuts are analyzed in the boundary layer and through the shock wave (Fig. 15 and 16, left). On Fig. 15 (right), it can be seen that the thickness of the boundary

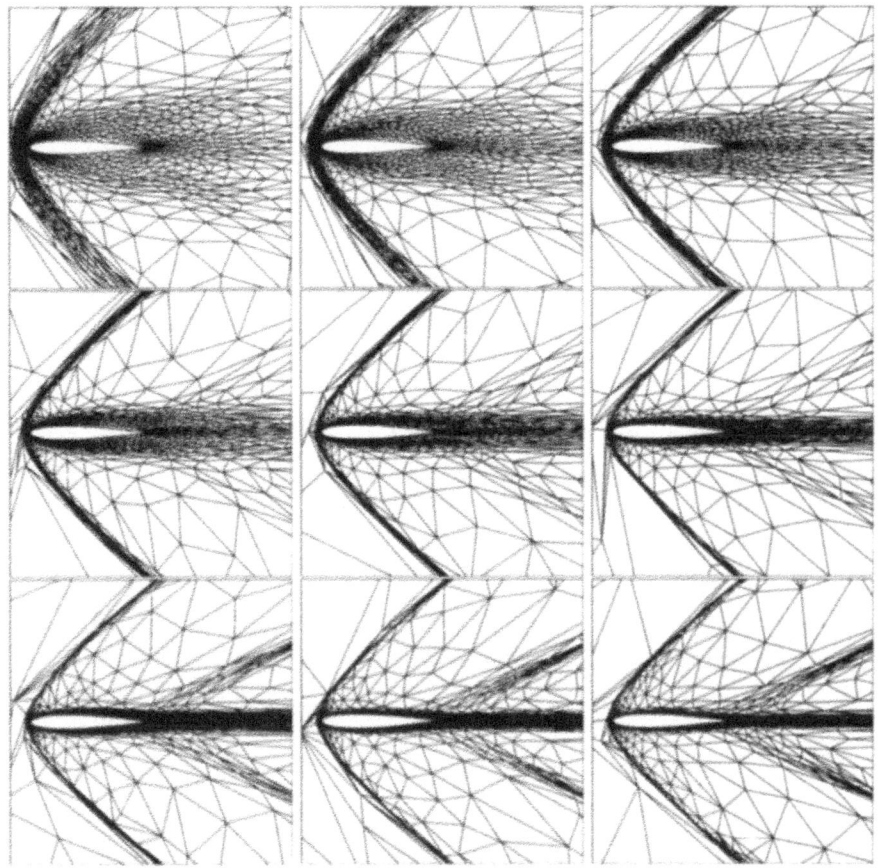

Figure 13.   Adapted meshes for flow at $Mach = 2.0$ and $Reynolds = 125, 250, 500,$ 1 000, 2 000, 4 000, 8 000, 16 000 and 32 000.

layer varies exactly as the square root of the Reynolds number and on Fig. 16 (right), the slope of the shock is exactly proportional to the Reynolds number. This is in perfect agreement with the laws of fluid mechanics and shows that, using adapted meshes, the numerical viscosity can be almost nil and the viscosity nearly equal to the physical one.

5. CONCLUSION

An anisotropic mesh adaptation method has been proposed and successfully implemented. The criterion is based on an estimate of the second derivatives of one of the field variables in the direction of the mesh edges. The adaptive step proceeds by successive local alterations of the previous mesh. The adaptive process and the coupling with the solver are both shown to

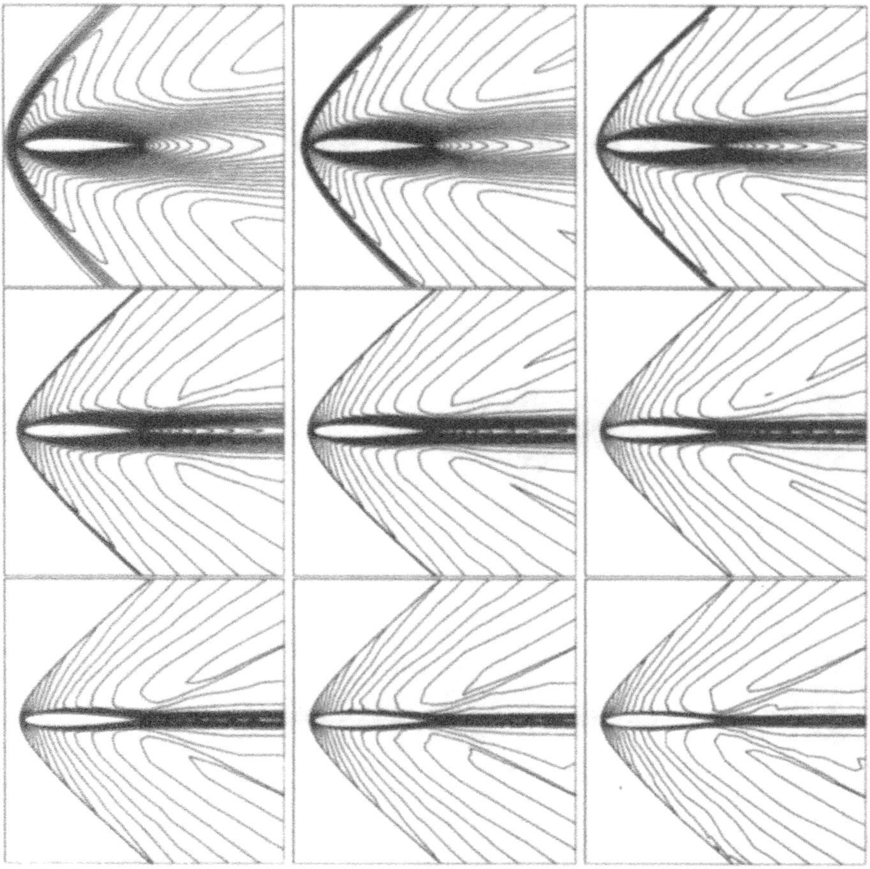

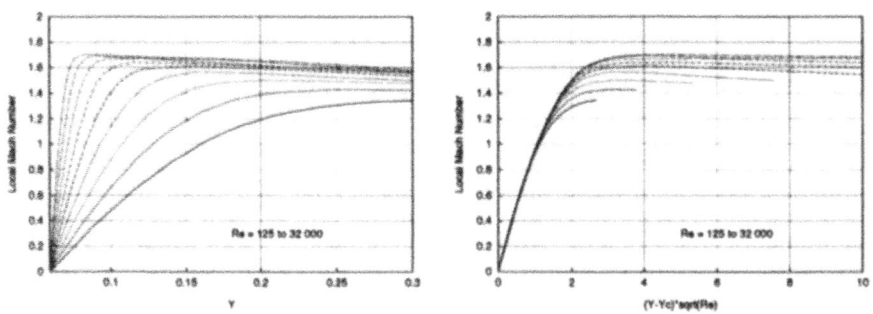

Figure 14.   Mach contours for flow at $Mach = 2.0$ and $Reynolds = 125, 250, 500, 1\,000,$ $2\,000, 4\,000, 8\,000.$ $16\,000$ and $32\,000$.

Figure 15.   Cut of the local Mach number through the boundary layer at $x = 0.3$ (left) and the same cut but non-dimensionalized (right).

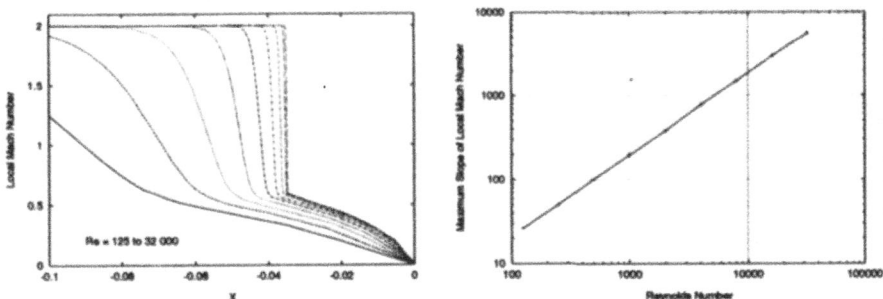

Figure 16. Cut of the local Mach number through the bow shock along $y = 0.0$ (left). Maximum slope of the local Mach number over the cut through the shock wave versus the prescribed Reynolds number (right).

converge.

Applications to laminar flows reveal two important benefits: the solutions are accurate even on coarse meshes, and the artificial viscosity can be reduced and often altogether eliminated.

The extension of the adaptive method to 3-D geometries has been done for structured grids (Sleiman *et al.*, 1996; Tam *et al.*, 1996). An unstructured tetrahedral mesh adaptation package is currently under development.

ACKNOWLEDGMENTS

The authors would like to thank NSERC and FCAR for Operating and Strategic grants and graduate scholarships under which this work was supported. Thanks are also due to Professor Sylvain Boivin of the Université du Québec à Chicoutimi for his permission to use the Navier-Stokes solver NS2D (Boivin, 1989; Boivin & Fortin, 1993; Boivin & Fortin, 1996) for the test case of § 4.2, and to Dr. Guido Baruzzi for the Euler solver (Baruzzi, 1995; Baruzzi *et al.*, 1995) included in the CFD code FENSAP2D from the CFD Lab for test case of § 4.1. The test case of § 3.1 has been computed using the code NSC2KE (Mohammadi, 1994; Mohammadi & Pironneau, 1994) developed at I.N.R.I.A. by Dr. Bijan Mohammadi and accessible through the WWW.

References

Ait-Ali-Yahia, D., Habashi, W.G., Tam, A., Vallet, M.-G., and Fortin, M., "A directionally adaptive methodology using an edge-based error estimate on quadrilateral grids," *Int. J. Num. Meth. Fluids*, Vol. 23, No. 7, 1996a, pp. 673-690.

Ait-Ali-Yahia, D. and Habashi, W.G., "A directionally-adaptive finite element method for high-speed flows," In *32nd AIAA/ASME/SAE/ASEE Joint Propulsion Conference*, AIAA-96-2553, Lake Buena Vista, FL, 1996b.

Baruzzi, G.S., "A Second Order Finite Element Method for the Solution of the Transonic Euler and Navier-Stokes Equations," Ph.D. thesis, Concordia University, Montréal, Qc, Canada, 1995.

Baruzzi, G.S., Habashi, W.G., Guèvremont, J.G., and Hafez, M.M., "A second order finite element method for the solution of the transonic Euler and Navier-Stokes equations," *Int. J. Num. Meth. Fluids*, Vol. 20, Nos. 8/9, 1995, pp. 671-693.

Boivin, S., "A numerical method for solving the compressible Navier-Stokes equations," *Impact Comp. Sc. Engng.*, Vol. 1, 1989, pp. 64-92.

Boivin, S. and Fortin, M., "A new artificial viscosity method for compressible viscous flow simulations by FEM," *Int. J. Comp. Fluid Dyn.*, Vol. 1, 1993, pp. 25-41.

Boivin, S. and Fortin, M., "A class of predictor-corrector schemes: Application to the resolution of the compressible Navier-Stokes equations by FEM," *Int. J. Num. Meth. Fluids*, 1996.

Brezzi, F. and Fortin, M., "Mixed and Hybrid Finite Element Methods," Springer Series in Computational Mathematics, Springer-Verlag, 1991.

Girault, V. and Raviart, P.-A., "Finite Element Methods for Navier-Stokes Equations: Theory and Algorithms," Springer-Verlag, 1980.

Gnoffo, P.A., "A vectorized finite-volume, adaptive grid algorithm for Navier-Stokes calculation," In *Numerical Grid Generation*, Thompson, J.F., ed., No. 9, New York, Elsevier North Holland, 1982, pp. 819-835.

Gnoffo, P.A., "A finite-volume, adaptive grid algorithm applied to planetary entry flowfields," *AIAA Journal*, Vol. 21, No. 9, 1983, pp. 1249-1254.

Mohammadi, B., "Fluid dynamics computation with NSC2KE, a user-guide, release 1.0," Technical Report RT-0164, Institut National de Recherche en Informatique et en Automatique, 1994.

Mohammadi, B. and Pironneau, O., "Analysis of the K-Epsilon Turbulence Model," Wiley & Sons and Masson, 1994.

Nakahashi, K. and Deiwert, G.S., "Three-dimensional adaptive grid method," *AIAA Journal*, Vol. 24, No. 6, 1985, pp. 948-954.

Nakahashi, K. and Deiwert, G.S., "Self-adaptive grid method with application to airfoil flow," *AIAA Journal*, Vol. 25, No. 4, 1987, pp. 513-520.

Palmerio, B., "An attraction-repulsion mesh adaption model for flow solution on unstructured grids," *Comp. and Fluids*, Vol. 23, No. 3, 1994, pp. 487-506.

Peraire, J., Vahdati, M., Morgan, K., and Zienkiewicz, O.C., "Adaptive remeshing for compressible flow computations," *J. Comp. Phys.*, Vol. 72, 1987, pp. 449-466.

Pironneau, O., "Méthode des éléments finis pour les fluides," Masson, 1988.

Pulliam, T.H. and Barton, J.T., "Euler computation of AGARD working group 07 airfoil test cases," In *AIAA 23rd Aerospace Sciences Meeting*, AIAA-85-0018, Reno, NV, 1985.

Sleiman, M., Tam, A., Robichaud, M.P., Peeters, M.F., Habashi, W.G., and Fortin, M., "Turbomachinery multistage simulation by a finite element adaptive approach," *41st ASME Gas Turbine and Aeroengine Congress*, ASME Paper 96-GT-418, Birmingham, U.K., 1996.

Tam, A., Habashi, W.G., Ait-Ali-Yahia, D., Robichaud, M.P., Sleiman, M., and Fortin, M., "A 3-D adaptive finite element method for turbomachinery," In *32nd AIAA/ASME/SAE/ASEE Joint Propulsion Conference*, AIAA-96-2659, Lake Buena Vista, FL, 1996.

Vallet, M.-G., "Génération de maillaes éléments finis anisotropes et adaptatifs," Ph.D. thesis, Université Pierre et Marie Curie, Paris VI, France, 1992.

Vallet, M.-G., Dompierre, J., Bourgault, Y., Fortin, M., and Habashi, W.G., "A directional error estimator for CFD," In *ASME Fluids Engineering Conference, FED-Vol. 238*, Vol. 3, San Diego, CA, 1996, pp. 209-215.

Warren, G.P., Anderson, W.K., Thomas, J.L., and Krist, S.L., "Grid convergence for adaptive methods," In *AIAA 29th Aerospace Sciences Meeting*, AIAA-91-1592-CP, Reno, NV, 1991.

Zienkiewicz, O.C. and Zhu, J.Z., "The superconvergent patch recovery and a posteriori error estimates. Part I: The recovery technique," *Int. J. Num. Meth. Engng.*, Vol. 33, No. 7, 1992, pp. 1331-1364.

ARTIFICIAL BOUNDARY CONDITIONS
FOR INFINITE-DOMAIN PROBLEMS*

SEMYON V. TSYNKOV[†]
NASA Langley Research Center
Mail Stop 128
Hampton, VA 23681–0001, U.S.A.

Abstract.
We present a new approach to constricting artificial boundary conditions for calculating three-dimensional external flows over finite bodies. The approach is based on application of the difference potentials method by V. S. Ryaben'kii and extends our previous technique developed for the two-dimensional case.

1. Introduction

General Remarks. External flows over finite bodies or configurations of bodies represent a wide class of important practical applications in fluid dynamics. To treat this type of problems numerically, one typically truncates the original infinite domain. The resulting truncated problem is obviously subdefinite unless supplemented by the proper closing procedure at the external computational boundary. The latter procedure is called the artificial boundary conditions (ABC's). In the ideal case, the ABC's would be specified so that the solution on the truncated domain coincides with the corresponding fragment of the original infinite-domain solution. The issue of ABC's is significant in CFD and in other areas of scientific computing; theoretical estimates and computational experiments by different authors show that the proper treatment of external boundaries has a profound impact on the overall performance of numerical algorithms and interpretation of the results.

---

*This paper was prepared while the author held a National Research Council Resident Research Associateship.
[†] s.v.tsynkov@larc.nasa.gov, http://fmad-www.larc.nasa.gov:80/~tsynkov/

V. Venkatakrishnan et al (eds.), Barriers and Challenges in Computational Fluid Dynamics, 119-137.
© 1998 *Kluwer Academic Publishers.*

Different ABC's methodologies have been studied extensively over the recent two decades. However, the construction of the ideal (i.e., exact) ABC's that would provide no error associated with the domain truncation and at the same time be computationally inexpensive, easy to implement, and geometrically universal, still remains a fairly remote possibility. Among the variety of approaches proposed to date only a very few can be regarded as the commonly used tools in CFD. As a rule, these approaches are based on the essential model simplifications (e.g., locally one-dimensional treatment near the external boundary) and therefore, often lack accuracy in computations. This, in turn, necessitates choosing the excessively large computational domains. On the other hand, these simple methods usually provide for local ABC's, and therefore, for cheap, geometrically universal, and algorithmically simple numerical procedures, which are attractive for practical use. There are, of course, methods of another kind, which typically provide for highly accurate and robust numerical algorithms. These methods, however, are not used routinely because the corresponding ABC's in most cases appear global. As a consequence, these ABC's may be relatively cumbersome and expensive; moreover, they can be derived easily only for the boundaries of regular shape. A survey of methods for setting the ABC's in different areas of scientific computing can be found in our recent work (Tsynkov, 1996a), as well as in the comprehensive reviews by Givoli (Givoli, 1991; Givoli, 1992).

In this paper, we concentrate on constructing the ABC's for the problems of computational aerodynamics. We, however, mention that this area constitutes a fraction of the possible range of applications for the different ABC's techniques. Besides the hydro- and aerodynamic problems (external flows, duct flows, boundary layers, free surfaces, etc.), the entire range includes the flows in porous media, filtration, MHD flows, plasma (e.g., solar wind), the problems of solid mechanics (in particular, elasticity and aeroelasticity), and the problems of wave propagation (electromagnetic, acoustic, seismic), just to name a few.

An Example of Exact Difference ABC's.   Before describing the actual flow problem, we provide here a simple model example that clarifies the nonlocal nature of exact ABC's. Consider a 2π-periodic in y formulation for the two-dimensional Yukawa equation driven by some compactly supported right-hand side:

$$\frac{\partial^2 u}{\partial x^2} + \frac{\partial^2 u}{\partial y^2} - \mu^2 u = f(x, y), \quad -\infty < x < \infty, \quad -\pi \leq y \leq \pi \qquad (1)$$

$$u(x, -\pi) = u(x, \pi)$$

$$u(x, y) \longrightarrow 0 \quad \text{as} \quad |x| \longrightarrow \infty$$

$$\text{supp} f(x, y) \subset \{(x, y) \mid x^2 + y^2 \leq 1\}$$

We discretize problem (1) on the Cartesian grid $(x_m, y_j) = (mh, -\pi + jh)$, $m = 0, \pm 1, \pm 2, \ldots, j = 0, \ldots, J - 1$, $J = 2\pi/h$, with the second order of accuracy using standard five-node stencil. Then, for $mh > 1$ the exact ABC at the linear artificial boundary $x = mh$ is given by

$$u_{m+1,j} = \sum_{l=0}^{J-1} \mathbf{T}_{j,l} u_{m,l}, \quad 0 \leq j \leq J - 1, \quad 0 \leq l \leq J - 1, \qquad (2)$$

where

$$\mathbf{T}_{j,l} = \frac{1}{J} \sum_{k=0}^{J-1} q_k^1 e^{ik(j-l)h}, \qquad (3)$$

and q_k^1 is the smaller root $(|q_k^1| < 1)$ of the equation

$$q^2 - 2(1 + 2\sin^2(kh/2) + \mu^2 h^2)q + 1 = 0. \qquad (4)$$

Boundary condition (2) is equivalent to the decay of the solution $u(x, y)$ as $x \longrightarrow \infty$; it was obtained by Fourier transforming problem (1) with respect to y and then selecting only the decaying mode (i.e., explicitly prohibiting the increasing one) for each wavenumber k. The boundary condition at $x = mh(mh < -1)$ that would be equivalent to the decay of $u(x, y)$ as $x \longrightarrow -\infty$ can be obtained in a similar way, it would require substituting the bigger root q_k^2 $(|q_k^2| > 1)$ of equation (4) into expression (3).

Boundary condition (2) is nonlocal, it is a matrix relation connecting the values of the solution at two consecutive grid rows along the entire artificial boundary. The nonlocality obviously originates from narrowing the variety of admissible solutions by selecting only some special (in this case decaying) modes in Fourier space. This situation is fairly general and arises in numerous applications (see (Tsynkov, 1996a)); analyzing the same example (1) on the PDE level we would have obtained the pseudodifferential relation connecting the solution $u(x, y)$ and its derivative with respect to x.

Main Objective and Approach. For external flow computations, our goal is to derive and practically implement the ABC's that would combine the advantages relevant to both global and local methodologies. Our approach

to constructing the ABC's is based on use of the finite-difference ana-
logues to Calderon's generalized potentials and boundary projections (see
the original work by Calderon (Calderon, 1963) and also work by Seeley (Seeley, 1966)). Ryaben'kii (see (Ryaben'kii, 1985; Ryaben'kii, 1987;
Ryaben'kii, 1996)) had modified the original construction and proposed a
numerical technique for the effective calculation of potentials and projections; this technique is known as the difference potentials method (DPM).

In work (Ryaben'kii & Tsynkov, 1995a) we describe the foundations
of the DPM-based approach to setting the ABC's for computation of
two-dimensional external viscous flows (Navier-Stokes equations). In work
(Tsynkov, 1995) we implement this approach along with the multigrid
Navier-Stokes algorithm by Swanson and Turkel (Swanson & Turkel, 1985;
Swanson & Turkel, 1987; Swanson & Turkel, 1996) and present some numerical results for subsonic and transonic laminar flows over single-element
airfoils. In work (Tsynkov et al., 1995) we show the results of subsequent numerical experiments and propose an approximate treatment of turbulence
in the far field. Our work (Ryaben'kii & Tsynkov, 1995b) delineates the
algorithm for solving one-dimensional systems of ordinary difference equations that arise when calculating difference potentials and the DPM-based
ABC's. In work (Tsynkov, 1996b) we extend the area of applications for
the DPM-based ABC's by analyzing two-dimensional flows that oscillate
in time; we also provide some solvability results for the linearized thin-
layer equations used for constructing the ABC's. In work (Ryaben'kii &
Tsynkov, 1996) we present a general survey of the DPM-based methodology
as applied to solving external problems in CFD, including parallel implementation of the algorithm, combined implementation of nonlocal ABC's
with multigrid, and entry-wise interpolation of the matrices of boundary
operators with respect to the Mach number and the angle of attack. Additionally, in work (Ryaben'kii & Tsynkov, 1996) one can find some new
theoretical results on the computation of generalized potentials, the construction of ABC's based on the direct implementation of boundary projections (thin-layer equations), and some numerical results for various airfoil
flows: laminar and turbulent, transonic and subsonic, including very low
Mach numbers (incompressible limit).

Below, we outline basic elements of the DPM-based ABC's for the case
of three-dimensional steady-state external viscous flows; this case is un-
doubtedly the one most demanded by the current computational practice.
Specifically, we address the flows around wing-shaped configurations. We
present some preliminary numerical results for the three-dimensional subsonic flow; for completeness we also reproduce some two-dimensional numerical results from our previous work. The results demonstrate clear superiority of the DPM-based ABC's over the standard existing methods.

2. External Flow Problem

Formulation.   We consider a steady-state flow of the viscous compressible perfect gas past a three-dimensional wing. The flow is uniform and subsonic at infinity; it is symmetric with respect to the Cartesian plane $z = 0$, which, in particular, implies that the free-stream velocity vector is parallel to this plane.

The flow equations are integrated on the grid generated around the wing. This grid actually defines the finite computational domain, the ABC's that would close the truncated problem should be set at the external co-ordinate surface of the grid. Let us designate this surface Γ; for one-block curvilinear boundary-fitted grid around ONERA M6 wing the schematic geometric setup is shown in Figure 1.

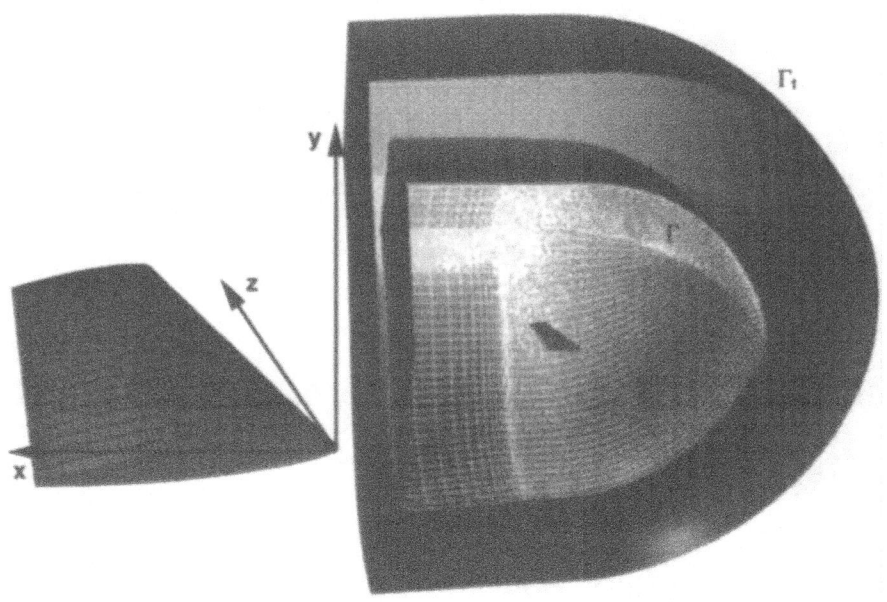

Figure 1.   Schematic geometric setup for the three-dimensional case. The wing on the left is enlarged.

The outermost coordinate surface of the grid is designated Γ_1 (see Figure 1); it represents the ghost nodes (or ghost cells for the finite-volume formulation). Clearly, when the stencil of the scheme used inside the computational domain is applied to any node from Γ, it generally requires some ghost cell data. Unless the required data are provided, the finite-difference system solved inside the computational domain appears subdefinite (i.e., it has fewer equations than unknowns). From the viewpoint of what the solution technique is, one can say that when some iterative solver is employed to integrate the flow equations inside the computational domain, the values of the solution at the ghost cells should be prescribed at each iteration in order to be able to advance the next "time" step.

Therefore, in the practical framework the closure of the discretized truncated problem means the specification of the solution values at the ghost cells. This will be done by means of the DPM-based ABC's so that the boundary data used for the closure admit an exterior complement that solves the problem outside the computational domain (see below).

First, we assume that the flow perturbations are small in the far field and linearize the problem outside the computational domain (i.e., outside Γ); we use the constant free-stream data as a background for the linearization. Clearly, for the very large computational domain the linearization is possible, as we approach the wing (the source of perturbations) the validity of linearization can always be verified a posteriori (see, e.g., our previous work (Tsynkov, 1995; Tsynkov et al., 1995; Ryaben'kii & Tsynkov, 1996)).

The linearized dimensionless thin-layer equations can be written as

$$\frac{\partial \rho}{\partial x} + \frac{\partial u}{\partial x} + \frac{\partial v}{\partial y} + \frac{\partial w}{\partial z} = 0, \tag{5a}$$

$$\frac{\partial u}{\partial x} + \frac{\partial p}{\partial x} - \frac{1}{Re}\left(\frac{\partial^2 u}{\partial y^2} + \frac{\partial^2 u}{\partial z^2}\right) = 0,$$

$$\frac{\partial v}{\partial x} + \frac{\partial p}{\partial y} - \frac{1}{Re}\left(\frac{4}{3}\frac{\partial^2 v}{\partial y^2} + \frac{\partial^2 v}{\partial z^2} + \frac{1}{3}\frac{\partial^2 w}{\partial y \partial z}\right) = 0,$$

$$\frac{\partial w}{\partial x} + \frac{\partial p}{\partial z} - \frac{1}{Re}\left(\frac{4}{3}\frac{\partial^2 w}{\partial z^2} + \frac{\partial^2 w}{\partial y^2} + \frac{1}{3}\frac{\partial^2 v}{\partial y \partial z}\right) = 0,$$

$$\frac{\partial p}{\partial x} - \frac{1}{M_0^2}\frac{\partial \rho}{\partial x} - \frac{\gamma}{Re\,Pr}\left[\left(\frac{\partial^2 p}{\partial y^2} + \frac{\partial^2 p}{\partial z^2}\right) - \frac{1}{\gamma M_0^2}\left(\frac{\partial^2 \rho}{\partial y^2} + \frac{\partial^2 \rho}{\partial z^2}\right)\right] = 0,$$

where ρ, u, v, w, and p are the perturbations of density, Cartesian velocity components, and pressure, respectively, M_0 is the free-stream Mach number, Re is the Reynolds number, Pr is the Prandtl number, and γ is the ratio of specific heats; the free-stream velocity is aligned with the positive x-direction. System (5a) is supplemented by the homogeneous boundary condition at infinity:

$$\mathbf{u} \equiv (\rho, u, v, w, p) \longrightarrow (0,0,0,0,0) \quad \text{as} \quad (x^2 + y^2 + z^2) \longrightarrow \infty, \tag{5b}$$

which corresponds to the free stream limit of the solution. Returning to the question of closing the discretized truncated system, we clarify that the boundary data provided by the DPM-based ABC's will admit an exterior complement that would solve the discrete counterpart of (5a) and meet boundary condition (5b) in a certain asymptotic sense.

DPM-based ABC's — Main Idea.   We discretize system (5a) on the auxiliary Cartesian grid with the second order of accuracy; we use first-order differences in x and second-order differences in y and z (see (Tsynkov, 1996b; Ryaben'kii & Tsynkov, 1996) for the details in two dimensions). The DPM will provide us with the *complete boundary classification* (in terms of the appropriate traces) of all those and only those exterior grid vector-functions that solve the discrete counterpart of (5a) outside the computational domain and satisfy boundary condition (5b) in some approximate sense. The foregoing boundary classification will be obtained as an image of the special projection operator, which can be considered as a discrete analogue of Calderon's pseudodifferential boundary projection. As we solve the Navier-Stokes equations inside the computational domain iteratively, every time we need to update the ghost cells we take a certain sufficient set of data from inside Γ (see below), project it onto the right manifold, i.e., onto the subspace in the space of boundary data that admit the correct exterior complement, and obtain the ghost cells values by calculating the trace of this complement on Γ_1.

Useful Analogy from the Theory of Cauchy Integral.   The following example (see (Ryaben'kii, 1985)) is not directly related to constructing the generalized potentials and projections for system (5a). It, however, provides a clear understanding of how the Calderon projection can be obtained for the most simple model case.

Let Γ be a smooth closed curve on the complex plane. Let $u(\zeta)$, $\zeta \in \Gamma$, be a complex-valued function satisfying Hölder condition. The Cauchy-type integral

$$P(z) = \frac{1}{2\pi i} \oint_\Gamma \frac{u(\zeta)}{\zeta - z} d\zeta \qquad (6)$$

has two limit values for $\zeta_0 \in \Gamma$ as we approach Γ from the different sides:

$$P^\pm(\zeta_0) = \text{p.v.} \frac{1}{2\pi i} \oint_\Gamma \frac{u(\zeta)}{\zeta - \zeta_0} d\zeta \pm \frac{1}{2} u(\zeta_0). \qquad (7)$$

Integral (6) becomes a Cauchy integral if and only if $P^-(\zeta) = 0$ (see (7)) for any $\zeta \in \Gamma$. This is a Sokhotskii-Plemelj condition, which can be rewritten as

$$u_\Gamma = \mathbf{P}_\Gamma u_\Gamma \;\Leftrightarrow\; u_\Gamma \in \mathrm{Im}\mathbf{P}_\Gamma, \qquad\qquad (8)$$

where

$$\mathbf{P}_\Gamma u_\Gamma \stackrel{def}{=} \left.\mathbf{P}u_\Gamma\right|_\Gamma \stackrel{def}{=} \left.\frac{\Theta_\Gamma}{2}u(z) + \mathrm{p.v.}\frac{1}{2\pi i}\oint_\Gamma \frac{u(\zeta)}{\zeta - z}d\zeta\right|_\Gamma. \qquad (9)$$

Θ_Γ in (9) is an indicator of the curve Γ. The operator $\mathbf{P}_\Gamma$ of (9) obviously satisfies $\mathbf{P}_\Gamma^2 = \mathbf{P}_\Gamma$, it can be regarded as Calderon's boundary projection for the Cauchy-Riemann system. Relation (8) selects all those and only those complex-valued functions on the curve Γ that are the traces of analytical functions on the domain.

DPM-based ABC's — Specific Implementation.   Let us split the nodes of the auxiliary Cartesian grid into two distinct groups: those that are inside Γ and those that are outside. Applying the stencil of the scheme for (5a) to each node of both groups, we consider the intersection of the grid sets swept by the stencil. This intersection is called the grid boundary γ, it is a multi-layered fringe of nodes of the auxiliary Cartesian grid located near Γ. Figure 2 shows an example of the grid boundary γ (several cross-sections in different directions).

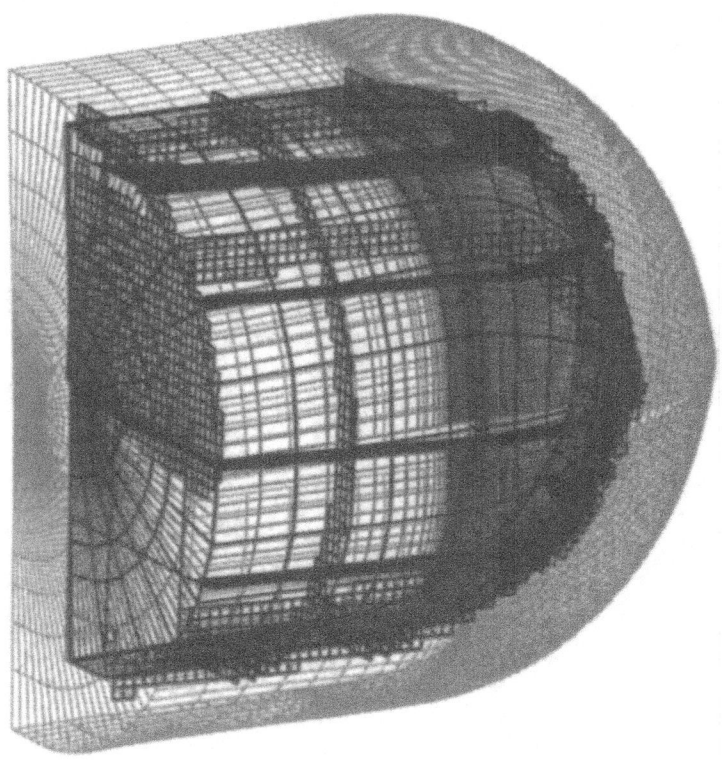

Figure 2. Grid sets for the three-dimensional case.

For any function $\mathbf{u}$ on the Cartesian grid we define its trace $\mathbf{Tr}_\gamma \mathbf{u}$ on γ as merely a contraction. For any grid function $\mathbf{u}_\gamma$ specified on γ we introduce the generalized potential $\mathbf{P}\,\mathbf{u}_\gamma$ with the density $\mathbf{u}_\gamma$, the generalized potential is defined on the auxiliary Cartesian grid on γ and outside it. The generalized potential is obtained as a solution of the special auxiliary problem (AP); solution of the AP replaces and extends the operation of convolution with the fundamental solution (compare to (6), (9)). The AP is driven by the right-hand side that depends on $\mathbf{u}_\gamma$, the formal construction of this right-hand side is the same in two- and three-dimensional cases and we refer the reader to our previous work (Ryaben'kii & Tsynkov, 1995a; Tsynkov, 1996b; Ryaben'kii & Tsynkov, 1996) for details. Boundary conditions for the AP should approximate boundary condition (5b) and at the same time ensure the finite-domain formulation for the AP. Therefore, we formulate the AP on a sufficiently large parallelepiped aligned with the Cartesian directions; the parallelepiped should fully contain Γ_1. We specify the periodicity boundary conditions in the y and z directions and also assume symmetry at $z = 0$. After the Fourier transform with respect to y and z the discrete counterpart to (5a) can be written as a family of one-dimensional difference equations:

$$\mathbf{A_k}\hat{\mathbf{u}}_{m,\mathbf{k}} + \mathbf{B_k}\hat{\mathbf{u}}_{m-1,\mathbf{k}} = \hat{\mathbf{f}}_{m-1/2,\mathbf{k}}, \tag{10}$$

$$m = 1,\ldots,M, \ \mathbf{k} \equiv (k_y, k_z),$$

$$k_y = 0,\ldots,J_y, \ k_z = 0,\ldots,J_z,$$

where $\mathbf{A_k}$ and $\mathbf{B_k}$ are the 5×5 matrices and $M + 1$, $2J_y + 1$, and $J_z + 1$ are the numbers of grid nodes in the x, y, and z directions, respectively (symmetry is taken into account, as well as the fact that $\mathbf{u}$ and $\mathbf{f}$ are real-valued). Boundary conditions in the x (streamwise) direction are specified separately for each pair of wavenumbers $\mathbf{k}$:

$$\mathbf{S}^-(\mathbf{k})\hat{\mathbf{u}}_{0,\mathbf{k}} = \mathbf{0}, \tag{11a}$$

$$\mathbf{S}^+(\mathbf{k})\hat{\mathbf{u}}_{M,\mathbf{k}} = \mathbf{0}, \tag{11b}$$

where

$$\mathbf{S}^-(\mathbf{k}) = \prod_{|\mu_s(\mathbf{k})|>1} (\mathbf{Q_k} - \mu_s(\mathbf{k})\mathbf{I}), \tag{12a}$$

$$\mathbf{S}^+(\mathbf{k}) = \prod_{|\mu_s(\mathbf{k})|\leq 1} (\mathbf{Q_k} - \mu_s(\mathbf{k})\mathbf{I}), \tag{12b}$$

$\mathbf{Q_k} = \mathbf{A_k}^{-1}\mathbf{B_k}$, and $\mu_s(\mathbf{k})$ are the eigenvalues of $\mathbf{Q_k}$. The semi-analytic boundary conditions (11a) and (11b) (the eigenvalues for (12) are calculated numerically) explicitly prohibit growing modes of the solution in the left and right directions, respectively, analogously to how it is done for the Yukawa equation example in Section 1, see (2). The periods in y and z should be chosen sufficiently large to ensure that the periodic solution considered near Γ and Γ_1 is sufficiently close to the theoretical non-periodic solution; the latter can be thought of as a limit when the periods approach infinity. The approximation of a non-periodic solution by the periodic one on a finite fixed interval as the period(s) increase(s) is discussed in our work (Ryaben'kii & Tsynkov, 1995a; Tsynkov, 1996b; Ryaben'kii & Tsynkov, 1996). In (Ryaben'kii & Tsynkov, 1996), we also discuss the possibility to replace the Fourier transforms by the non-unitary transforms. The latter appear when the grid in y or z is stretched (which provides for a drastic cost reduction) and the corresponding eigenfunctions consequently form a skew basis.

The foregoing AP allows us to calculate the generalized difference potential $\mathbf{P}\,\mathbf{u}_\gamma$ for any grid density $\mathbf{u}_\gamma$ specified on γ. The composition of the operators $\mathbf{Tr}_\gamma$ and $\mathbf{P}$, $\mathbf{P}_\gamma \equiv \mathbf{Tr}_\gamma\mathbf{P}$, is a projection, $\mathbf{P}_\gamma^2 = \mathbf{P}_\gamma$, it is a discrete counterpart of Calderon's boundary projection for system (5a). As in the preceding Cauchy-Riemann example, the image of this projection, $\mathrm{Im}\mathbf{P}_\gamma$, contains all those and only those $\mathbf{u}_\gamma$'s that are the traces of some exterior difference solution to (5a) that satisfies the boundary conditions of the AP (periodicity in y and z and boundary conditions (11) in x). These boundary conditions, in turn, approximate (5b).

Having constructed the procedure for calculating the potentials and projections for the discrete version of (5a), we can now close the system inside the computational domain, i.e., obtain the ABC's. First, we take $\mathbf{u}$ and $\partial\mathbf{u}/\partial n$ on Γ, n is the normal, (these data are available from inside the computational domain) and, using interpolation $\mathbf{R}_\Gamma$ along Γ and the first two terms of the Taylor expansion (denoted π_γ), obtain $\mathbf{u}_\gamma$:

$$\mathbf{u}_\gamma = \pi_\gamma \mathbf{R}_\Gamma \left(\mathbf{u}, \frac{\partial \mathbf{u}}{\partial n} \right)\bigg|_\Gamma. \tag{13}$$

Then, we need to calculate the potential $\mathbf{P}\,\mathbf{v}_\gamma$ for the density $\mathbf{v}_\gamma = \mathbf{P}_\gamma \mathbf{u}_\gamma$ and interpolate it to the nodes Γ_1:

$$\mathbf{u}\big|_{\Gamma_1} = \mathbf{R}_{\Gamma_1} \mathbf{P}\,\mathbf{v}_\gamma \equiv \mathbf{R}_{\Gamma_1} \mathbf{P}\,\mathbf{u}_\gamma. \tag{14}$$

Finally, the ABC's are obtained in the operator form

$$\mathbf{u}\big|_{\Gamma_1} = \mathbf{T} \left(\mathbf{u}, \frac{\partial \mathbf{u}}{\partial n} \right)\bigg|_\Gamma, \tag{15}$$

where $\mathbf{T}$ is composed of the operations (13) and (14). Boundary condition (15) is applied every time we need to update the ghost cells values in the course of the iteration process. The implementation of ABC's (15) can either be direct or involve preliminary calculation of the matrix $\mathbf{T}$. In the latter case, the runtime implementation of the ABC's (15) is reduced to a matrix-vector multiplication.

3. Numerical Experiments

Two-Dimensional Case.   Here, we reproduce some numerical results from our previous work (see, e.g., (Ryaben'kii & Tsynkov, 1996)) for two-dimensional flow past RAE2822 airfoil.

The computational domain is formed by the C-type curvilinear grid generated around the airfoil. On this grid, the Navier-Stokes equations are integrated using the code **FLOMG** by Swanson and Turkel (Swanson & Turkel, 1985; Swanson & Turkel, 1987; Swanson & Turkel, 1996). This code is based on the central-difference finite-volume discretization in space with the first- and third-order artificial dissipation. Pseudo-time iterations are used for obtaining the steady-state solution; the integration in time is done

by the five-stage Runge-Kutta algorithm (with the Courant number calculated locally) supplemented by the residual smoothing. For the purpose of
accelerating the convergence, the multigrid methodology is implemented;
in our computations we used three subsequent grid levels with W cycles;
the full multigrid preconditioning (FMG) could be employed as well. The
DPM-based ABC's were set only on the finest level of multigrid on the
final FMG stage, boundary data for the coarser levels were provided by the
coarsening procedure. The standard treatment of the external boundary
in the code FLOMG is based on the locally one-dimensional characteristics
analysis, which may or may not be supplemented by the point-vortex (p.-
v.) correction (Thomas & Salas, 1985); the corresponding results serve as
a reference point for the comparison.

In Table 1, we present the numerical results obtained with the two types
of ABC's for the supercritical flow around RAE2822 airfoil.

TABLE 1. RAE2822: $M_0 = 0.73$; $Re_0 = 6.5 \cdot 10^6$; $\alpha = 2.79°$; basic grid 640×128 nodes; normal grid spacing near the surface $0.5 \cdot 10^{-5}$.

Domain "radius"	2.5 chords		8 chords		50 chords	
Grid	600×104		608×112		640×128	
Type of ABC'	p.-v.	DPM	p.- v.	DPM	p.-v.	DPM
Wave lift, C_l	0.8653	0.8591	0.8624	0.8589	0.8603	0.8593
Relative error	0.58%	0.02%	0.24%	0.04%	0%	0%
Wave drag, $C_d \times 10$	0.1203	0.1263	0.1209	0.1261	0.1255	0.1260
Relative error	4.14%	0.24%	3.67%	0.08%	0%	0%
Full drag, $C_D \times 10$	0.1755	0.1816	0.1762	0.1815	0.1810	0.1815
Relative error	3.04%	0.05%	2.65%	0%	0%	0%

Each consecutive smaller grid in Table 1 is obtained by peeling off several
external rows of cells of the preceding bigger grid; this is done to eliminate
the influence that the grid change near the airfoil may exert on the solution.

From Table 1, one can see that the corresponding asymptotic values
of the force coefficients (those obtained for the 50 chords computational
domain) are very close to one another for both types of the ABC's. As,
however, the artificial boundary approaches the airfoil the discrepancy between the corresponding values increases: the force coefficients obtained on
the basis of boundary conditions (15) deviate from their asymptotic values
much slighter than the coefficients obtained using local ABC's do. In other
words, the nonlocal DPM-based ABC's allow one to use much smaller computational domains than the standard boundary conditions do and to still
maintain high accuracy of computations. Moreover, unlike the ABC's (15),
which perform well for all coefficients, the point-vortex boundary conditions perform much better for the lift than they do for the drag coefficient.

This behavior seems reasonable since the point-vortex model is a lift-based treatment.

In Table 2 we compare the results obtained with the two aforementioned types of ABC's on the different grids. One can see that boundary conditions (15) outperform the standard ABC's in this case as well.

TABLE 2. RAE2822: $M_0 = 0.73$; $Re_0 = 6.5 \cdot 10^6$; $\alpha = 2.79°$; normal grid spacing near the surface $0.5 \cdot 10^{-5}$.

Domain "radius"	2.5 chords		50 chords			
Grid	320×64		320×64		640×128	
Type of ABC's	p.-v.	DPM	p.-v.	DPM	p.-v.	DPM
Wave lift, C_l	0.8688	0.8560	0.8504	0.8492	0.8603	0.8593
Relative error	2.15%	0.38%	1.15%	1.17%	0%	0%
Wave drag, $C_d \times 10$	0.1123	0.1259	0.1260	0.1265	0.1255	0.1260
Relative error	10.5%	0.07%	0.40%	0.39%	0%	0%
Skin friction, $C_f \times 100$	0.5469	0.5492	0.5478	0.5480	0.5543	0.5544
Relative error	1.34%	0.94%	1.17%	1.15%	0%	0%
Full drag, $C_D \times 10$	0.1670	0.1808	0.1808	0.1814	0.1810	0.1815
Relative error	7.73%	0.39%	0.11%	0.05%	0%	0%

We also note that in certain cases the DPM-based ABC's may noticeably speed up (by up to a factor of three) the convergence of the multigrid iterations, see (Ryaben'kii & Tsynkov, 1995a; Tsynkov, 1995; Tsynkov et al., 1995). The discussion on combined implementation of the DPM-based ABC's with multigrid is contained in (Ryaben'kii & Tsynkov, 1996).

Three-Dimensional Case.   Here, we demonstrate some new results on computation of the external flows around ONERA M6 wing.

We use the code TLNS3D by Vatsa, et al. (Vatsa *et al.*, 1993) to integrate the thin-layer equations on the curvilinear grid (see Figures 1 and 2) generated around the wing. The type of the spatial discretization and the pseudo-time iteration mechanism in TLNS3D are similar to those in FLOMG. In addition, we use the preconditioning technique of (Turkel *et al.*, 1996) to improve the convergence to steady state. Unlike the two-dimensional case, here we implement the DPM-based ABC's only on the first and the last Runge-Kutta stages, which seems to make very little difference compared to the implementation on all five stages; the boundary data for the three intermediate stages are provided from the DPM-based ABC's on the first stage. Another important difference compared to the two-dimensional case is that the standard treatment of the external boundary in three dimensions is based merely on the locally one-dimensional characteristics analysis and extrapolation (the point-vortex model is not applicable).

In Table 3, we present the numerical results obtained with the two types of ABC's for the subcritical flow around ONERA M6 wing. As one can clearly see, the DPM-based ABC's in three space dimensions are capable of producing accurate results on the small computational domains, whereas the performance of the standard technique deteriorates as the domain shrinks.

TABLE 3.   ONERA M6: $M_0 = 0.5$; $Re_0 = 11.7 \cdot 10^6$; $\alpha = 3.06°$.

Domain "radius"	1.25 root chords		2 root chords		10 root chords	
Grid			$197 \times 49 \times 33$			
Type of ABC's	standard	DPM	standard	DPM	standard	DPM
Full lift, C_L	0.2218	0.2065	0.2185	0.2065	0.2081	0.2072
Relative error	6.58%	0.34%	5.0%	0.34%	0%	0%
Full drag, $C_D \times 100$	0.817	0.791	0.793	0.791	0.787	0.788
Relative error	3.8%	0.38%	0.76%	0.38%	0%	0%

Let us note that for this series of computations, the dimensions of all three grids are the same, smaller in size grids are obtained by scaling down the original 10 chords grid. The concentration of nodes near the wing that results from the down-scaling obviously contributes to the general improvement in the accuracy of the solution.

We have also conducted some three-dimensional computations for the standard transonic case: $M_0 = 0.84$, other parameters are as in Table 3. The corresponding results will be presented in a future paper. Here we only note that besides accuracy improvement the application of the DPM-based ABC's to the transonic flow resulted in a higher convergence rate (the rate of decrease of residual) for the small (3 chords) computational domain, as well as in a much faster convergence of other parameters, including those deemed as sensitive, e.g., the number of supersonic points in the domain.

Miscellaneous Issues.   Here, we will primarily discuss the computational cost of the DPM-based ABC's and possible ways for its reduction.

The cost of the DPM-based ABC's in two space dimensions is not high, the ABC's may add about 10% to the cost of the original integration procedure. In three space dimensions, the situation is so far less favorable, the DPM-based ABC's typically add from 25% to 30% of extra computer time. Since the three-dimensional work is currently on its initial stage, we will hopefully be able to find the proper algorithm changes that would reduce the aforementioned cost. Apart from the algorithm changes there are some "indirect" ways for reducing the cost of the DPM-based ABC's.

Parallel implementation of the ABC's on the multi-processor machines is a relatively simple task because the solution algorithm for the linear AP is easily parallelizable. For two space dimensions, an up to five times speedup compared to the single-processor version has been achieved on an eight-processor CRAY Y-MP.

In three space dimensions, carrying out multiple computations on the same grid is very often the case (many different flow regimes in one geometric setting). In so doing, the external geometry that influences the ABC's is fixed and the boundary conditions depend only on the aerodynamic parameters, e.g., Mach number M_0 and the angle of attack α. Therefore, the entry-wise interpolation of the matrices of boundary operators T with respect to these parameters has a substantial promise in this case. Indeed, after a noticeable startup expense for calculating the "reference" T's for some selected values of the parameters, any other matrix needed for any specific regime within the prescribed range of the parameters can be obtained for virtually no extra cost by means of the interpolation. This methodology has been tested for the two-dimensional case; the results, which seem quite encouraging, are summarized in Table 4.

TABLE 4. RAE2822: $M_0 = 0.73$; $Re_0 = 6.5 \cdot 10^6$; $\alpha = 2.79°$; normal grid spacing near the surface $0.5 \cdot 10^{-5}$. Entry-wise interpolation of the matrices of boundary operators T.

Domain "radius"		2.5 chords			50 chords	
Grid		600×104			640×128	
Type of ABC's	p.-v.	DPM			p.-v.	DPM
		genuine	Int. M_0	Int. α		
Wave lift, C_l	0.8653	0.8591	0.8593	0.8587	0.8603	0.8593
Relative error	0.58%	0.02%	0.0%	0.07%	0%	0%
Wave drag, $C_d \times 10$	0.1203	0.1263	0.1257	0.1252	0.1255	0.1260
Relative error	4.14%	0.24%	0.24%	0.6%	0%	0%
Full drag, $C_D \times 10$	0.1755	0.1816	0.1811	0.1805	0.1810	0.1815
Relative error	3.04%	0.05%	0.22%	0.55%	0%	0%

Finally, we mention that the nonlocality of the DPM-based ABC's is expressed by the fact that the matrix T is dense. However, this matrix is typically structured. From the standpoint of physics, the structure of T reflects the simple consideration that each specific node influences its close neighbors stronger than it influences the remote points. For system (5a), the operator T from (15) is composed of 5×5 blocks as shown in Figure 3a. The qualitative meaning of each block in Figure 3a is how one variable of the corresponding pair influences another one at the boundary. Stronger

$\rho - \rho$	u - ρ	v - ρ	w - ρ	p - ρ
$\rho - u$	u - u	v - u	w - u	p - u
$\rho - v$	u - v	v - v	w - v	p - v
$\rho - w$	u - w	v - w	w - w	w - p
$\rho - \rho$	u - p	v - p	w - p	p - p

Figure 3a. Qualitative block structure of the boundary operator **T** for three-dimensional flow computations.

interdependence between closer nodes results in the fact that the near-diagonal entries for each block are considerably larger than the off-diagonal ones. The latter observation has, in fact, been corroborated computationally (for the two-dimensional case).

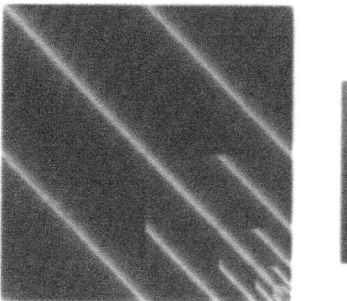

Figure 3b.   Multiresolution representation of the boundary operator for the two-dimensional Yukawa equation.

To effectively calculate matrix-vector products for the structured matrices of the foregoing type, one can use multiresolution-based data compression algorithms, e.g., the one proposed by Harten and Yad-Shalom in (Harten & Yad-Shalom, 1994). As an example, we apply this algorithm to the matrix **T** from (3). Of course, from the rigorous standpoint the operator **T** of (3) has not that much in common with the operator **T** that corresponds to system (5a) and is schematically represented in Figure 3a. Qualitatively, however, one can talk about some resemblance between the operator **T** of (3) and each block of the operator **T** in Figure 3a. Indeed, the resemblance is caused by the aforementioned physical reason: closer the two nodes are stronger their influence on one another is. The nested multiresolution representation (Harten & Yad-Shalom, 1994) (logarithmic scale) for the boundary operator $\mathbf{T}$ of (3), $J = 256$, is shown in Figure 3b.

We can see that all the nonzero entries are concentrated near the main diagonal of each nested block; as we recede the diagonal the values rapidly fall below the double-precision machine accuracy (10^{-13}) threshold. Therefore (Harten & Yad-Shalom, 1994), this nested structure can be used for the efficient calculation of matrix-vector products with consecutively improving resolution. The concentration of nonzero entries near the main diagonal also implies effective localization of the boundary conditions. This feature may have certain promise for the future use in the view of the multi-block grids.

4. Time-Dependent Problems.

Time-dependent problems are the next major challenge from the standpoint of treating numerically the infinite-domain formulations. As mentioned in the beginning, the issue of ABC's for the time-dependent problems is significant not only in CFD but also in other areas of scientific computing, e.g., those originating from acoustics and electrodynamics.

The primary difficulty that distinguishes the time-dependent problems from the steady-state ones is that here the exact ABC's are nonlocal not only in space but also in time (except in the simplest one-dimensional problems). As concerns the approximate approaches, the same basic trend that is relevant to the steady-state problems also holds for the time-dependent ones: higher the accuracy demands for the boundary procedure are more of the nonlocal nature of the ABC's should somehow be taken into account.

Of course, the nonlocality of the ABC's in time presents a very serious computational obstacle, because as the solution evolves such boundary conditions would become more and more expensive from the standpoints of both memory and computer time requirements. Therefore, the most crucial numerical issue in constructing the time-dependent ABC's is how to effectively restrict the nonlocality of the boundary conditions in time.

In this paper, we neither present our work on the time-dependent ABC's nor review thoroughly the approaches by other authors. We just briefly mention several recent developments that are promising in our opinion.

The particular case of time-periodic problems (in the framework of two-dimensional external flows) has been analyzed by Tsynkov in (Tsynkov, 1996b); the analysis is based on the Fourier transform in time, after which the problem is reduced to a family of the steady-state systems. This approach cannot be carried on to the general time-dependent problems in a straightforward manner.

Sofronov in (Sofronov, 1993; Sofronov, 1995; Sofronov, 1996) analyzes the wave equation outside a sphere and the Euler equations in a three-dimensional axially symmetric duct. His approach to the localization in time is based on the recursive calculation of convolutions with the special kernels (sums of exponents) that arise when representing boundary conditions is an certain integral form.

Grote and Keller in (Grote & Keller, 1995) also analyze the wave equation outside a sphere, their approach is based on reducing the problem to the one-dimensional wave equation after separating the variables in the spherical coordinates. Both this work and the aforementioned work by Sofronov have some geometric limitations.

Berenger in (Berenger, 1994) proposes the so-called perfectly matched layers for the Maxwell equations. Those are the layers of a special model medium that envelop the actual computational domain and rapidly damp the outgoing waves. The equations of this model medium have no direct physical meaning, which gives rise to certain concerns. In particular, Abarbanel and Gottlieb have recently shown in (Abarbanel & Gottlieb, 1997) that the perfectly matched layers by Berenger may yield weak instabilities.

Ryaben'kii in (Ryaben'kii, 1990) suggests exploiting the lacunas that exist in the solutions of the linear hyperbolic systems when the number of space dimensions is odd. Provided that the basic numerical method used for computations is capable of clear capturing these lacunas (i.e., clear capturing the rear fronts of the waves), this approach may give a key to constructing effective ABC's in many time-dependent problems.

References

Abarbanel, S. and Gottlieb, D., "A Mathematical Analysis of the PML Method," submitted for publication, 1997.

Berenger, J.-P., "A Perfectly Matched Layer for the Absorption of Electromagnetic Waves," *Journal of Computational Physics*, Vol. 114, 1994, pp. 185–200.

Calderon, A. P., "Boundary-Value Problems for Elliptic Equations," In *Proceedings of the Soviet-American Conference on Partial Differential Equations at Novosibirsk*, Fizmatgiz, Moscow, 1963, pp. 303–304.

Givoli, D., "Non-reflecting Boundary Conditions," *Journal of Computational Physics*, Vol. 94, 1991, pp. 1–29.

Givoli, D., *Numerical Methods for Problems in Infinite Domains*, Elsevier, Amsterdam, 1992.

Grote, M. J. and Keller, J. B., "Exact Nonreflecting Boundary Conditions for the Time Dependent Wave Equation," *SIAM Journal on Applied Mathematics*, Vol. 55, 1995, pp. 280–297.

Harten, A. and Yad-Shalom, I., "Fast Multiresolution Algorithms for Matrix-Vector Multiplication," *SIAM Journal on Numerical Analysis*, Vol. 31, 1994, pp. 1191–1218.

Ryaben'kii, V. S., "Boundary Equations with Projections," *Russian Mathematical Surveys*, Vol. 40, 1985, pp. 147–183.

Ryaben'kii, V. S., *Difference Potentials Method for Some Problems of Continuous Media Mechanics*, Nauka, Moscow, 1987 (in Russian).

Ryaben'kii, V. S., "Exact Transfer of Boundary Conditions," *Computer Mechanics of Solids*, Vol. 1, 1990, pp. 129–145 (in Russian).

Ryaben'kii, V. S., "Difference Potentials Method and its Applications," *Math. Nachr.*, Vol. 177, 1996, pp. 251–264.

Ryaben'kii, V. S. and Tsynkov, S. V., "Artificial Boundary Conditions for the Numerical Solution of External Viscous Flow Problems," *SIAM Journal on Numerical Analysis*, Vol. 32, 1995, pp. 1355–1389.

Ryaben'kii, V. S. and Tsynkov, S. V., "An Effective Numerical Technique for Solving a Special Class of Ordinary Difference Equations," *Applied Numerical Mathematics*, Vol. 18, 1995, pp. 489–501.

Ryaben'kii, V.S. and Tsynkov, S. V., "An Application of the Difference Potentials Method to Solving External Problems in CFD," to appear in *CFD Review 1996*, Hafez, M. and Oshima, K., eds.

Seeley, R. T., "Singular Integrals and Boundary Value Problems," *American Journal of Mathematics*, Vol. 88, 1966, pp. 781–809.

Sofronov, I. L., "Conditions for Complete Transparency on the Sphere for the Three-Dimensional Wave Equation," *Russian Academy of Sciences Doklady, Mathematics*, Vol. 46, 1993, pp. 397–401.

Sofronov, I. L., "Generation of 2D and 3D Artificial Boundary Conditions Transparent for Waves Outgoing to Infinity," University of Stuttgart, Mathematical Institute A, Preprint No. 96-9, 1996.

Sofronov, I. L., "Transparent Boundary Conditions for Unsteady Transonic Flow Problems in Wind Tunnel," University of Stuttgart, Mathematical Institute A, Preprint No. 95-21, 1995.

Swanson, R. C. and Turkel, E., "A Multistage Time-Stepping Scheme for the Navier-Stokes Equations," *AIAA Paper No. 85-0035*, January 1985.

Swanson, R. C. and Turkel, E., "Artificial Dissipation and Central Difference Schemes for the Euler and Navier-Stokes Equations," *AIAA Paper No. 87-1107*, June 1987.

Swanson, R. C. and Turkel, E., "Multistage Schemes with Multigrid for the Euler and Navier-Stokes Equations. Volume I: Components and Analysis," *NASA Technical Paper No. 3631*, Langley Research Center, 1996.

Thomas, J. L. and Salas, M. D., "Far-Field Boundary Conditions for Transonic Lifting Solutions to the Euler Equations," *AIAA Paper No. 85-0020*, 1985.

Tsynkov, S. V., "An Application of Nonlocal External Conditions to Viscous Flow Computations," *Journal of Computational Physics*, Vol. 116, 1995, pp. 212–225.

Tsynkov, S. V., "Artificial Boundary Conditions Based on the Difference Potentials Method," *NASA Technical Memorandum No. 110265*, Langley Research Center, July 1996a.

Tsynkov, S. V., "Artificial Boundary Conditions for Computation of Oscillating External Flows," to appear in *SIAM Journal on Scientific Computing*; also *NASA Technical Memorandum No. 4714*, Langley Research Center, August 1996b.

Tsynkov, S. V., Turkel, E., and Abarbanel, S., "External Flow Computations Using Global Boundary Conditions," *AIAA Journal*, Vol. 34, 1996, pp. 700–706; also *AIAA Paper No. 95-0564*, January 1995.

Turkel, E., Vatsa, V. N., and Radespiel, R., "Preconditioning Methods for Low-Speed Flows," *AIAA Paper No. 96-2460-CP*, June 1996.

Vatsa, V. N., Sanetrik, M. D., and Parlette, E. B., "Development of a Flexible and Efficient Multigrid-Based Multiblock Flow Solver," *AIAA Paper No. 93-0677*, January 1993.

ISSUES AND STRATEGIES FOR HYPERBOLIC PROBLEMS WITH STIFF SOURCE TERMS

M. ARORA AND P. L. ROE
W.M. Keck Foundation Laboratory
for Computational Fluid Dynamics
Department of Aerospace Engineering
University of Michigan
Ann Arbor, MI 48109-2118, USA

1. Introduction

Fluid flow becomes much more complicated when the physical, thermodynamic, or chemical state of the fluid is influenced by events in the flow. Typically, there is an equilibrium state toward which the fluid tends, and the governing equations have the form

$$\mathbf{u}_t + \mathbf{F}_x = \mathbf{Q}(\mathbf{u})/\tau, \tag{1}$$

where $\mathbf{u}$ is an extended set of variables incorporating additional degrees of freedom, $\mathbf{F}(\mathbf{u})$ is a flux function, $\mathbf{Q}(\mathbf{u})$ is a *source term* that vanishes in the equilibrium state, and τ is a *relaxation time*. In practice, there may be several such times and these could be of similar or disparate magnitudes. It is relatively straightforward to incorporate the source term(s) into standard methods for solving the homogeneous problem, unless τ happens to be much smaller than the residence time $\Delta x/u$ of a particle in a computational cell. The relaxation is then too fast to be followed accurately on the time scale of the flow calculation, and the problem is said to be *under-resolved* or *stiff*. Stiff problems frequently arise in Computational Fluid Dynamics when attempting to model rapid chemical reactions or strong departure from thermodynamic equilibrium, but as yet there has been little by way of systematic method development directed toward them.

   If we really need to see the detailed behaviour of the relaxation processes, there is probably no alternative to taking time steps that are of order τ (and spatial intervals of order $\lambda\tau$, where λ is a typical wavespeed). There may, however, be regions of the computation where relaxation is

139

unimportant, or only its asymptotic effects are required. We would then like to take timesteps limited only by the large-scale behaviour, but this could introduce large errors due to the source terms. A method for which this does not happen is said to be robust with respect to the stiffness.

In practice, reactive or relaxing flows are often computed using some form of the *operator splitting* procedure deriving from Yanenko (Yanenko, 1971) and Strang (Strang, 1968). Typically, solutions to the homogeneous problem ($\mathbf{Q} = 0$) are alternated with solutions to the ordinary differential equations ($\mathbf{F} = 0$). There are some philosophical objections to this procedure, especially if the homogeneous problems are solved using the modern high-resolution schemes that are often preferred when the source terms are truly absent, rather than being temporarily neglected. A strong objection is that the high-resolution schemes use Riemann solvers that assume 'frozen' wavespeeds, whereas the flow may actually, in many regions, be closer to equilibrium physics. The alternating processes may each be physically inappropriate. The loss of accuracy resulting from this procedure has been found in (Pember, 1993; Jin, 1995) and is confirmed below even in a case where the individual operators are solved exactly.

In Section 2 of this paper we exhibit and analyse what seems to be the simplest linear model to capture the essence of the difficulties; a rather complete analysis is possible. In Section 2.1 we review results from (Liu, 1987) and elaborate on these in the remainder of the Section. We give a rather full discussion here because we believe that the model clearly illuminates some issues of code design. In Section 3, we discuss how this linear problem might be treated using the method of characteristics. It is clear that any difficulties which remain after such drastic simplification will be very hard to dispel in a more complex setting. In particular, we find that the simple and popular strategy of operator splitting degenerates to very poor behaviour in the stiff limit, even though its component parts in this case are actually exact. Section 4 describes a strategy for nonlinear problems on noncharacteristic grids, motivated by but not closely reproducing, the analysis in Section 3. Finally, in Section 5, we show how the strategy performs on a simplified version of nonequilibrium gasdynamics, the Broadwell equations.

2. The Model Problem

A 2×2 problem of the type we wish to consider is (see (Liu, 1987))

$$u_t + v_x = 0, \tag{2}$$

$$v_t + a_F^2 u_x = -\frac{1}{\tau}(v - a_E u). \tag{3}$$

Eqn. (2) might stand for a set of conservation laws governing the overall mass, momentum and energy of a fluid comprising several phases, states or chemical components. Eqn (3) could stand for supplementary laws describing how the flux v is driven toward a limiting value equal to $a_E u$. In this section we will review the analytical properties of these equations.

For very high frequency waves, the differentiated terms on the left of (3) dominate the undifferentiated terms on the right, and we have in effect a homogeneous hyperbolic system with wave speeds $\pm a_F$. This is the *frozen flow limit*. For very low frequencies the domination is reversed, and the second equation enforces $v = a_E u$, so that (2) becomes

$$u_t + a_E u_x = 0, \tag{4}$$

a simple advection equation that holds in the *equilibrium limit*, sometimes called the *hydrodynamic limit*.

2.1. ASYMPTOTIC ANALYSIS

The near-equilibrium case can be treated by supposing that

$$v = a_E u + v_1,$$

where v_1 is small but not rapidly changing. It is not an early time transient, but a slowly decaying perturbation of the long-time equilibrium flow. Under these circumstances, differentiating does not change the orders of magnitude, and we have $v_t \simeq a_E u_t$, and hence from (2), $v_t \simeq -a_E v_x \simeq -a_E^2 u_x$. Inserting this into (3) gives

$$v_1 = -\tau (a_F^2 - a_E^2) u_x, \tag{5}$$

and so from (2) that

$$u_t + a_E u_x = \tau (a_F^2 - a_E^2) u_{xx}. \tag{6}$$

This is the near-equilibrium *asymptotic limit*, stable if $a_E^2 < a_F^2$.

2.2. GENERAL SOLUTION

The general case can be analysed by introducing a potential $\phi(x,t)$ such that $\phi_x = -u$, $\phi_t = v$. Then (2) is trivially satisfied, and (3) is satisfied if

$$\phi_{tt} - a_F^2 \phi_{xx} = -\frac{1}{\tau}(\phi_t + a_E \phi_x). \tag{7}$$

Since u, v are derivatives of ϕ they also obey this scalar equation.

2.2.1. *Dispersion Analysis*

A solution $\phi = \exp i(\omega t - \xi x)$ of (7) satisfies the *dispersion relationship*

$$\omega^2 - a_F^2 \xi^2 - \frac{i}{\tau}(\omega - a_E\xi) = 0. \tag{8}$$

To treat the initial-value problem we set ξ real and $\omega = \omega_R + i\omega_I$, where ω_R/ξ will be the propagation speed of the wave and ω_I/ξ the damping rate. Eliminating ω_R between the real and imaginary parts of (8) gives

$$4\,\tau^3\omega_I{}^4 - 8\,\tau^2\omega_I{}^3 + \left(5\,\tau + 4\,a_F{}^2\xi^2\tau^3\right)\omega_I{}^2$$
$$- \left(4\,a_F{}^2\xi^2\tau^2 + 1\right)\omega_I + (a_F^2 - a_E^2)\xi^2\tau \;=\; 0. \tag{9}$$

This equation has coefficients of alternating sign, and hence only positive roots, if $a_F^2 \geq a_E^2$ and $\tau \geq 0$, which is the condition for stability and gives positive dissipation in (6). If we introduce a dimensionless damping $\Omega = \tau\omega_I$, and a dimensionless wavenumber $\Xi = \tau a_F\xi$, this becomes

$$4\Omega^4 - 8\Omega^3 + (4\Xi^2 + 5)\Omega^2 - (4\Xi^2 + 1)\Omega + \Xi^2\left(1 - \frac{a_E^2}{a_F^2}\right) = 0. \tag{10}$$

We also introduce a dimensionless wavespeed $A = \omega_R/(\xi a_F)$ and find (by eliminating ω_I from (8)) that

$$4\Xi^2 A^4 - (4\Xi^2 - 1)A^2 - \frac{a_E^2}{a_F^2} = 0 \tag{11}$$

Plots of these quantities are given in Fig 1 for the cases $a_E = 0, a_E = a_F/2$. In general there are two distinct wave modes. If a_E is a substantial fraction of a_F, the wave whose phase speed has the opposite sign to a_E is more strongly damped, and the disappearance of this wave could be said to mark the onset of the asymptotic regime. In Fig 1 we indicate the asymptotic region with a letter 'a'. In this region, it can be determined that

$$A \;=\; \frac{a_E}{a_F}\left[1 + 2(1 - \frac{a_E^2}{a_F^2})\Xi^2 + \mathcal{O}(\Xi^4)\right], \tag{12}$$

$$\Omega \;=\; 1 - (1 - \frac{a_E^2}{a_F^2})\Xi^2 + \mathcal{O}(\Xi^4). \tag{13}$$

We would like to build this behaviour of the surviving wave into a numerical scheme. Probably, it will not be important what happens to the other wave at large times, so long as it is strongly damped.

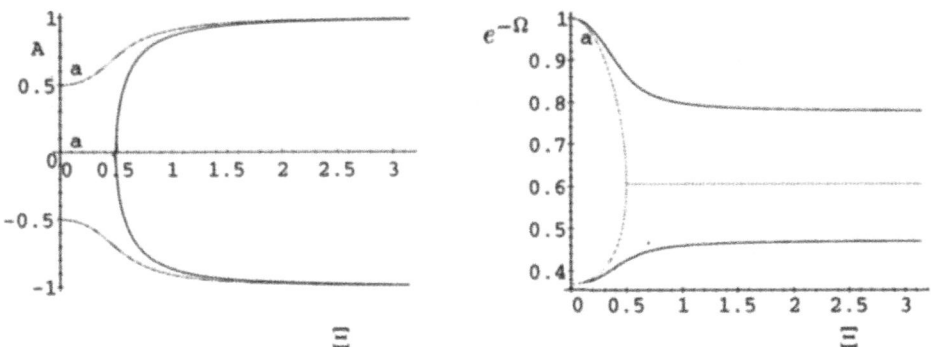

Figure 1. Dimensionless phase speed (left) and dissipation (right) versus dimensionless wavenumber for $a_E/a_F=0$ (these plots bifurcate at $\Xi = 0.5$) and for $a_E/a_F=0.5$.

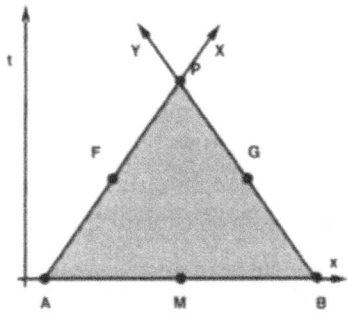

Figure 2. Characteristic coordinates, domain of dependence of P, and grid points used for characteristic stencil.

There are two interesting special cases. When a_E is small, both modes are moderately damped at high frequencies, but only one of them survives into the asymptotic regime. The case $a_E = 0$ (which has physical significance as the *Hyperbolic Heat Equation* (Cattaneo, 1958; Wiggert, 1977)) is included in Fig 1. The other case is if $|a_E| = a_F$, when it is easy to show that the wavespeeds are always $\pm a_F$ for all wavenumbers, so that $A = \pm 1$, and that Ω is always zero (for the wave whose speed has the same sign as a_E) or unity.

This analysis opens to question upwind schemes based on the frozen wavespeed a_F, since equilibrium waves may have a very different speed.

2.2.2. *Characteristic Analysis, Riemann's Function and the IVP*

Introduce coordinates X, Y such that $\partial_X = \partial_t + a_F \partial_x$ and $\partial_Y = \partial_t - a_F \partial_x$. (See Fig 2.) We obtain characteristic equations

$$a_F u_X + v_X = -\frac{1}{\tau}(v - a_E u), \quad a_F u_Y - v_Y = \frac{1}{\tau}(v - a_E u), \quad (14)$$

and the potential equation becomes

$$\phi_{XY} + a\phi_X + b\phi_Y = 0, \quad (15)$$

where

$$a = \frac{1}{2\tau}\left(1 + \frac{a_E}{a_F}\right), \quad b = \frac{1}{2\tau}\left(1 - \frac{a_E}{a_F}\right). \quad (16)$$

Eqn (15) is a special case of Riemann's differential equation (Courant & Hilbert, 1953) (pp 449-460), for which there is a rather general method of solution. One introduces a 'Riemann function', satisfying the *adjoint equation*,

$$R_{XY} - aR_X - bR_Y = 0, \quad (17)$$

and boundary conditions

$$R_X = bR \quad \text{on} \quad Y = 0,$$

$$R_Y = aR \quad \text{on} \quad X = 0.$$

It can be shown that the solution of this adjoint problem is

$$R = e^{(bX + aY)} I_0 \left(4abXY\right)^{\frac{1}{2}}. \quad (18)$$

An explicit solution to the initial boundary-value problem is then given by (see (Courant & Hilbert, 1953))

$$\phi_P = \frac{1}{2}[\phi_A R_A + \phi_B R_B] + \frac{1}{2a_F}\int_{AB} R\left(\frac{\phi}{\tau} + \phi_t\right) - R_t \phi \, dx. \quad (19)$$

Since u is also governed by (7), we can replace ϕ by u in (19) and then use (2) to eliminate u_t. The resulting solution for u is

$$u_P = \frac{1}{2}\left[e^{-bt}u_A + e^{-at}u_B\right] - \frac{1}{2a_F}\int_{AB} Rv_x \, dx + \frac{1}{2a_F}\int_{AB}\left(\frac{R}{\tau} - R_t\right) u \, dx \quad (20)$$

where we have also used the simple form of R along AP, BP. Similarly, if we solve for v_t using (3), then

$$v_P = \frac{1}{2}\left[e^{-bt}v_A + e^{-at}v_B\right] - \frac{a_F}{2}\int_{AB} Ru_x \, dx + \frac{a_F}{2}\int_{AB}\left(\frac{a_E Ru}{\tau} - R_t v\right) dx. \quad (21)$$

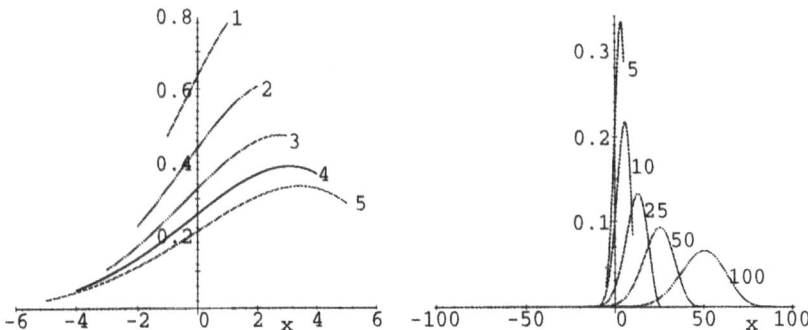

Figure 3.   The fundamental solution with $a_F = 1.0, a_E = 0.5$ at times $t/\tau = 1, 2, 3, 4,$
5 (left) and $t/\tau = 5, 10, 25, 50, 100$ (right).

The forms of these expressions are interesting; we see terms that clearly
represent propagation along the characteristics, and terms obtained by integration over the domain of dependence. These integrals can be thought
of as representing interaction between the two wave families.

2.3. A FUNDAMENTAL SOLUTION

The only difference between the potential equation (15) and the adjoint.
equation (17) is the sign of τ, so by reversing the signs of a, b in (18) we
have a solution of the potential equation,

$$\phi = e^{-(bX+aY)} I_0(4abXY)^{\frac{1}{2}}, \tag{22}$$

where it is understood that ϕ vanishes for $x^2 > a_F^2 t^2$. The evolution of this
solution is shown in Figure 3 for parameters $a_F = 1.0, a_E = 0.5$ at $t/\tau =$
1, 2, 3, 4, 5,10, 25, 50, 100. It can be regarded as solving an initial value
problem for which $\phi(x,0) = 0$, $\phi_t(x,0) = \delta(x)$. One can see that at large
times the solution approaches an advection-diffusion limit. Indeed, using
the asymptotic relationship $I_0(z) \simeq e^z/\sqrt{2\pi z}$, we find

$$\phi = \exp\left\{-\frac{(x - a_E t)^2}{4\tau a_F^2 t}\right\} \left[\frac{\pi^2 t^2}{\tau^2}\left(1 - \frac{a_E^2}{a_F^2}\right)\right]^{-\frac{1}{4}}. \tag{23}$$

We will now discuss several features of this solution that typify the
various regimes that arise in general.

2.3.1. *Evolution of the Fundamental Solution*
The first noticeable event in Fig 3 is the rapid disappearance of the left-
going shock. This actually decays like e^{-at} which is the decay rate for all

high-frequency left-going waves (see Fig 1(right)). The right-going shock also decays, but more slowly, like e^{-bt}. It is no longer visible at $t = 25$. Soon afterward, the solution becomes that of the typical advection diffusion equation (6). To within 1%, this has taken place by $t = 50\tau$. A robust scheme would reproduce accurately both the early and late time behaviour.

For numerical work involving shockwaves, it might only be important to ensure the correct wavespeed at large times, since the internal structure resulting from dissipation is not usually resolved anyway. A rarefaction wave, however, has its corners rounded by the effective diffusion (Jones, 1964). For this or other purposes it might be important to represent the diffusion accurately also. To achieve second-order accuracy in the asymptotic regime, the error in A must be $\mathcal{O}(\Xi^2)$, and the error in Ω must be $\mathcal{O}(\Xi^3)$.

3. Numerical Schemes for the Model

We will discuss in some detail two plausible ways to discretise the model equations on a grid composed of the frozen characteristics ($\Delta x = a_F \Delta t$). Without source terms, this is normally a very satisfactory way to solve 2×2 hyperbolic problems; in the linear case it is even exact. Difficulties that turn up in this context must be expected to recur in any more complex setting.

3.1. AN OPERATOR-SPLITTING SCHEME

In the commonly employed strategy of operator splitting (Yanenko, 1971; Strang, 1968), we take a half-time step with the o.d.e system

$$u_t = 0, \tag{24}$$

$$v_t = -\frac{1}{\tau}(v - a_E u), \tag{25}$$

(done here analytically), next a full step with the homogeneous equations

$$u_t + v_x = 0, \tag{26}$$

$$v_t + a_F^2 u_x = 0, \tag{27}$$

(another exact process on a characteristic grid) and then a second half-step with the o.d.e.s. The outcome of combining these stages on the stencil shown in Fig 2 is

$$u_P = \frac{1}{2}(u_A + u_B) + \frac{(1 - e^{-k})a_E}{2a_F}(u_A - u_B) + \frac{e^{-k}}{2a_F}(v_A - v_B) \tag{28}$$

$$v_P = \frac{e^{-2k}}{2}(v_A + v_B) + \frac{(1 - e^{-2k})a_E}{2}(u_A + u_B) \tag{29}$$

$$+ \frac{e^{-k}a_F^2 + (1 - e^{-k})^2 a_E^2}{2a_F}(u_A - u_B) + \frac{e^{-k}(1 - e^{-k})a_E}{2a_F}(v_A - v_B),$$

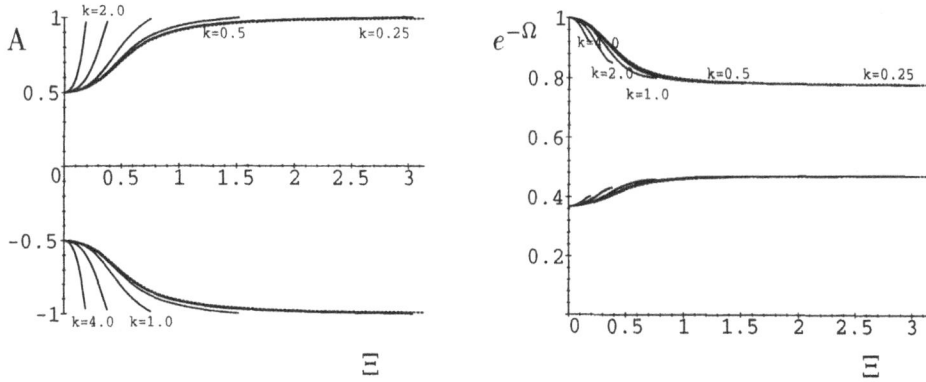

Figure 4. Dimensionless phase speed (left) and dissipation (right) versus dimensionless wavenumber for the operator splitting scheme $a_E/a_F=0.5$.

where $k = \Delta t/(2\tau)$ is a measure of the *numerical stiffness*.

Let the discrete solution be defined by

$$u(j\Delta x, n\Delta t) \;=\; Z^n \exp ij\xi\Delta x \;=\; Z^n \exp 2ijk\Xi. \qquad (30)$$

We then find a surprisingly simple *discrete dispersion relationship*

$$Z^2 - (1 + e^{-2k}) \cos(2k\Xi)Z + e^{-2k} - \frac{ia_E}{a_F}(1 - e^{-2k}) \sin(2k\Xi) = 0, \qquad (31)$$

which is plotted in Fig 4. It is of course no longer a function only of the dimensionless wavenumber Ξ but also of the stiffness parameter k. To any finite value of k corresponds a maximum wavenumber resolvable on that grid. These are the '$2\Delta x$' waves, for which $\Xi = \pi/(4k)$. A *well-resolved wave* is one for which Ξ is in some sense small compared with this limit. Provided also that k is small (say $k < 0.25$) we see that the discrete and analytical dispersion relationships match well. For larger k, there are serious errors in both the phase speed and dissipation. This implies that the waves need to be well resolved both in space and time.

At any k, the correct wavespeed is recovered in the equilibrium limit $\Xi \to 0$, but the '$2\Delta x$' waves always propagate with the frozen speeds, even if the grid is so coarse ($k \gg 1, \Xi_{max} \ll 1$) that they ought to be close to equilibrium behaviour. The behaviour of the scheme for large k can be understood by neglecting all of the exponentially small terms (e^{-k}) in (28,29), which then simplify to

$$u_P \;=\; \frac{1}{2}(u_A + u_B) + \frac{a_E}{2a_F}(u_A - u_B), \qquad (32)$$

$$v_P \;=\; a_E u_P. \qquad (33)$$

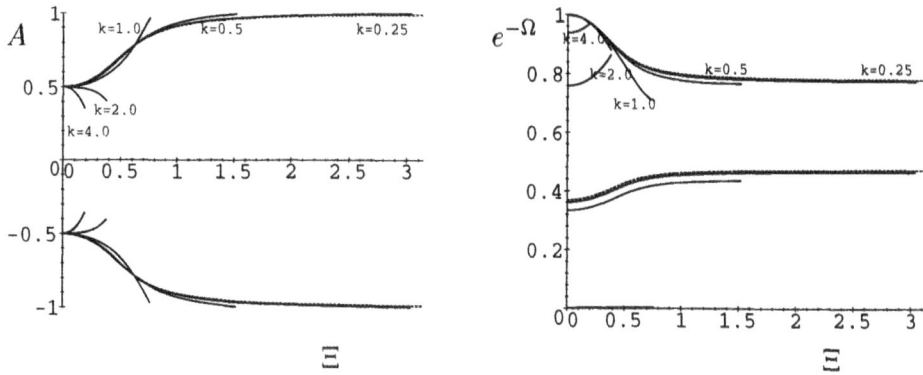

Figure 5. Dimensionless phase speed (left) and dissipation (right) versus dimensionless wavenumber for the point implicit scheme $a_E/a_F = 0.5$. Note in the dissipation plot how the upper and lower branches merge when k is large and Ξ is small.

Thus, the solution is forced into exact equilibrium, and the advection equation (4) for u is solved by the first-order Lax-Friedrichs scheme (since in this limit a_E/a_F can be identified with a Courant number). Recall that this scheme contains the most dissipation that can be added to any advection problem without creating instability. Thus, in the equilibrium limit, the operator-splitting scheme will generate solutions in which the dissipation is completely dominated by numerical effects, rather than the physical dissipation described by (6). It can also be expected to generate very dispersed shockwaves. Similar conclusions regarding the equilibrium behaviour of operator-splitting schemes were reached by Pember (Pember, 1993) and Jin (Jin, 1995).

Moreover, even the enforcement of equilibrium is something of an illusion. If the initial data actually is in equilibrium ($v_A = a_E u_A, v_B = a_E u_B$), the first half step of the ordinary differential equation will not disturb this, but the solution of the homogeneous problem will. Equilibrium has to be reimposed by the second ODE step. Thus the clean dispersion relation of the overall process actually conceals a strongly oscillatory behaviour that cannot be trusted to balance out so well in more complex problems.

3.2. A POINT-IMPLICIT SCHEME

Consider the characteristic equations (14). Each may be discretized by taking a simple difference on the LHS and evaluating the RHS with weights μ at P and $(1 - \mu)$ at A. Because each equation contains both u and v at the new time level, a small 2×2 set of algebraic equations must be solved. Therefore, we call this method *point implicit*. Here it is simple to just solve the equations and make the scheme explicit, as (taking $\mu = \frac{1}{2}$)

$$u_P = \frac{1}{2}(u_A + u_B) + \frac{ka_E}{2a_F}(u_A - u_B) + \frac{1-k}{2a_F}(v_A - v_B), \qquad (34)$$

$$v_P = \frac{1-k}{2(1+k)}(v_A + v_B) + \frac{ka_E}{1+k}(u_A + u_B)$$
$$+ \frac{a_F^2 + k^2 a_E^2}{2(1+k)a_F}(u_A - u_B) + \frac{k(1-k)a_E}{2(1+k)a_F}(v_A - v_B). \qquad (35)$$

This method was applied to practical problems of hyperbolic heat conduction in (Wiggert, 1977). The dispersion relationship is now

$$Z^2 - \frac{2}{1+k}\cos(2k\Xi)Z + \frac{1-k}{1+k} - \frac{2ika_E}{a_F(1+k)}\sin(2k\Xi)Z = 0. \qquad (36)$$

This is plotted in Fig 5 and is quite different from the results for operator splitting. For example, it is no longer true that '$2\Delta x$' waves travel at the frozen speeds. Their speeds are still wrong, obviously, but much closer to the proper ones. The other notable difference is that the dispersion in the asymptotic region is much more accurate. In fact, very remarkably, it can be shown that the dispersion relationship for small Ξ matches the exact relationship (10) with error $O(\Xi^4)$ for any k. That is, the point implicit method remains uniformly second-order accurate, for any degree of stiffness, for the solution mode that should survive.

This is the good news. However, the mode that should be damped out in the equilibrium limit shows no dissipation whatsoever ($\Omega(\Xi, k) \to 1$ for both waves). This is due to the factor $(1-k)/(1+k)$ in the first term of (35) which is a terrible approximation to e^{-2k} if k is large. Another dubious feature is that several coefficients of the scheme (34,35) become large if k is large. So the good results depend on cancellation between large terms, which is again unlikely to happen for more realistic problems.

For this scheme some intriguing simplifications occur when $k = 1$. The update of u is again done by a Lax-Friedrichs scheme, but this time the inherent dissipation in this happens to be physically correct.

3.3. OTHER METHODS

Numerically, a finite difference scheme of arbitrarily high order could be obtained by reconstructing $u(x), v(x)$ within AB using neighbouring information and then carrying out the integration of the exact formulae (20,21). To avoid numerical integration, we would like to integrate analytically the product of the Riemann function or its time-derivative with arbitrary polynomials. In (Roe & Arora, 1993), the formulae were given for the case $a_E = 0$.

The simplest such method would involve linear interpolation on the interval AB. It seems intuitive that such a method ought to be optimal within the class of all schemes predicting the solution at P from data at A, B and therefore superior to the two schemes just discussed. We have not repeated the analysis for $a_E \neq 0$, but present below the formula for $a_E = 0$.

$$u_P = \frac{1}{2}(u_A + u_B) + \frac{1 - e^{-2k}}{4ka_F}(v_A - v_B) \qquad (37)$$

$$v_P = \frac{e^{-2k}}{2}(v_A + v_B) + \frac{a_F(1 - e^{-2k})}{4k}(u_A - u_B) \qquad (38)$$

Surprisingly, we found that this was not such a good method. Either by examining the dispersion relationship or by numerical testing it came out very similar to the operator-splitting method. A partial explanation seems to be that under-resolved waves are poorly represented by a linear variation, and that any scheme that succeeds in evolving them accurately (to whatever extent) from such a coarse approximation, must owe a good deal to dumb luck. A much better (third-order) scheme resulted from fitting a parabola to the data at A, M, B and then taking the integrals, but that scheme was too complex to form a suitable basis for further work. We formed the impression that some of the improvement from using the data at M, apart from the mathematical gain of one order in accuracy, was due to propagating information along paths other than the frozen characteristics.

Without claiming to have exhausted all the possibilities, we have devised a compromise scheme that seems fairly robust. In the point implicit scheme we weight the two integration points by chosing a value of μ that depends on k in such a way that we move from $\mu = \frac{1}{2}$ (most accurate) for small k, to $\mu = 1$ (most robust) for large k. A semi-empirical choice (Arora, 1996) is

$$\mu = \frac{1}{1 - e^{-2k}} - \frac{1}{2k}.$$

Although degrading the accuracy of the point-implicit method on the model problem, this 'PI$_\mu$' scheme remained superior to the operator-splitting method and proved distinctly more robust in non-linear settings (Arora, 1996).

4. A MUSCL-type Algorithm

Consider a finite-volume scheme on the grid shown in Fig 6. The update formula will be

$$U_j^{n+1} - U_j^n = -\frac{\Delta t}{\Delta x}\left[F_{j+\frac{1}{2}}^{n+\frac{1}{2}} - F_{j-\frac{1}{2}}^{n+\frac{1}{2}}\right] + \Delta t Q_j^{n+\frac{1}{2}} \qquad (39)$$

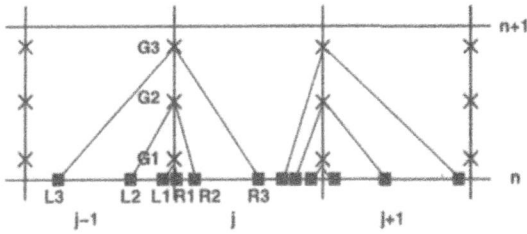

Figure 6. Part of a one-dimensional finite-volume mesh, showing the Gauss points on the interface and the characteristics traced back from them.

where $\mathbf{F}_{j+\frac{1}{2}}^{n+\frac{1}{2}}$ approximates the average flux along interface $j+\frac{1}{2}$ and $\mathbf{Q}_j^{n+\frac{1}{2}}$ approximates the mean source within the cell j. Our strategy for evaluating these terms is as follows.

Along the interface, some components of the solution will vary strongly, perhaps with rapid exponential decay. To capture this, we integrate as accurately as possible, using Gaussian quadrature. After some experimentation we settled on three-point quadrature at the points G_1, G_2, G_3 defined by $(n + \frac{1}{2})\Delta t, (n + \frac{1}{2}(1 \pm \sqrt{\frac{3}{5}}))\Delta t$ with weights $w_{1,3} = \frac{5}{18}, w_2 = \frac{4}{9}$. The flux at each G_k was determined by tracing back the characteristics to points L_k, R_k in the initial data. We used the PPM reconstruction (Colella & Woodward, 1984) to define the data at these points. The solution at G_k was found from the data at L_k, R_k by applying the PI_μ scheme. If there were more than two independent variables we drew as many frozen characteristics as required and applied the PI_μ scheme to all of them. It was found that using the point-implicit scheme was prone to instability, and the operator-splitting approach created additional diffusion. These observations are consistent with the analysis of the model problem. The source term $\mathbf{Q}_j^{n+\frac{1}{2}}$ was evaluated for simplicity at level $n + 1$, in a point implicit manner. For further details see (Arora, 1996)

5. An Application

The Broadwell equations describe a discrete-velocity gas whose particles can move only with speeds $0, \pm 1$. They are commonly used to study aspects of the transition from particle behaviour to fluid behaviour. In terms of density ρ, momentum m, and 'pressure' z, the equations are

$$\rho_t + m_x = 0, \tag{40}$$

$$m_t + z_x = 0, \tag{41}$$

$$z_t + m_x = \frac{m^2 + \rho^2 - 2\rho z}{2\tau}. \tag{42}$$

In the 'hydrodynamic limit', $z = \frac{1}{2}(m^2/\rho + \rho)$ and (41) takes on a fairly familiar appearance as a momentum equation. We have computed a problem studied also in (Jin, 1995), with initial data $(\rho, m, z)_L = (1, 1, 1)$, $(\rho, m, z)_R = (1, 0.13962, 1)$ which should evolve at large times into a right-moving shock. At early times, there will be waves moving with the speed of the frozen characteristics, $dx/dt = 0, \pm 1$. To reveal the transitions, we compute to a single final time $(t = 0.5)$, but with three different relaxation times. The results in Fig 7 (top left) assume $\tau = 1$ and test the capability of the algorithm in the frozen limit. The results at top right assume $\tau = 10^{-8}$ and represent the equilibrium limit. Those at bottom left assume $\tau = 10^{-2}$. In comparison with the fully relaxed results they show a more diffuse shock and some residual disturbances near the initial discontinuity at $x = 0.2$. Finally, at lower right, we show results for different data $\{(\rho, m, z)_L = (1, 0, 1), (\rho, m, z)_R = (0.2, 0, 1)\}$, assuming $\tau = 10^{-8}$. For this data the fully relaxed solution features a left-moving rarefaction as well as a right-moving shock.

All four solutions appear satisfactory, although we have no exact solutions against which to compare, and the discontinuities are not quite so crisp as conventional high-resolution schemes would provide. In (Arora, 1996), the method was also applied to a variety of linear and non-linear test problems taken from (Pember, 1993; Jin, 1995; Jin & Levermore, 1996), to the Euler equations for flow in a conducting pipe, and to a 10-moment model of a relaxing gas.

6. Conclusions

We feel that code design for stiff problems presents a number of subtle challenges and are sceptical of the theoretical foundations on which current practical codes are based. We especially distrust operator splitting, but make no claim to have discovered the best alternative strategy. We have found that analysis of a simple model problem is revealing, but suspect that additional information remains to be mined from it. Rational design of stiff solvers for PDEs is still in its infancy, and although our infant seems to be as robust as anyone else's, we cannot prove its superiority. It does operate within a standard finite-volume framework, as does the proposal of Pember (Pember, 1993) but our results seem to be slightly less oscillatory, perhaps because of our more elaborate flux integration. We join with him in feeling that an upwind method acknowledging both frozen and equilibrium wavepaths would be of benefit. As far as we can tell the computational cost of our method is less than that of the semidiscrete solvers put forward in (Jin, 1995; Jin & Levermore, 1996). The results, judged rather subjectively, seem to be at least as good. We hope to report in the future on further gains in accuracy or simplicity.

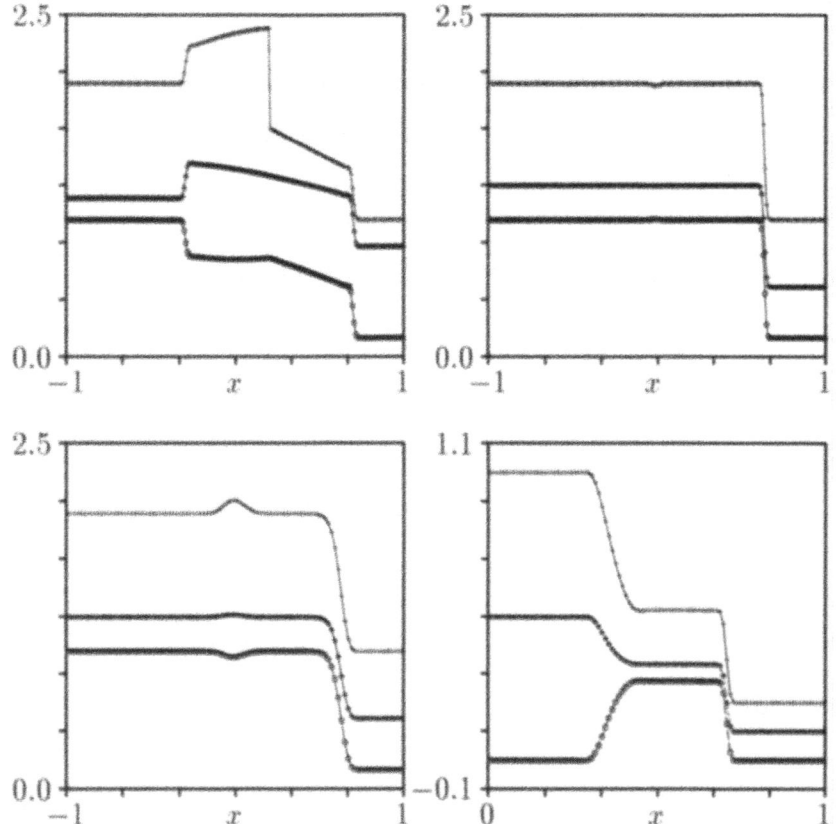

Figure 7. Solutions for the conserved variables density ρ (+), momentum m (o) and z (◇). All grids have 200 intervals, except for the solution at bottom left, which uses 100 intervals.

References

Arora, M., "Explicit characteristic-based high-resolution algorithms for hyperbolic conservation laws with stiff source terms," *Ph. D. Thesis*, Department of Aerospace Engineering, University of Michigan, 1996.

Cattaneo, C., "A form of heat conduction equation which eliminates the paradox of instantaneous propagation," *Comptes Rendues*, Vol. 247, 1958, pp. 431-433.

Colella, P. and Woodward, P., "The piecewise-parabolic method (PPM) for gas-dynamical simulations," *J. Comput. Phys.*, Vol. 54, 1984, pp. 174-201.

Courant, R. and Hilbert, D., *Methods of Mathematical Physics*, Interscience Publishers, New York, 1953.

Jin, S., "Runge-Kutta methods for hyperbolic conservation laws with stiff source terms," *J. Comput. Phys.* Vol. 122, 1995, pp. 51-67.

Jin, S. and Levermore, C. D., "Numerical schemes for hyperbolic conservation laws with stiff relaxation terms," *J. Comput. Phys.*, Vol. 126, 1996, pp. 373-389.

Jones, J. G., "One the near-equilibrium and near-frozen regions in an expansion wave in a relaxing gas," *J. Fluid Mech.*, Vol. 19, 1964, pp. 81-102.

Liu, T.-P., "Hyperbolic conservation laws with relaxation," *Comm. Math. Phys.*, Vol.

108, 1987, pp. 153-175.

Pember, R. B., "Numerical methods for hyperbolic conservation laws with stiff relaxation," *SIAM JSC*, Vol. 14, 1993, pp. 824-859.

Roe, P. L. and Arora, M., "Characteristic-based schemes for dispersive waves I. The method of characteristics for smooth solutions," *Num. Meth. PDEs*, Vol. 9, 1993, pp. 459-505.

Strang, G., "On the construction and comparison of difference schemes," *SINUM*, Vol. 5, 1968, pp. 506-517.

Wiggert, D. C., "Analysis of early-time transient heat conduction by method of characteristics," *J. Heat Transfer*, Vol. 99, 1977, pp. 35-40.

Yanenko, N. N., "The method of fractional steps; the solution of problems of mathematical physics in several variables," Springer-Verlag, Berlin, New York, 1971.

NUMERICAL METHODS FOR A ONE-DIMENSIONAL INTERFACE SEPARATING COMPRESSIBLE AND INCOMPRESSIBLE FLOWS

R. FEDKIW, B. MERRIMAN AND S. OSHER [0]

UCLA, Department of Mathematics,

405 Hilgard Ave., LA, CA 90095

Abstract.

We develop 1D numerical methods for treating an interface separating a liquid drop and a high speed gas flow. The droplet is an incompressible Navier-Stokes fluid. The gas is a compressible, multi-species, chemically reactive Navier-Stokes fluid (Fedkiw *et al.*, 1996; Fedkiw, 1996). The interface is followed with a marker particle, although the level set method will be used for the eventual 2D extension (Sussman, 1995). Away from the interface, we solve the equations with TVD Runge Kutta schemes in time and conservative finite difference ENO schemes in space (Shu and Osher, 1988). Near the interface, we cannot apply this discretization, since the equations differ in both number and type across the interface. Instead we use the interface location for domain decomposition, and apply a moving control volume formulation nearby. This is done in a conservative framework, compatible with the outer finite difference scheme. Full details are given for a simple forward Euler time stepping scheme, and this has direct, although algorithmically complicated, extensions to 2nd and 3rd order Runge Kutta methods. Future work will focus on the extension to 2D, and simplifications of the higher order time stepping algorithms.

1. Introduction

The level set method has been used to model the interface between different fluids. In this procedure, each numerical flux function is constructed by staying on the appropriate side of the level set. These fluxes are then

[0]Research supported in part by ARPA URI-ONR-N00014-92-J-1890, NSF #DMS 94-04942, and ARO DAAH04-95-1-0155.

V. Venkatakrishnan et al (eds.), Barriers and Challenges in Computational Fluid Dynamics, 155-194.

© 1998 *Kluwer Academic Publishers.*

differenced to update the conserved variables. This method works well for most of the numerical domain. However, it does not work as well for points which are adjacent to the level set. A point which is adjacent to the level set will be updated using a flux from each side of it. Using these "straddled" fluxes can lead to problems.

If the media on opposite sides of the interface are very similar, then one is able to update a point adjacent to the level set by using "straddled" fluxes. In (Mulder *et al.*, 1991) the authors modeled two distinct gas phases (e.g. helium and air) and had little trouble with "straddled" fluxes. This is because the two gas phases are fairly similar. In (Sussman, 1995), dissimilar fluids are treated, such as water and air. In order to use "straddled" fluxes, conditions are imposed to make the fluids similar: both are treated as incompressible fluids, and their physical properties are averaged together over a few points (around 2 or 3) near the interface in order to smooth out the discrepancies.

There are times when the fluids on opposite sides of the interface are quite different and we are not free to simply average their properties. Even worse, the equations on opposite sides of the level set may differ in number and type. In these cases it does not make mathematical sense to use "straddled" fluxes; a new method must be employed. We construct a flux located on the moving interface, using sub-grid information. Then this *interface flux* can be utilized in a conservative way to update the points adjacent to the interface, on one or both sides as required by the equations posed on each side. In this way the different fluids do not incorrectly interact with each other through interior flux points. They interact with each other through the *interface flux* using the appropriate mathematical model designed for the interface.

We will study the 1D case of a liquid and gas, such as water and air. The liquid is an incompressible Navier-Stokes fluid. The gas is a compressible, multi-species, chemically reactive Navier-Stokes fluid (Fedkiw *et al.*, 1996; Fedkiw, 1996). The interface is followed with a marker particle instead of a level set. In 1D there is little advantage to using a level set. In 2D we will use the level set method, since it is the most efficient technique (Mulder *et al.*, 1991; Sussman, 1995). Away from the interface, we solve the equations with TVD Runge Kutta schemes in time and ENO schemes in space (Shu and Osher, 1988). The ENO schemes are implemented using Marquina's Jacobian method (Fedkiw *et al.*, 1996). Near the interface, we cannot employ standard methods, since the equations differ in both number and type. Instead we use the interface location for domain decomposition. A moving flux function is constructed for the interface in such a way that it is valid for both domains. We use this moving flux function in a conservative framework to update the conserved variables together with forward Euler time stepping. With some added algorithmic complexity, we can also directly extend this time stepping technique to methods based on a composite of Euler time steps, such as 2nd and 3rd order TVD Runge Kutta methods (Shu and Osher, 1988).

2. Equations

2.1. GAS PHASE MODEL EQUATIONS

The 1D Navier-Stokes equations for multi-species flow with chemical reactions are

$$\vec{U}_t + [\vec{F}(\vec{U})]_x = [\vec{F}_v(\vec{U})]_x + \vec{S} \tag{1}$$

$$E = -p + \frac{\rho u^2}{2} + \rho h \tag{2}$$

where

$$\vec{U} = \begin{pmatrix} \rho \\ \rho u \\ E \\ \rho Y_1 \\ \vdots \\ \rho Y_{N-1} \end{pmatrix}, \quad \vec{F}(\vec{U}) = \begin{pmatrix} \rho u \\ \rho u^2 + p \\ (E+p)u \\ \rho u Y_1 \\ \vdots \\ \rho u Y_{N-1} \end{pmatrix} \tag{3}$$

$$F_v(U) = \begin{pmatrix} 0 \\ \tau_{11} \\ u\tau_{11} + Q_1 \\ \rho D_{1,m}(Y_1)_x \\ \vdots \\ \rho D_{N-1,m}(Y_{N-1})_x \end{pmatrix}, \quad \vec{S} = \begin{pmatrix} 0 \\ 0 \\ 0 \\ \dot{\omega}_1 \\ \vdots \\ \dot{\omega}_{N-1} \end{pmatrix} \tag{4}$$

$$\tau_{11} = \frac{4}{3}\mu u_x, \quad Q_1 = kT_x + \rho \sum_{i=1}^{N} h_i D_{i,m}(Y_i)_x \tag{5}$$

$$\dot{\omega}_i = \dot{\omega}_i(T, p, Y_1, \ldots, Y_{N-1}) \tag{6}$$

where t is time, x is the spatial dimension, ρ is the density, u is the velocity, E is the energy per unit volume, h is enthalpy per unit mass, p is the pressure, N is the number of species being considered, Y_i is the mass fraction of species i, μ is the mixture viscosity, k is the mixture thermal conductivity, $D_{i,m}$ is the mass diffusivity of species i into the mixture, and $\dot{\omega}_i$ is the mass production rate of species i. (Fedkiw et al., 1996) Note that

$$Y_N = 1 - \sum_{i=1}^{N-1} Y_i \tag{7}$$

defines the last mass fraction.

2.1.1. Energy and Enthalpy
We write the enthalpy per unit mass as

$$h = e + \frac{p}{\rho} = \sum_{i=1}^{N} Y_i e_i + \frac{\sum_{i=1}^{N} p_i}{\rho} = \sum_{i=1}^{N} Y_i \left(e_i + \frac{p_i}{\rho Y_i} \right) = \sum_{i=1}^{N} Y_i h_i \tag{8}$$

where e_i, p_i, and h_i are the internal energy, partial pressure, and the enthalpy per unit mass of the ith gas respectively. Using the first equality in 8 gives an alternate form for equation 2

$$E = \rho e + \frac{\rho u^2}{2} \tag{9}$$

where e is the internal energy per unit mass.

In a *perfect gas*, the internal energy, enthalpy, and specific heats are functions of the temperature only. In this case we can write

$$h_i = h_i(T) \qquad e_i = e_i(T) \tag{10}$$

$$c_{p_i} = c_{p_i}(T) \qquad c_{v_i} = c_{v_i}(T) \tag{11}$$

$$dh_i(T) = c_{p_i}(T)dT \qquad de_i(T) = c_{v_i}(T)dT \tag{12}$$

where c_{p_i} is the specific heat at constant pressure of the ith species, and c_{v_i} is the specific heat at constant volume of the ith species (Anderson, 1989).

We can integrate the first equation in 12 starting from $T = 298K$ to get

$$h_i(T) = h_i^{298} + \int_{298}^{T} c_{p_i}(s)ds \tag{13}$$

where h_i^{298} is the enthalpy per unit mass at 298K for species i. This is also sometimes called the heat of formation at 298K, which is a standard constant that can be found in the JANAF Thermochemical Tables (Stall and Prophet, 1971). To speed up the actual implementation, we construct a table of $h_i(T)$ for each species including every integer temperature between $0K$ and $4800K$. We approximate the integral to desired accuracy, using the CHEMKIN code (Kee *et al.*, 1986) to give us the values of $c_{p_i}(T)$ when needed. This is done once at the beginning of the code. During computation, if we need $h_i(T)$ for a non-integral value of the temperature, we interpolate linearly. (Fedkiw *et al.*, 1996)

A *calorically perfect* gas has c_{p_i} constant, while a *thermally perfect* gas has $c_{p_i}(T)$ a function of the temperature (Anderson, 1989). In regimes where CHEMKIN does not have data for $c_{p_i}(T)$, we usually extrapolate with a calorically perfect assumption.

2.1.2. *Equation of State, Specific Heats, and Gamma*
The equation of state for multi-species flow is

$$p = \rho RT \qquad (14)$$

where

$$R = \frac{R_u}{W} \qquad (15)$$

where $R_u = 8314$ J/(kmol K) is the universal gas constant, and W is the molecular weight of the mixture given by

$$W = \frac{1}{\sum_{i=1}^{N} \frac{Y_i}{W_i}} \qquad (16)$$

where W_i is the molecular weight of the ith species.

Gamma for the mixture is

$$\gamma = \frac{c_p}{c_p - \frac{R_u}{W}} \qquad (17)$$

where c_p is the specific heat at constant pressure of the mixture given by

$$c_p = \sum_{i=1}^{N} Y_i c_{p_i} \qquad (18)$$

(Fedkiw *et al.*, 1996).

2.1.3. *Temperature*
Since various thermodynamic quantities are functions of temperature, it is necessary to have a means of computing temperature from the primary conserved variables (mass, momentum and energy.or enthalpy) evolved during the calculation.

We get an expression for the temperature by combining equation 2 with equation 14 to get

$$T = C_1 + C_2 h(T) \tag{19}$$

where C_1 and C_2 are constants if the conserved variables are fixed. This equation is implicit for temperature.

We rewrite equation 19 as,

$$f(T) = T - C_1 - C_2 h(T) = 0 \tag{20}$$

and note that

$$\frac{df(T)}{dT} = 1 - C_2 \frac{dh(T)}{dT} = 1 - C_2 c_p(T) = 1 - \frac{c_p(T)}{R} = \frac{-1}{\gamma(T) - 1} \tag{21}$$

where $\gamma(T)$ is a function of temperature. Since $\gamma(T)$ is always greater than one, this shows that $f(T)$ is a strictly decreasing function of temperature. We solve equation 20 with Newton-Raphson iteration. (Fedkiw et al., 1996)

2.1.4. Eigensystem

The ENO spatial discretization is based on upwind discretization of the characteristic fields, and so requires full characteristic data for the convective part of the system of conservation laws, i.e. the eigensystem of $\vec{F}'(\vec{U})$. See appendix A for details.

2.2. LIQUID PHASE MODEL EQUATIONS

The incompressible liquid phase satisfies the general incompressibility condition $\nabla \cdot u = 0$, which in 1D reduces to the trivial form

$$u_x = 0 \tag{22}$$

with the density also a constant. For water,

$$\rho = 1000 \frac{kg}{m^3} \tag{23}$$

We will also assume the thermal conductivity is a constant, which is reasonable in the liquid phase. For water in the temperature range from range from 298K to 373K, the thermal conductivity is fixed at

$$k = .65 \frac{J}{mK\,sec} \tag{24}$$

instead of a varying function of temperature.

Using these simplifications, the 1D Navier-Stokes equations for a non-reacting, incompressible flow reduce to

$$\vec{U}_t + [\vec{F}(\vec{U})]_x = [\vec{F}_v(\vec{U})]_x \tag{25}$$

where

$$\vec{U} = \begin{pmatrix} \rho u \\ E \end{pmatrix}, \quad \vec{F}(\vec{U}) = \begin{pmatrix} p \\ (E+p)u \end{pmatrix}, \quad F_v(U) = \begin{pmatrix} 0 \\ kT_x \end{pmatrix} \tag{26}$$

where

$$E = \rho e + \frac{\rho u^2}{2} \tag{27}$$

which is in the form of equation 9. We use this form rather than the enthalpy form to avoid use of the pressure, since there is no local equation of state available for pressure anyway.

2.2.1. Comments on the Equations

The first equation in equation 25 can be written as

$$u_t = \frac{-p_x}{\rho} \tag{28}$$

and by taking an x-derivative of both sides of this equation we get

$$p_{xx} = 0 \tag{29}$$

implying that the pressure profile is linear.

Substituting equation 27 into the second equation in equation 25 yields

$$\rho e_t + \rho u u_t + u \rho e_x + u p_x = kT_{xx} \tag{30}$$

which can be simplified with the use of equation 28 to get

$$\rho e_t + u \rho e_x = kT_{xx} \tag{31}$$

or in conservation form as

$$(\rho e)_t + (u \rho e)_x = (kT_x)_x \tag{32}$$

for the convection and diffusion of internal energy per unit volume. It is interesting to note that equation 28 determines both the momentum and the kinetic energy. Thus, equation 28 cancels out the kinetic energy in equation 30 leaving only a conservation equation for internal energy, equation 32.

2.2.2. *Temperature*

As for the gas phase, we need to compute the liquid temperature from the primary variables (momentum and energy). However, the incompressible liquid does not have a local equation of state such as the gas law 14. Thus we cannot follow the temperature calculation used for the gas phase. Here we derive an appropriate relation for the temperature of the liquid phase.

Consider equation 8 for the vapor phase of the liquid,

$$h = e + \frac{p}{\rho} \tag{33}$$

The vapor phase is a gas and thus satisfies the gas law, 14. Using this yields

$$h = e + \left(\frac{R_u}{W}\right) T \tag{34}$$

where W is the molecular weight of the liquid molecules. Thus the internal energy of the vapor phase is

$$e(T) = h(T) - \left(\frac{R_u}{W}\right) T \tag{35}$$

and we can write the internal energy of liquid phase as that of the vapor minus the heat of vaporization,

$$e(T) = h(T) - \left(\frac{R_u}{W}\right) T - H_{vap}(T) \tag{36}$$

where $H_{vap}(T)$ is the latent heat of vaporization of the liquid as a function of temperature. This quantity is tabulated for water and many other liquids.

To speed up the actual implementation, we construct a table of $e(T)$ for the liquid, and handle it in the same way as the enthalpy tables constructed for each species. In order to construct this table consistent with the gas, we first construct a table of enthalpy $h(T)$ for liquid vapor. Then we construct as much of a table as possible for the latent heat of vaporization $H_{vap}(T)$ (this obviously isn't tabulated for very high or low values). Then we can use equation 36 to construct a partial table of internal energies $e(T)$ for the liquid. We need to complete the table outside the range of tabulated latent heat of vaporization $H_{vap}(T)$. To do this, we assume that the liquid is calorically perfect outside the temperature range of tabulated $H_{vap}(T)$. We integrate the second equation in 12 twice. First starting from $T = T_{min}$ and then again starting from $T = T_{max}$

$$e(T) = e(T_{min}) + c_v(T_{min})(T - T_{min}) \tag{37}$$

$$e(T) = e(T_{max}) + c_v(T_{max})(T - T_{max}) \tag{38}$$

where T_{min} and T_{max} were the lowest and highest temperature in the already completed portion of the table of $e(T)$. Note that $c_v(T_{min})$ and $c_v(T_{max})$ are two different constants. We than use equation 37 to fill in the low temperature portion of the table, and equation 38 to fill in the high temperature portion of the table.

We arrive at an implicit expression for the temperature from 27, which can be written as

$$f(T) = e(T) + C = 0 \tag{39}$$

where C is a constant if the conserved variables are fixed. Note that

$$\frac{df(T)}{dT} = \frac{de(T)}{dT} = c_v(T) \tag{40}$$

shows that $f(T)$ is a strictly increasing function of temperature. We then solve equation 39 for the temperature via Newton-Raphson iteration.

3. Numerical Methods

We attempt to construct numerical flux functions at the midpoint of every pair of grid nodes. They are constructed using 3rd order conservative finite difference ENO (Shu and Osher, 1988) with Marquina's Jacobian (Fedkiw *et al.*, 1996) on the convection terms along with conservative central differencing on the diffusive terms. These fluxes are then differenced to update the conserved variables. This method works well for most of the numerical domain. However, it does not work as well for points which are adjacent to the interface. A point which is adjacent to the interface would be updated using a flux from each side of the interface. Since our equations differ in number and type, it does not make sense to use these "straddled" fluxes. Instead, we construct a flux located on the moving interface. This *interface flux* is constructed directly from the model equations valid at the interface. Then the *interface flux* is used in a conservative way to update the points adjacent to the interface.

3.1. CONSERVATION METHOD

Our information is stored as point values on the grid. For normal nodes— those not adjacent to the interface—we construct ENO numerical flux functions which update the point values to the desired order of accuracy. In a cell which contains the interface, we do not construct a numerical flux function. Instead, we construct a physical flux at the interface.

We wish to use moving physical fluxes to update the points adjacent to the interface. For this purpose, we use the cell average framework. Thus, a grid point which is adjacent to the interface has two stored values. The point value is used to find numerical flux functions for updating the interior domain, away from the interface. The cell average value is used to update the grid point itself.

Consider updating the cell average for a point adjacent to the interface. We do this with two physical fluxes. One of them is obviously the moving physical flux on the interface. The other should be constructed away from the interface, where a numerical flux function already exists. To maintain conservation and compatibility with the outer finite difference ENO scheme, we use the pre-existing numerical flux function as an approximation to the desired physical flux function at this point. Since the numerical flux is only a second order accurate approximation to the physical flux, this degrades the local order of our numerical method. Here a choice must be made between order and conservation. We could get higher order by violating conservation, but we favor preserving discrete conservation to capture unresolved phenomena that may arise from abrupt interface motion.

Since the interface moves at the speed of the fluid which it separates, the CFL condition limits its travel to one grid cell per time step. We will eliminate the case where the interface lies exactly on a grid point. We can always move it off the grid point by some small amount based on the order of the method or on machine precision. Thus, there are two cases to consider:

1. The interface does not cross a grid point during the course of an Euler time step. It ends up in the the same cell in which it started.

2. The interface crosses a grid point during the course of an Euler time step. It appears in an adjacent cell at the end of the time step.

For the first case construction, refer to Case 1 in Figure 1 for illustration. In this case the interface remains between the grid points x_i and x_{i+1}. The pointwise values of the conserved variables here are $\vec{U}_i$ and $\vec{U}_{i+1}$, while the cell average values are $(\vec{U}_i)_{ave}$ and $(\vec{U}_{i+1})_{ave}$. We do not construct a numerical flux function $\vec{F}_{i+\frac{1}{2}}$ in the cell which contains the interface. Instead, we construct an interface flux, $\vec{F}_I$. We still construct numerical fluxes away from the interface, including $\vec{F}_{i-\frac{1}{2}}$ and $\vec{F}_{i+\frac{3}{2}}$. At time n, we define h_i^n and h_{i+1}^n as the sizes (lengths) of the cells over which the cell averages $(\vec{U}_i^n)_{ave}$ and $(\vec{U}_{i+1}^n)_{ave}$ are respectively defined. At time $n+1$, we define h_i^{n+1} and h_{i+1}^{n+1} as the sizes of the cells over which the cell averages $(\vec{U}_i^{n+1})_{ave}$ and $(\vec{U}_{i+1}^{n+1})_{ave}$ are respectively defined.

Since the interface moves at the speed of the fluid, we know that the velocity profile is continuous at the interface, although the derivatives of the velocity are not necessarily continuous. At time n, we can interpolate to find the interface velocity v_I^n. For our specific problem, equation 22 tells us that the velocity profile in the liquid is constant. Thus, v_I^n is identical to the liquid velocity at time n, and the interface moves a distance of $v_I^n dt$ during an Euler time step. So it is possible to find the future location of the interface, before updating the conserved variables. Note that a level set representation of the interface will have the same property in multiple dimensions. This ability is essential for the numerical method presented here.

We write the numerical scheme as

$$(\vec{U}_i^{n+1})_{ave} h_i^{n+1} = (\vec{U}_i^n)_{ave} h_i^n + dt(\vec{F}_{i-\frac{1}{2}} - \vec{F}_I) \tag{41}$$

$$(\vec{U}_{i+1}^{n+1})_{ave} h_{i+1}^{n+1} = (\vec{U}_{i+1}^n)_{ave} h_{i+1}^n + dt(\vec{F}_I - \vec{F}_{i+\frac{3}{2}}) \tag{42}$$

which both can be read as: the new "stuff" equals the old "stuff", plus the flux in from the right, minus the flux out to the left. We can solve equations 41 and 42 for $(\vec{U}_i^{n+1})_{ave}$ and $(\vec{U}_{i+1}^{n+1})_{ave}$ by dividing by h_i^{n+1} and h_{i+1}^{n+1} respectively. Remember that h_i^{n+1} and h_{i+1}^{n+1} are known, since we are able to move the interface first. Once we find $(\vec{U}_i^{n+1})_{ave}$ and $(\vec{U}_{i+1}^{n+1})_{ave}$ we still need to find the pointwise values $\vec{U}_i^{n+1}$ and $\vec{U}_{i+1}^{n+1}$.

Consider the second case where the interface moves into an adjacent cell, and refer to Case 2 in Figure 1 for illustration. Since moving to the left and to the right are symmetric, we only consider the case where the interface moves into the cell to the right. It starts between x_i and x_{i+1} and ends up between x_{i+1} and x_{i+2}. The pointwise values of the conserved variables here are $\vec{U}_i$, $\vec{U}_{i+1}$, and $\vec{U}_{i+2}$, while the cell average values are $(\vec{U}_i)_{ave}$, $(\vec{U}_{i+1})_{ave}$, and $(\vec{U}_{i+2})_{ave}$. Note that we do not know $(\vec{U}_{i+2})_{ave}$ and must construct it. Once again, we construct an interface flux, $\vec{F}_I$, to replace the numerical flux $\vec{F}_{i+\frac{1}{2}}$. Also, we still construct fluxes away from the interface, including $\vec{F}_{i-\frac{1}{2}}$, $\vec{F}_{i+\frac{3}{2}}$, and $\vec{F}_{i+\frac{5}{2}}$. At time n, we have h_i^n, h_{i+1}^n, and dx as the size of the cells over which the cell averages $(\vec{U}_i^n)_{ave}$, $(\vec{U}_{i+1}^n)_{ave}$, and $(\vec{U}_{i+2}^n)_{ave}$ are respectively defined. At time $n+1$, we have dx, h_i^{n+1}, and h_{i+2}^{n+1} as the size of the cells over which the cell averages $(\vec{U}_i^{n+1})_{ave}$, $(\vec{U}_{i+1}^{n+1})_{ave}$, and $(\vec{U}_{i+2}^{n+1})_{ave}$ are respectively defined.

We write the numerical scheme as

$$(\vec{U}_i^{n+1})_{ave} dx + (\vec{U}_{i+1}^{n+1})_{ave} h_{i+1}^{n+1} = (\vec{U}_i^n)_{ave} h_i^n + dt(\vec{F}_{i-\frac{1}{2}} - \vec{F}_I) \tag{43}$$

$$(\vec{U}_{i+2}^{n+1})_{ave} h_{i+2}^{n+1} = (\vec{U}_{i+1}^n)_{ave} h_{i+1}^n + (\vec{U}_{i+2}^n)_{ave} dx + dt(\vec{F}_I - \vec{F}_{i+\frac{5}{2}}) \tag{44}$$

which can both be read as: the new "stuff" equals the old "stuff", plus the flux in from the right, minus the flux out to the left. In equation 43, we need to find $(\vec{U}_i^{n+1})_{ave}$ and $(\vec{U}_{i+1}^{n+1})_{ave}$ along with their pointwise values $\vec{U}_i^{n+1}$ and $\vec{U}_{i+1}^{n+1}$. In equation 44, we find $(\vec{U}_{i+2}^{n+1})_{ave}$ by dividing by h_{i+2}^{n+1}. After this, we still need to find the pointwise value $\vec{U}_{i+2}^{n+1}$.

A note on the CFL condition. The size of the cells adjacent to the interface will not necessarily be dx, like the cells in the rest of the domain. However, it is easy to see that their size will always be larger than $\frac{dx}{2}$ and smaller than $\frac{3dx}{2}$. Thus, the minimum cell size for use in the CFL conditions $\frac{dx}{2}$, and this determines the largest stable time step.

3.2. FINDING UNKNOWN CELL AVERAGES AND POINT VALUES

In equation 44, we need $(\vec{U}_{i+2}^n)_{ave}$. Since the cell averages are only known for the cells adjacent to the interface, we need to construct $(\vec{U}_{i+2}^n)_{ave}$. We use a 2nd order, piecewise linear construction given by

$$(\vec{U}_{i+2}^n)_{ave} = \frac{\vec{U}_{i+1}^n + 6\vec{U}_{i+2}^n + \vec{U}_{i+3}^n}{8} \tag{45}$$

which is depicted graphically in Figure 2.

In equations 41, 42, and 44 we can solve for $(\vec{U}_i^{n+1})_{ave}$, $(\vec{U}_{i+1}^{n+1})_{ave}$, and $(\vec{U}_{i+2}^{n+1})_{ave}$. From these we have to reconstruct the point values $\vec{U}_i^{n+1}$, $\vec{U}_{i+1}^{n+1}$, and $\vec{U}_{i+2}^{n+1}$ respectively. We only show how to reconstruct the point value $\vec{U}_i^{n+1}$ from the cell average $(\vec{U}_i^{n+1})_{ave}$, since the other two reconstructions are symmetric to this one. We use a 2nd order, linear reconstruction which is consistent with the cell average,

$$\vec{U}_i^{n+1} = \left(\frac{(\vec{U}_i^{n+1})_{ave} - \vec{U}_{i-1}^{n+1}}{\frac{dx}{2} + \frac{h_i^{n+1}}{2}} \right) dx + \vec{U}_{i-1}^{n+1} \tag{46}$$

which is depicted graphically in Figure 2. Since these reconstructions could give non-physical values, we find the pointwise value at the interface using the same reconstruction

$$\vec{U}_I = \left(\frac{(\vec{U}_i^{n+1})_{ave} - \vec{U}_{i-1}^{n+1}}{\frac{dx}{2} + \frac{h_i^{n+1}}{2}} \right) \left(\frac{dx}{2} + h_i^{n+1} \right) + \vec{U}_{i-1}^{n+1} \tag{47}$$

and check to be sure this value is within the physically allowed range (positive densities, etc). If not, then we replace equation 46 with

$$\vec{U}_i^{n+1} = (\vec{U}_i^{n+1})_{ave} \tag{48}$$

which is a 1st order accurate reconstruction.

From equation 43, we need to find $(\vec{U}_i^{n+1})_{ave}$ and $(\vec{U}_{i+1}^{n+1})_{ave}$ along with their pointwise values $\vec{U}_i^{n+1}$ and $\vec{U}_{i+1}^{n+1}$. To do this, we we define the total cell average through

$$(\vec{U}_T)_{ave}(dx + h_{i+1}^{n+1}) = (\vec{U}_i^{n+1})_{ave}dx + (\vec{U}_{i+1}^{n+1})_{ave}h_{i+1}^{n+1} \tag{49}$$

where $(\vec{U}_T)_{ave}$ is the total cell average. We can combine equations 43 and 49 to get

$$(\vec{U}_T)_{ave}(dx + h_{i+1}^{n+1}) = (\vec{U}_i^n)_{ave}h_i^n + dt(\vec{F}_{i-\frac{1}{2}} - \vec{F}_I) \tag{50}$$

and then solve for $(\vec{U}_T)_{ave}$ by dividing by $dx + h_{i+1}^{n+1}$. We use a 2nd order, linear reconstruction which is consistent with the total cell average,

$$\vec{U}_i^{n+1} = \left(\frac{(\vec{U}_T)_{ave} - \vec{U}_{i-1}^{n+1}}{\frac{dx}{2} + \frac{dx+h_{i+1}^{n+1}}{2}}\right) dx + \vec{U}_{i-1}^{n+1} \tag{51}$$

$$\vec{U}_{i+1}^{n+1} = 2\vec{U}_i^{n+1} - \vec{U}_{i-1}^{n+1} \tag{52}$$

$$(\vec{U}_i^{n+1})_{ave} = \vec{U}_i^{n+1} \tag{53}$$

$$(\vec{U}_{i+1}^{n+1})_{ave} = \frac{(\vec{U}_T)_{ave}(dx + h_{i+1}^{n+1}) - (\vec{U}_i^{n+1})_{ave}dx}{h_{i+1}^{n+1}} \tag{54}$$

which is depicted graphically in Figure 2. Note that equation 54 comes from a rearrangement of equation 49. Since these reconstructions could give non-physical values, we find the pointwise value at the interface using the same reconstruction

$$\vec{U}_I = \left(\frac{(\vec{U}_T)_{ave} - \vec{U}_{i-1}^{n+1}}{\frac{dx}{2} + \frac{dx+h_{i+1}^{n+1}}{2}}\right)\left(\frac{3dx}{2} + h_{i+1}^{n+1}\right) + \vec{U}_{i-1}^{n+1} \tag{55}$$

and check to be sure this value is within the physically allowed range. If not, then we replace equations 51, 52, 53, and 54 with

$$\vec{U}_i^{n+1} = \vec{U}_{i+1}^{n+1} = (\vec{U}_i^{n+1})_{ave} = (\vec{U}_{i+1}^{n+1})_{ave} = (\vec{U}_T)_{ave} \tag{56}$$

which is a 1st order accurate reconstruction.

3.3. INTERFACE FLUX

We have discussed how to implement the *interface flux*, F_I, in the numerical method. Next we will consider its construction. The interface flux will model the physics of the interface. The general physical flux looks like,

$$F_I = \vec{F}(\vec{U}) - \vec{F}_v(\vec{U}) = \begin{pmatrix} \rho u \\ \rho u^2 + p - \tau_{11} \\ (E+p)u - u\tau_{11} - Q_1 \\ \rho u Y_1 - \rho D_{1,m}(Y_1)_x \\ \vdots \\ \rho u Y_{N-1} - \rho D_{N-1,m}(Y_{N-1})_x \end{pmatrix} \tag{57}$$

but this can be greatly simplified.

Since the interface moves with the speed of the fluid, and it separates two fluids, there are no particles crossing the interface. Thus, there is no convection of mass, momentum, or energy across the interface. Likewise, there is no mass diffusion across the interface. So, equation 57 reduces to

$$F_I = \begin{pmatrix} 0 \\ p - \tau_{11} \\ u(p - \tau_{11}) - Q_1 \\ 0 \\ \vdots \\ 0 \end{pmatrix} = \begin{pmatrix} 0 \\ p - \frac{4}{3}\mu u_x \\ u(p - \frac{4}{3}\mu u_x) - kT_x \\ 0 \\ \vdots \\ 0 \end{pmatrix} \tag{58}$$

where $p - \frac{4}{3}\mu u_x$ is the net force on the interface. This force changes both the momentum and the kinetic energy. Also, $-kT_x$ is the diffusive heat flux across the interface.

Since we have a conservative scheme, the interface flux must have some consistency with both sides of the interface. It makes physical sense that both temperature and velocity are continuous. In addition to this, we note two principles of the interface which will help us:

1. The net stress on the interface is zero.
2. Heat which flows into the interface flows back out.

These are equivalent to

$$\left(p - \frac{4}{3}\mu u_x\right)_{gas} = \left(p - \frac{4}{3}\mu u_x\right)_{liquid} \tag{59}$$

$$(kT_x)_{gas} = (kT_x)_{liquid} \tag{60}$$

and obviously also give jump conditions for the interface. With these conditions, we see the F_I in equation 58 is consistent with both sides of the interface. Note that in our case equation 59 reduces to

$$\left(p - \frac{4}{3}\mu u_x\right)_{gas} = p_{liquid} \tag{61}$$

since the velocity in the liquid is constant.

3.4. EVALUATION OF THE INTERFACE FLUX

We will consider Figure 1 with gas on the left and liquid on the right. The opposite case is symmetric.

3.4.1. *Mass Fractions*
In order to evaluate the viscosity and thermal conductivity on the gas side of the interface, we need to know the mass fractions. The first $N - 1$ mass fractions are interpolated from the gas, since they do not exist in the liquid. The Nth mass fraction is formed from those using equation 7. We interpolate to high order, but reduce the order if non-physical values are predicted at the interface, i.e. any negative mass fraction. Mass fractions may also be predicted to be greater than one, but equation 7 shows that we cannot have a mass fraction greater than one without another mass fraction being negative. The third, second, and first order interpolations for the mass fractions are $\vec{Y} = (Y_1, \cdots, Y_{N-1})$ are

$$\begin{aligned}
\vec{Y}_I &= \left(\frac{\vec{Y}_i - 2\vec{Y}_{i-1} + \vec{Y}_{i-2}}{2dx^2}\right)\left(h_i^n - \frac{dx}{2}\right)^2 + \\
&\quad \left(\frac{3\vec{Y}_i - 4\vec{Y}_{i-1} + \vec{Y}_{i-2}}{2dx}\right)\left(h_i^n - \frac{dx}{2}\right) + \vec{Y}_i
\end{aligned} \tag{62}$$

$$\vec{Y}_I = \left(\frac{\vec{Y}_i - \vec{Y}_{i-1}}{dx}\right)\left(h_i^n - \frac{dx}{2}\right) + \vec{Y}_i \tag{63}$$

$$\vec{Y}_I = \vec{Y}_i \tag{64}$$

respectively.

3.4.2. *Temperature and Thermal Conductivity*
Given the interface temperature T_I, we can compute discrete approximations to T_x in both the liquid and gas phases using one sided differences.

Note we should not difference across the interface, since T is not necessarily smooth across the interface.

Thus the discretized form of the energy flux equation 60 yields an implicit equation for T_I,

$$f(T_I) = (kT_x)_{gas} - (kT_x)_{liquid} = 0 \qquad (65)$$

(the thermal conductivities are functions of T_I as well, in general, and also of the known mass fractions at the interface in the gas phase).

In equation 65, we replace T_x on the gas side of the interface with the following third order discretization,

$$A_1 = \frac{T_I - T_i}{h_i^n - \frac{dx}{2}}, \qquad A_2 = \frac{T_I - T_{i-1}}{h_i^n + \frac{dx}{2}}, \qquad A_3 = \frac{T_I - T_{i-2}}{h_i^n + \frac{3dx}{2}}$$

$$A_4 = \frac{\left(h_i^n + \frac{dx}{2}\right) A_1 - \left(h_i^n - \frac{dx}{2}\right) A_2}{dx}, A_5 = \frac{\left(h_i^n + \frac{3dx}{2}\right) A_1 - \left(h_i^n - \frac{dx}{2}\right) A_3}{2dx}$$

$$T_x = \frac{\left(h_i^n + \frac{3dx}{2}\right) A_4 - \left(h_i^n + \frac{dx}{2}\right) A_5}{dx} \qquad (66)$$

where T_x is a function of T_I. We replace T_x on the liquid side of the interface using the symmetric, third order equivalent to equation 66.

Finally, we solve equation 65 for T_I using Newton-Raphson iteration. As an initial guess for the interface temperature we use the average

$$(T_I)_o = \frac{T_i + T_{i+1}}{2} \qquad (67)$$

although a k-weighted harmonic average would yield and even better guess.

The resulting T_I is used to evaluate the kT_x term in the interface flux. It is important to use the exact same numerical discretization of T_x to evaluate the interface flux as was used in the Newton-Raphson iteration, i.e. equation 66. If we used a different discretization for the interface flux evaluation, say fourth order, than there is no guarantee that equation 60 holds in the discretized form. If equation 60 does not hold in discretized form, there could be problems with both accuracy and conservation, depending on how the interface flux is treated.

3.4.3. *Velocity and Viscosity*

Now that we know the mass fractions and the temperature at the interface, we can compute the viscosity on the gas side of the interface using Chemkin (Kee *et al.*, 1986).

The velocity will be continuous at the interface, although its derivative is usually not. Since the velocity in the liquid is constant, the interface velocity, u_I, will be identical to the liquid velocity. Also, $u_x = 0$ on the liquid side of the interface. To compute u_x on the gas side of the interface, we use the known interface velocity, u_I, to interpolate u_x to third order

$$A_1 = \frac{u_I - u_i}{h_i^n - \frac{dx}{2}}, \qquad A_2 = \frac{u_I - u_{i-1}}{h_i^n + \frac{dx}{2}}, \qquad A_3 = \frac{u_I - u_{i-2}}{h_i^n + \frac{3dx}{2}}$$

$$A_4 = \frac{\left(h_i^n + \frac{dx}{2}\right) A_1 - \left(h_i^n - \frac{dx}{2}\right) A_2}{dx}, A_5 = \frac{\left(h_i^n + \frac{3dx}{2}\right) A_1 - \left(h_i^n - \frac{dx}{2}\right) A_3}{2dx}$$

$$u_x = \frac{\left(h_i^n + \frac{3dx}{2}\right) A_4 - \left(h_i^n + \frac{dx}{2}\right) A_5}{dx} \tag{68}$$

3.4.4. *Pressure*
The pressure at the interface is interpolated from the gas phase, since we do not have a local equation of state for the pressure in the liquid. We interpolate to high order, but reduce the order if non-physical values are predicted at the interface, i.e. if $p \leq 0$. The third, second, and first order interpolations for the pressure are

$$p_I = \left(\frac{p_i - 2p_{i-1} + p_{i-2}}{2dx^2}\right) \left(h_i^n - \frac{dx}{2}\right)^2 + \left(\frac{3p_i - 4p_{i-1} + p_{i-2}}{2dx}\right) \left(h_i^n - \frac{dx}{2}\right) + p_i \tag{69}$$

$$p_I = \left(\frac{p_i - p_{i-1}}{dx}\right) \left(h_i^n - \frac{dx}{2}\right) + p_i \tag{70}$$

$$p_I = p_i \tag{71}$$

respectively. Note that this is the same as that for the mass fractions.

Since we know p, μ, and u_x in the gas, we can use equation 61 to find the pressure on the liquid side of the interface. In 2D this would also include any pressure jump due to surface tension forces. This in turn provides the pressure boundary condition needed to solve the global equation for pressure inside the incompressible liquid.

Note that our procedure of interpolating pressure in from the gas phase, and using the pressure balance across the interface to deduce pressure on the liquid side, makes the implicit assumption that a pressure shock does not exist at the interface. If one does, a brief transient error may result during the time it takes for the shock to be transmitted one cell.

3.5. PRACTICAL USE OF THE INTERFACE FLUX

Once the interface flux has been constructed, it has a single value which is used on both sides of the interface. Most of its terms are identically zero, but we do have a non-zero flux for both momentum and energy.

If momentum, ρu, and total energy, E, are included in the conserved variables on both sides of the interface, then we just apply the interface flux to update the conserved variables according to equations 41 and 42 or equations 43 and 44, depending on which are relevant. In equation 1, for the gas side of the interface, $\vec{U}$ contains momentum and total energy as conserved variables, as can be seen in $\vec{U}$ in equation 3. In equation 25, for the liquid side of the interface, $\vec{U}$ contains momentum and total energy as conserved variables, as can be seen in $\vec{U}$ in equation 26. Thus, if equations 1 and 25 are our conservation equations for gas and liquid, then no special treatment is needed.

Suppose that we used equations 28 and 32 for the liquid. Then one has to be careful when updating. Equation 28 can be rewritten as

$$(\rho u)_t = -p_x \tag{72}$$

since ρ is a constant. Then, we can update the momentum using the interface flux. This is straight forward. It is not as easy to see how to handle equation 32. In fact, there are two choices:

1. We can maintain conservation of *total energy*, E, which will maintain global conservation of energy.
2. We can maintain conservation of *internal energy*, ρe, which is dictated by the special incompressible assumptions leading to equation 32. This will not maintain global conservation of energy.

3.5.1. *Conservation of Internal Energy in the Liquid*
The kinetic energy is defined as

$$KE = \frac{\rho u^2}{2} \tag{73}$$

where ρ is a constant in the liquid. Thus u uniquely determines both the momentum and the kinetic energy. Therefore, the entire interface flux for energy is not needed. On the liquid side, we rewrite the interface flux as

$$F_I = \begin{pmatrix} 0 \\ p_{liquid} \\ (-kT_x)_{liquid} \\ 0 \\ \vdots \\ 0 \end{pmatrix} \tag{74}$$

with the kinetic energy portion of the energy flux deleted.

One should be careful when using this method since it is not conservative. Using the unaltered interface flux yields a change in kinetic energy of

$$\triangle KE = dt(up) \tag{75}$$

across the interface. Using the altered interface flux in equation 74, the change in velocity is

$$\triangle u = dt \left(\frac{p}{\rho} \right) \tag{76}$$

across the interface. The corresponding change in kinetic energy is

$$\triangle KE = \frac{\rho(u + \triangle u)^2}{2} - \frac{\rho u^2}{2} = dt(up) + dt^2 \left(\frac{p^2}{2\rho} \right) \tag{77}$$

across the interface. Note that equation 75 gives the conservative update of the kinetic energy, while equation 77 differs from the conservative update in the second order term. These updates were done for Euler's Method. For third order Runge-Kutta, the error in conservation would be fourth order. This method is only conservative up to the order of the temporal scheme.

3.5.2. *Global Conservation of Total Energy*
To keep the scheme entirely conservative, we have to move the truncation error seen in equation 77 to a different location. This is done by using the same value of the interface flux on both sides of the interface. Again, we only need modifications on the liquid side.

The velocity is used to compute the kinetic energy which is then added to the internal energy to get the total energy,

$$E^n = \frac{\rho(u^n)^2}{2} + (\rho e)^n \tag{78}$$

which is updated to E^{n+1} using the interface flux. We also update the momentum with the interface flux and find the new velocity. Then the internal energy is reconstructed as

$$(\rho e)^{n+1} = E^{n+1} - \frac{\rho (u^{n+1})^2}{2} \qquad (79)$$

to finish the conservative scheme. The change in internal energy across the interface is

$$\triangle (\rho e) = \triangle E - \left(\frac{\rho (u^{n+1})^2}{2} - \frac{\rho (u^n)^2}{2} \right) = -kT_x - dt^2 \left(\frac{p^2}{2\rho} \right) \qquad (80)$$

when it should only have been $-kT_x$. Here the second order error is in the non-physical conversion of kinetic energy to internal energy, which is not dictated by equation 32.

3.6. DISCRETIZATION IN THE LIQUID

We use equations 1 and 25 as our conservation equations for gas and liquid, so that momentum and total energy are included in the conserved variables on both sides of the interface. Thus, we use the same value of the interface flux on both sides of the interface. The gas equations are easily discretized (Fedkiw *et al.*, 1996), but the liquid equations require some thought.

Consider the the first row in equation 25,

$$(\rho u)_t + p_x = 0 \qquad (81)$$

which can be updated once we know the pressure. Since the interface evaluation process gave us the pressure on both sides of the droplet, we can compute p_x as a constant since the pressure profile in the droplet is known to be linear from equation 29. Thus p_x is just a source term. Since fluxes are needed to properly do the interface portion of the calculation, we decompose this spatially constant source term into fluxes. Each flux is assigned the known pressure value at the flux point. Then, differencing the fluxes gives us back our spatially constant source term p_x.

Consider the last row in equation 25,

$$E_t + (uE + up)_x = (kT_x)_x \qquad (82)$$

which can be simplified using the fact that density is constant in the droplet. Constant density, along with $u_x = 0$ from equation 22 implies that $(KE)_x = 0$ in the droplet. These considerations simplify equation 82 to

$$E_t + (u\rho e)_x + up_x = (kT_x)_x \qquad (83)$$

where up_x is a constant, since velocity is constant in the liquid, and pressure is linear. Thus up_x is a spatially constant source term. Once again we decompose the source term up_x into fluxes, where each flux is evaluated as the known value of up. Differencing the fluxes gives us back our spatially constant source term up_x, since u is constant. The diffusion term kT_x is discretized with conservative central differencing. The convection term $(upe)_x$ is just the convection of internal energy and can be discretized with ENO, using u as the upwind direction.

In equation 81 and equation 83, we have written all the terms defining the fluxes at the cell walls. This allows us to easily apply our interface method.

3.7. SPECIAL STABILITY CONSIDERATIONS

Th numerical method designed above generates a small amount of noise which propagates into the surrounding gas. To stop this, we lower the accuracy of the flux adjacent to the interface so that it is only second order. This single flux acts as some sort of a filter. The reasons for this mild instability are still unclear at this time.

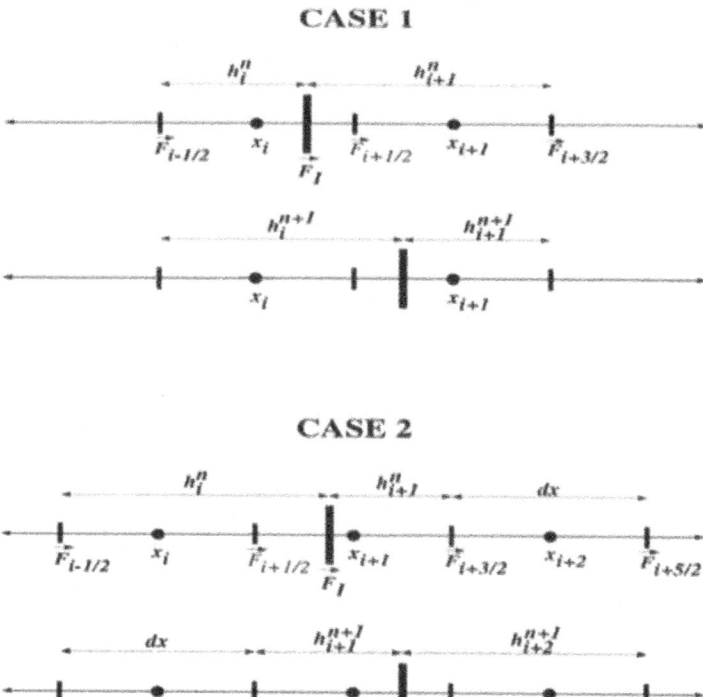

Figure 1.   In case 1, the interface stays between grid nodes. In case 2, the interface crosses a grid node and ends up in an adjacent cell.

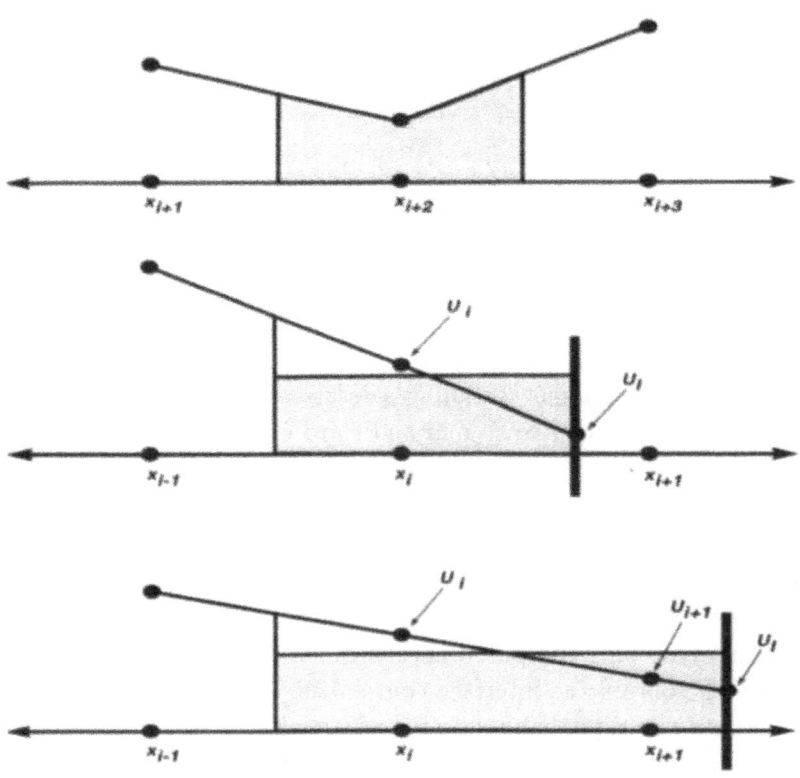

Figure 2. Geometric representation of the transition back and forth from cell averages to point values.

4. Runge-Kutta Methods

For time integration of an equation of the form

$$\vec{U}_t = \vec{f}(\vec{U}) \tag{84}$$

we use the TVD Runge Kutta methods (Shu and Osher, 1988). First order TVD Runge Kutta is forward Euler,

$$\vec{U}^{n+1} = \vec{U}^n + \Delta t \vec{f}(\vec{U}^n) \tag{85}$$

Second order TVD Runge Kutta is Heun's predictor-corrector method,

$$\vec{U}^\star = \vec{U}^n + \Delta t \vec{f}(\vec{U}^n) \tag{86}$$

$$\vec{U}^{n+1} = \vec{U}^n + \Delta t \left(\frac{1}{2} \vec{f}(\vec{U}^n) + \frac{1}{2} \vec{f}(\vec{U}^\star) \right) \qquad (87)$$

A third order TVD Runge Kutta method is given by,

$$\vec{U}^\star = \vec{U}^n + \Delta t \vec{f}(\vec{U}^n) \qquad (88)$$

$$\vec{U}^{\star\star} = \vec{U}^n + \Delta t \left(\frac{1}{4} \vec{f}(\vec{U}^n) + \frac{1}{4} \vec{f}(\vec{U}^\star) \right) \qquad (89)$$

$$\vec{U}^{n+1} = \vec{U}^n + \Delta t \left(\frac{1}{6} \vec{f}(\vec{U}^n) + \frac{1}{6} \vec{f}(\vec{U}^\star) + \frac{2}{3} \vec{f}(\vec{U}^{\star\star}) \right) \qquad (90)$$

The scheme in the previous section was described using Euler's method. Our scheme can be extended to second and third order TVD Runge Kutta as well. Each partial Runge Kutta step is seen as an "Euler type" step where the right hand side is the average of the right hand sides from the appropriate time levels. For points "away" from the interface, the right hand side is evaluated in the standard way. However, one must be careful for points "near" the interface.

Normal fluxes are only computed in cells which do not contain the interface. Cells which contain the interface contain the special moving interface flux. To compute the right hand side for a partial step of a Runge Kutta method, we intersect the normal fluxes across all time levels that are used in the right hand side of the partial step. Then we use these intersected fluxes to find the convective derivative on the right hand side for each time level. Then the convective derivatives are averaged in the appropriate way.

There is a grid point or two left which are not updated. These are updated using the interface method with the new averaged interface flux and its nearest averaged neighbor which lives in the intersection described above. Essentially, these points are updated in a conservative way using some of the interpolation described in the previous section. We have employed second and third order Runge Kutta methods and prefer third order. We will not spell out the details of the higher order Runge Kutta here, since there are many cases and it is quite lengthy. (More to come on higher order Runge Kutta in a future report.)

5. Numerical Examples

In the figures, we do not show the actual values of density and pressure for the water droplet. Density is not shown, since it is is off the scale reasonable for the gas. Pressure is not shown, since it is not computed in the 1D water model. It has a linear profile and must be calculated using the pressure in the neighboring air. We use "place holder" values for the density and the pressure, just to show the location of the water droplet. However, the values for the velocity and the temperature are unaltered.

5.1. EXAMPLE 1

We take a 1-D domain of length $10cm$ and a grid with 100 points. We have solid wall boundaries on both sides of the domain. Inside the domain we have an incompressible water droplet and a compressible, thermally perfect gas. We will assume that they are both inviscid for this calculation.

The water droplet is five grid cells long and starts at rest near the left hand side of the domain with an initial temperature of $298K$. The rest of the chamber is filled with argon gas which also starts at rest with a temperature of $1000K$. The gas to the right of the drop is at standard atmospheric pressure, while the gas to the left of the drop has a pressure over five times higher than atmospheric pressure.

We expect the high pressure air to drive the droplet to the right. The air on the left will expand and cool, while the air on the right will be compressed and heated. Eventually the droplet will compress the air to the right enough to allow a net force to the left which will slow the droplet and reverse its direction. This process will repeat and the droplet will bounce back and forth in a spring like fashion.

Third order ENO with Marquina's Jacobian is used to discretize the relevant fluxes. Then our phase interface method is coupled to third order TVD Runge Kutta in the appropriate way.

Figure 3 shows the droplet after a short time. It is being driven to the left by the pressure difference. The droplet continues to the right. The air to the left is decompressed and cooled, while the air to the right is compressed and heated. This can be seen in Figure 4 where the velocity of the droplet is over $30\frac{m}{s}$. Note that the pressure on the right now exceeds that on the left, and the net force is now to the left slowing down the droplet. The droplet will keep slowing down until it comes to a complete rest for an brief instant, before it starts moving in the opposite direction. Figure 5 shows the droplet almost at rest, just before it turns around in the other direction. Figure 6 shows the droplet moving in the opposite direction Now it is decompressing and cooling the air on the right while it compresses and heats the air on the left. This process continues, mimicking a spring-like phenomena.

Note that we conserve total energy. Thus there will be an error in the computation of the internal energy in the water. Equation 80 shows that the internal energy will decrease in a nonphysical way proportional to $\frac{p^2}{2\rho}$. Since $\frac{p^2}{2\rho}$ is larger on the right hand side of the droplet then on the left, for most of our computation, we would expect the temperature on the right of the droplet to be lower than that on the left. This can be seen in figures 4, 5, and 6.

5.2. EXAMPLE 2

We take a 1-D domain of length $10cm$ and a grid with 100 points. We have solid wall boundaries on both sides of the domain. Inside the domain we have an incompressible water droplet and a compressible, thermally perfect gas. We will assume that they are both inviscid for this calculation.

The water droplet is five grid cells long and starts at rest near the right hand side of the domain with an initial temperature of $298K$. The rest of the chamber is filled with argon gas which also starts at rest with a temperature of $1000K$. Near the center of the domain, we impose a pressure jump. The gas to the right of this jump is at standard atmospheric pressure, while the gas to the left of this jump has a pressure over five times higher than atmospheric pressure. This is a "Sod" shock tube problem with a water droplet included in the tube.

We expect the pressure jump to split into a shock, a contact discontinuity, and a rarefaction. The shock will travel to the right of the domain and collide with the water droplet, reflecting off in the opposite direction.

Third order ENO with Marquina's Jacobian is used to discretize the relevant fluxes. Then our phase interface method is coupled to third order TVD Runge Kutta in the appropriate way.

Figure 7 shows the shock traveling to the right toward the water droplet. It hits the droplet, reflects off, and then travels in the opposite direction as shown in figure 8. Note that the shock will impart momentum to the droplet forcing it to the right with high pressure. This can be seen slightly in figure 8. Figure 9 shows the solution at a later time, where it is more apparent that the droplet is moving to the right. Here, the velocity of the droplet is positive and the velocity of the air to the right of the droplet has a linear profile.

Figures 10 and 11 show the same example with 400 points, just to illustrate convergence. Note that there is a peak in the density and a dip in the temperature by the droplet after reflection. This is the standard effect due to loss of room to interpolate.

5.3. EXAMPLE 3

We repeat example 1, except this time we allow the flow to be fully viscous.

Once again the droplet is driven to the left, slows down, and reverses direction. Figure 12 shows the droplet after reversing direction. Comparing this to figure 6, we can see that the cold droplet is absorbing heat from the hot gas. The gas temperature profile near the droplet is falling, while the corresponding density profile is increasing.

Figure 13 shows the drop after reversing direction once again, now traveling to the right.

R. FEDKIW ET AL.

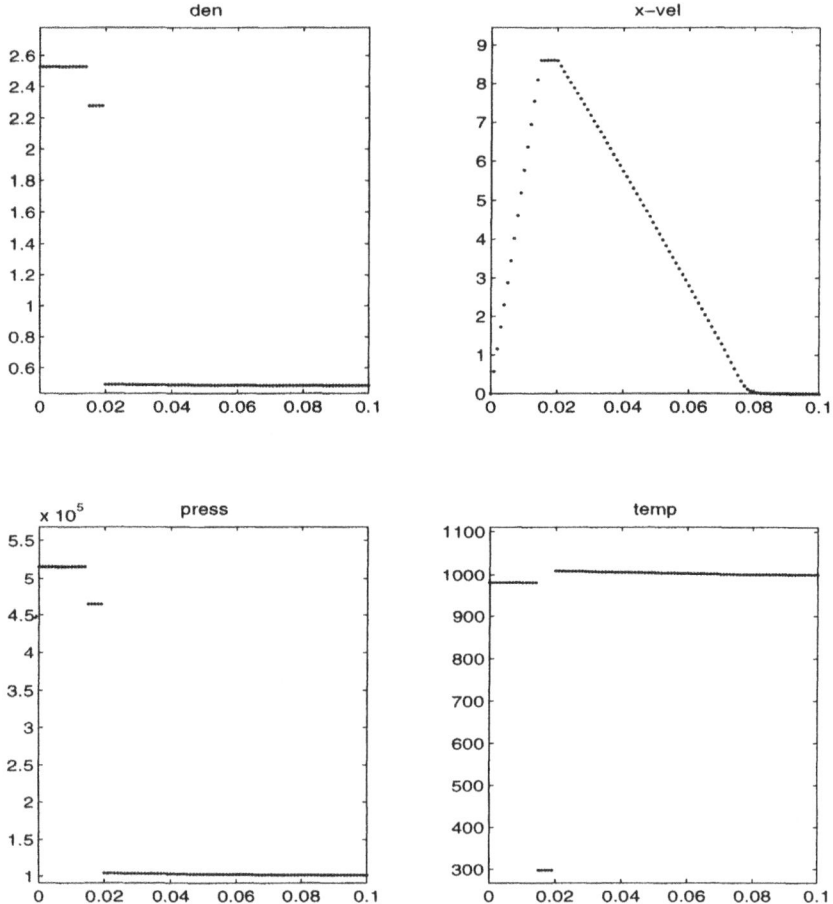

Figure 3.   The water droplet is initially driven to the right from the pressure difference in the air. Inviscid case.

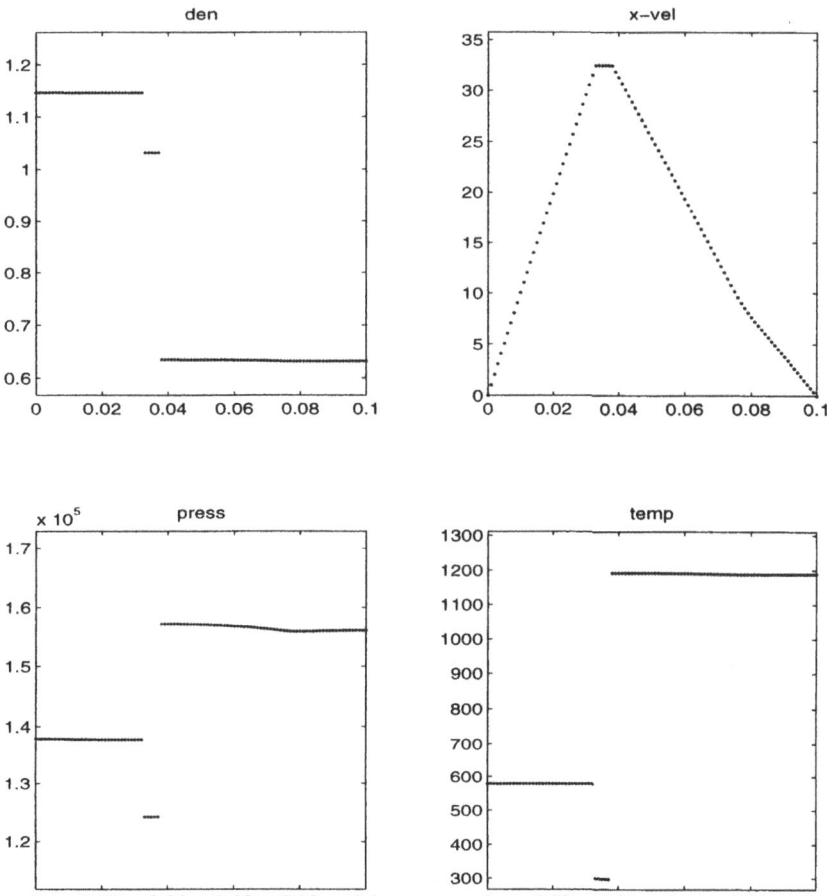

Figure 4. The gas on the left has been decompressed and cooled, while the gas on the right has been compressed and heated. Now, the net force is to the left and the droplet will start to slow down. Inviscid case.

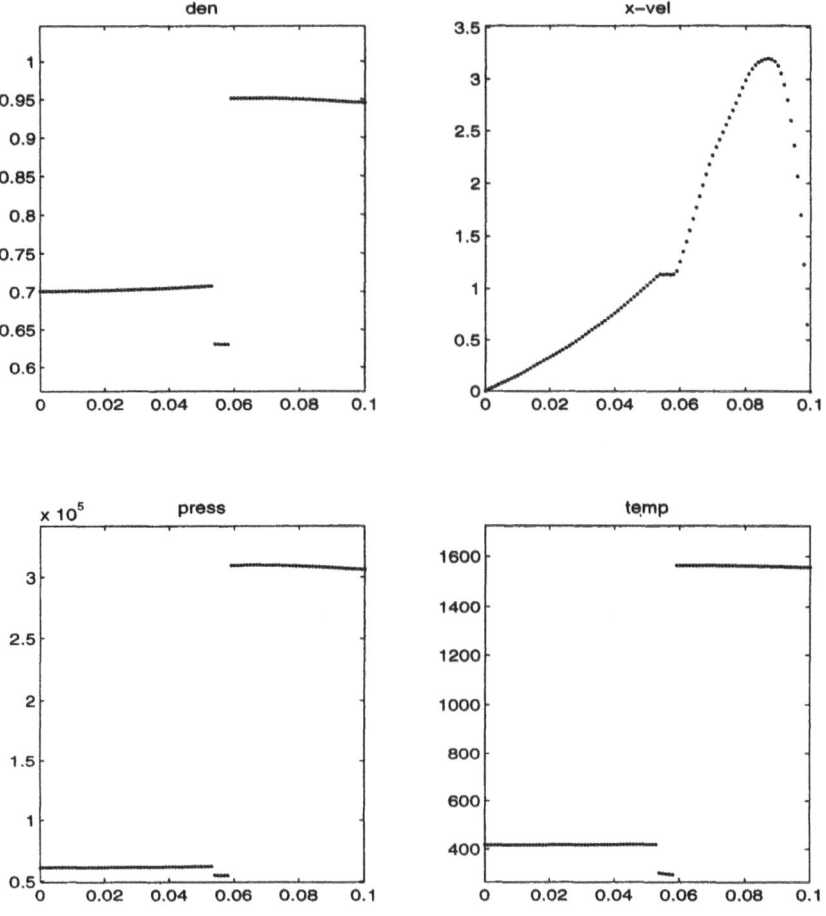

Figure 5. The droplet has almost come to a complete rest, after which it will begin to move in the opposite direction. Inviscid case.

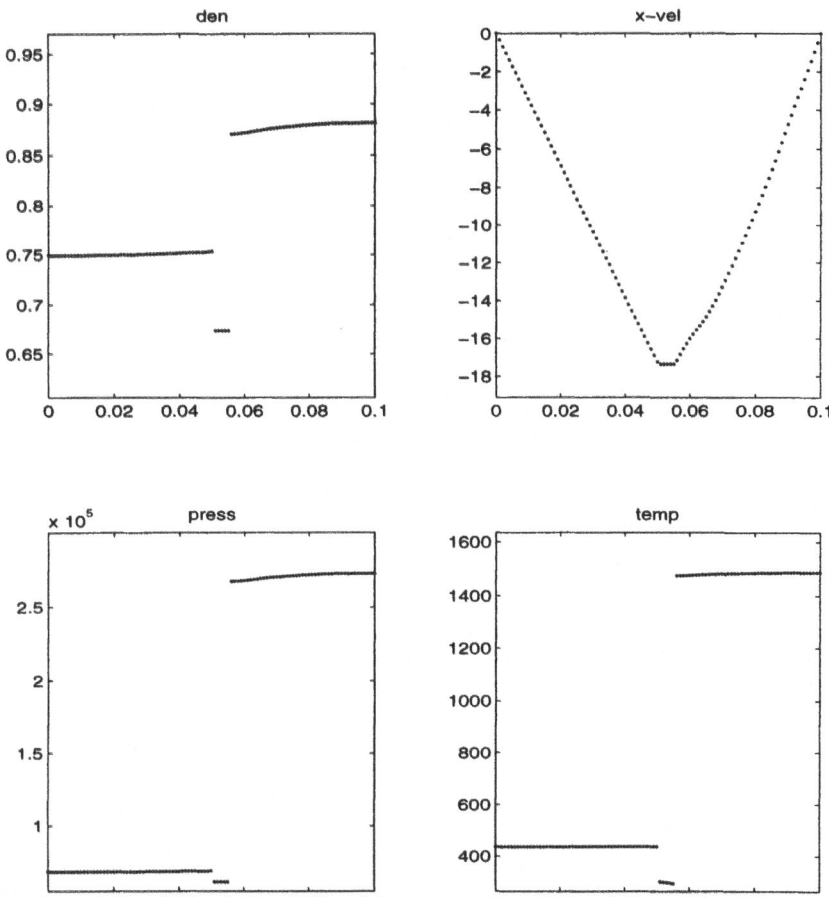

Figure 6. Now, the droplet moves to the left. It decompresses and cools the gas to the left, while it compresses and heats the gas to the right. Inviscid case.

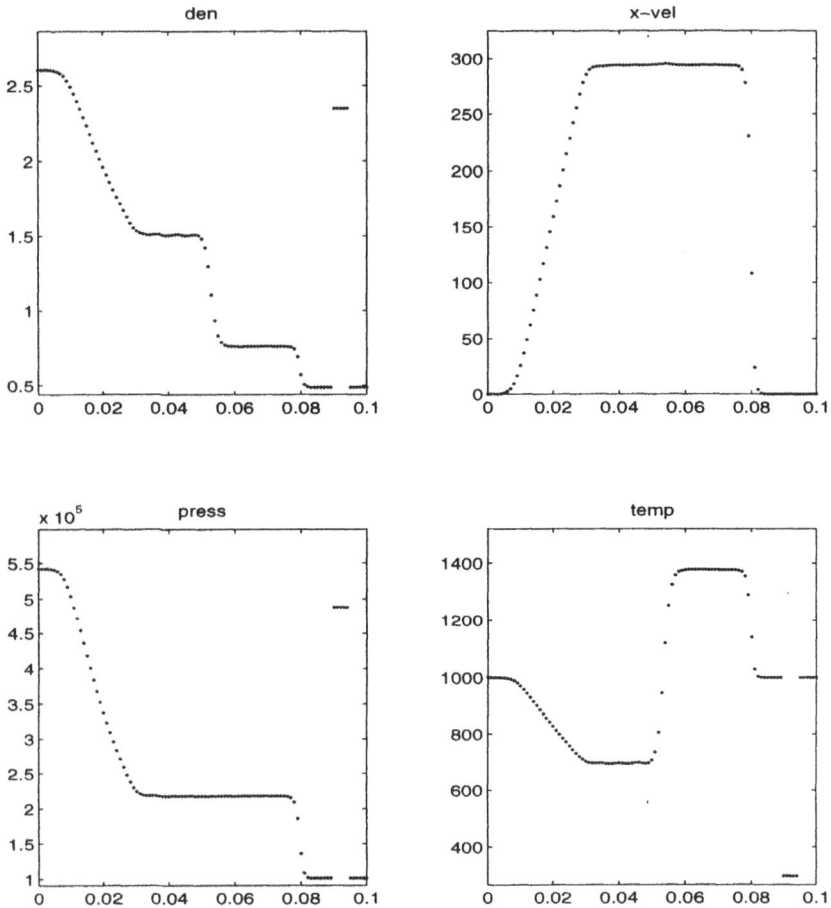

Figure 7.   The shock is traveling to the right and is about to collide with the water droplet. 100 grid points.

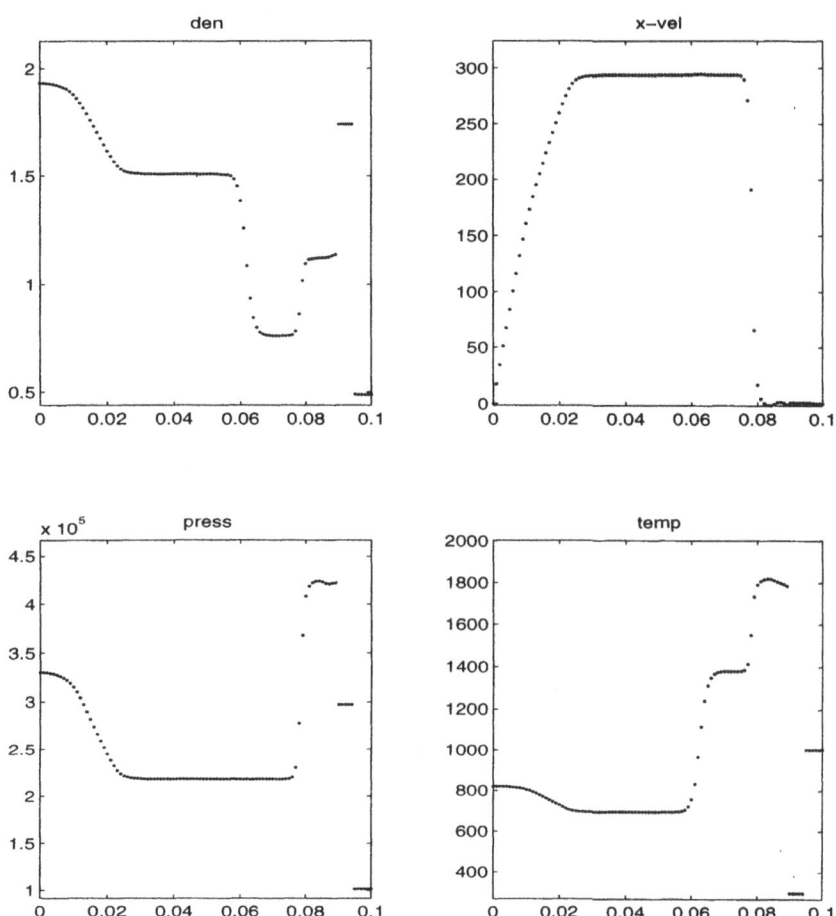

Figure 8. The shock has reflected off the water droplet and is now traveling to the left. 100 grid points.

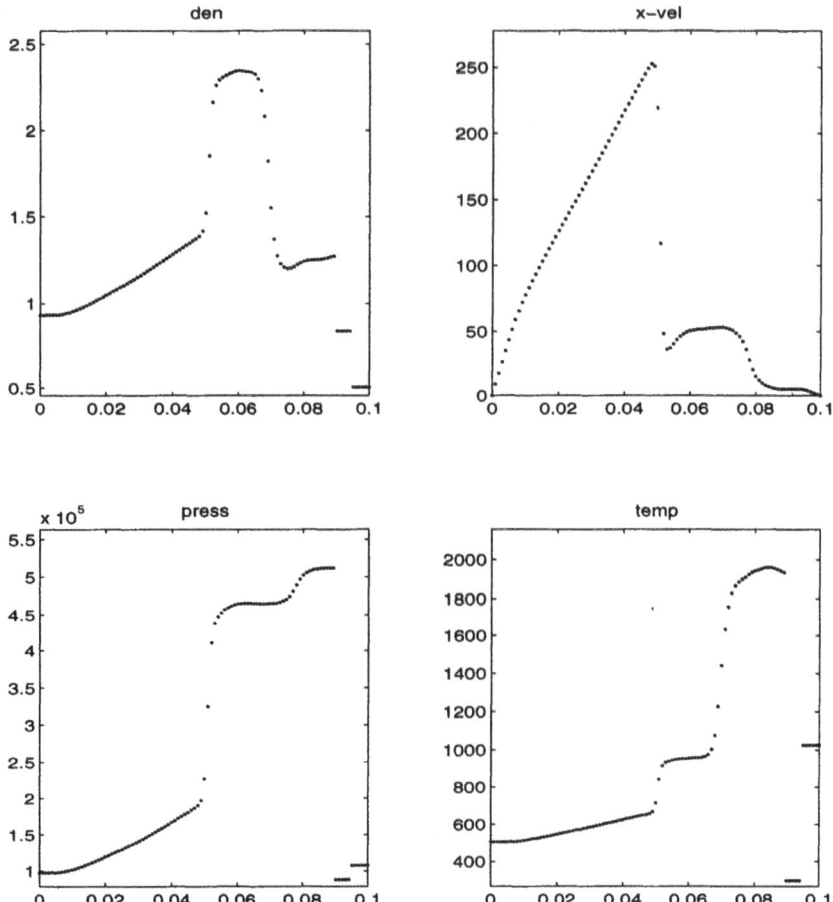

Figure 9. The droplet has a positive velocity and is moving slowly to the right. The air to the right of the droplet has a linear velocity profile. 100 grid points.

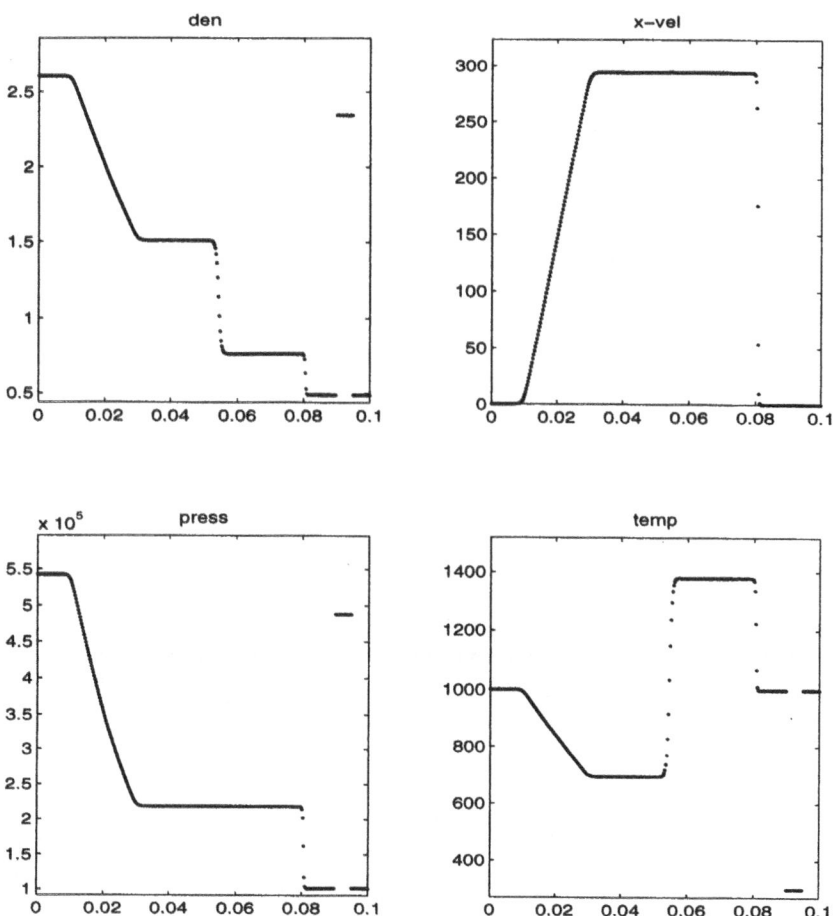

Figure 10. The shock is traveling to the right and is about to collide with the water droplet. 400 grid points.

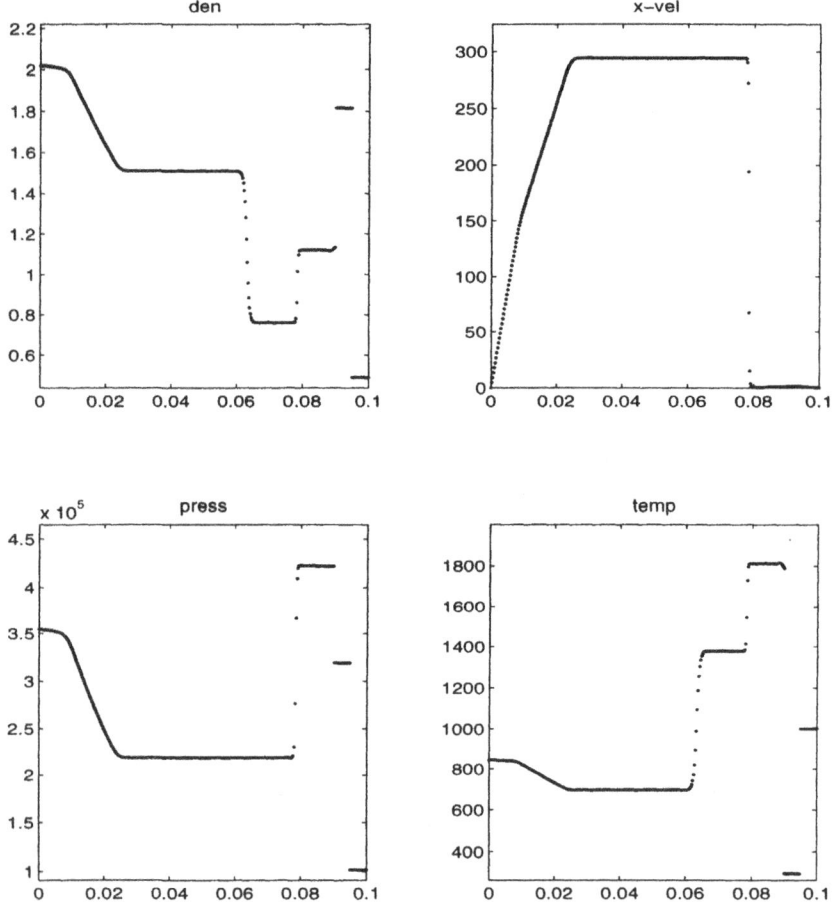

Figure 11. The shock has reflected off the water droplet and is now traveling to the left. 400 grid points.

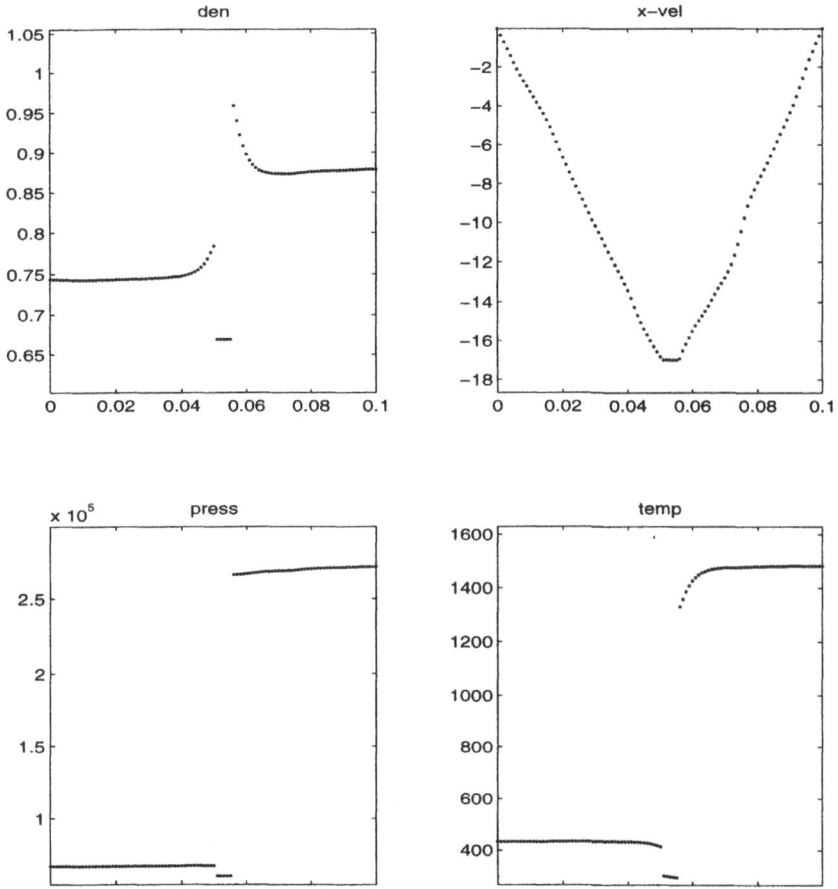

Figure 12. Now, the droplet moves to the left. It decompresses and cools the gas to the left, while it compresses and heats the gas to the right. Viscous case.

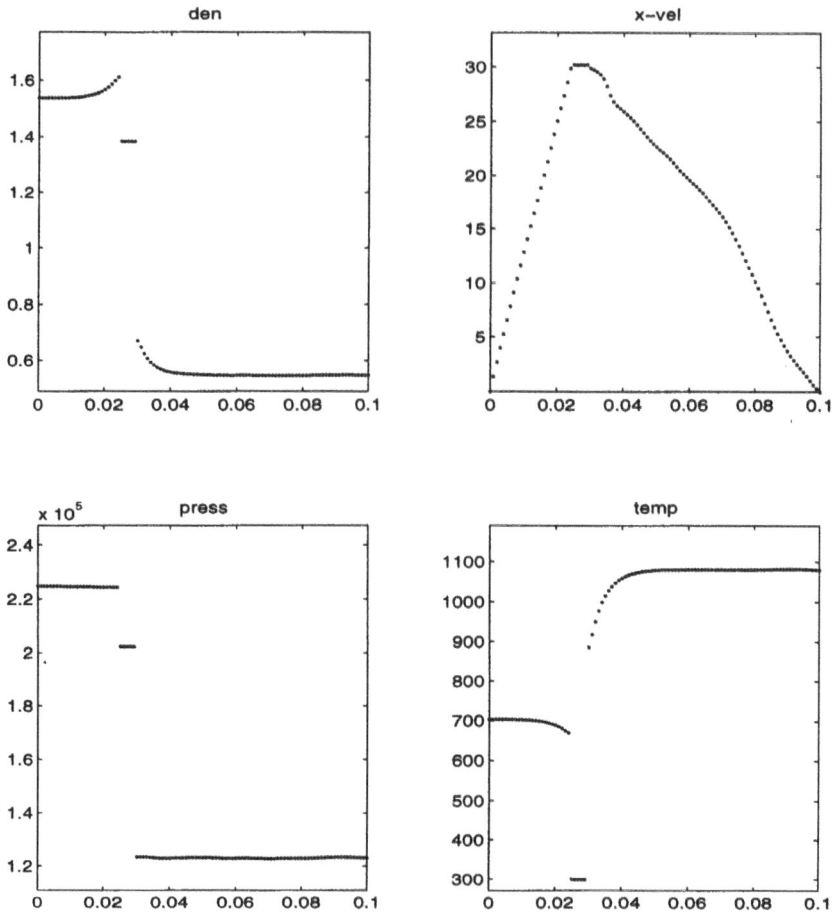

Figure 13.   Once again, the droplet changes direction. It is now move to the right. Viscous case.

6. Conclusion

We have developed a new approach to multiphase flow involving a liquid droplet hit by a high speed gas flow, or more generally for any mixed compressible-incompressible flows. Our approach is based on decomposing the domain into distinct regions where the respective, different, model equations apply, and discretizing each region with the most appropriate techniques. Proper boundary conditions are imposed on the internal interface boundary, mainly in terms of continuity or conservation conditions. In order to maintain discrete conservation and use finite difference ENO methods in the gas phase, we developed a novel formulation using a Lagrangian control volume method near the interface. Novel time stepping schemes are also developed for nodes near the interface, since a given node will change its unknowns and governing equations as the interface passes over it.

For basic proof of principle and to investigate the details of of the novel spatial discretization and time stepping required near the interface, we did this initial theory and computations in 1D, Our experiments showed good results for both smooth flows and flows with shocks reflecting from the liquid surface, Future work will extend the method to 2D, using a level set representation of the interface.

References

Anderson, John D., "Hypersonic and High Temperature Gas Dynamics," McGraw-Hill, 1989.

Atkinson, Kendall E., "An Introduction to Numerical Analysis," Wiley, 1989.

Fedkiw, Ronald P., "A Survey of Chemically Reacting, Compressible Flows," UCLA (Dissertation), 1996.

Fedkiw, R., Merriman, B., and Osher, S., "Numerical methods for a mixture of thermally perfect and/or calorically perfect gaseous species with chemical reactions," *UCLA CAM Report 96-1*, January 1996.

Fedkiw, R., Merriman, B., Donat, R., and Osher, S., "The Penultimate Scheme for Systems of Conservation Laws: Finite Difference ENO with Marquina's Flux Splitting," *UCLA CAM Report 96-18*, July 1996.

Kee, Miller and Jefferson, CHEMKIN: A General Purpose Problem Independent, Transportable Fortran Chemical Kinetics Code Package," *SAND 80-8003*, Sandia National Laboratories, March 1986.

Mulder, W., Osher, S., and Sethian, J. A., "Computing Interface Motion in Compressible Gas Dynamics," *J. Numer. Analysis*, Vol. 30, 1991, pp. 542.

Shu, C.-W. and Osher, S., "Efficient Implementation of Essentially Non-Oscillatory Shock Capturing Schemes II (two)," *UCLA CAM Report 88-12*, April 1988.

Stall, D.R. and Prophet, H., "JANAF Thermochemical Tables," National Standard Reference Data Series, 1971.

Sussman, M., "A Level Set Approach to Computing Incompressible Two Phase Flow," Ph.D. Thesis, UCLA Department of Mathematics, 1995.

A. Eigensystem

Consider $\vec{F}(\vec{U})$ defined in equation 3.

The eigenvalues of the Jacobian matrix of $\vec{F}(\vec{U})$ are

$$\lambda^1 = u - c \tag{91}$$

$$\lambda^2 = \cdots = \lambda^{N+2} = u \tag{92}$$

$$\lambda^{N+3} = u + c \tag{93}$$

The left eigenvectors $\vec{L}^p$, are the rows of the following matrix.

$$
\begin{pmatrix}
\frac{b_2}{2} + \frac{u}{2c} + \frac{b_3}{2} & -\frac{b_1 u}{2} - \frac{1}{2c} & \frac{b_1}{2} & \frac{-b_1 z_1}{2} & \cdots & \frac{-b_1 z_{N-1}}{2} \\
1 - b_2 - b_3 & b_1 u & -b_1 & b_1 z_1 & \cdots & b_1 z_{N-1} \\
-Y_1 & 0 & 0 & & & \\
\vdots & \vdots & \vdots & & I & \\
-Y_{N-1} & 0 & 0 & & & \\
\frac{b_2}{2} - \frac{u}{2c} + \frac{b_3}{2} & -\frac{b_1 u}{2} + \frac{1}{2c} & \frac{b_1}{2} & \frac{-b_1 z_1}{2} & \cdots & \frac{-b_1 z_{N-1}}{2}
\end{pmatrix}
\tag{94}
$$

The right eigenvectors $\vec{R}^p$, are the columns of the following matrix.

$$
\begin{pmatrix}
1 & 1 & 0 & \cdots & 0 & 1 \\
u - c & u & 0 & \cdots & 0 & u + c \\
H - uc & H - \frac{1}{b_1} & z_1 & \cdots & z_{N-1} & H + uc \\
Y_1 & Y_1 & & & & Y_1 \\
\vdots & \vdots & & I & & \vdots \\
Y_{N-1} & Y_{N-1} & & & & Y_{N-1}
\end{pmatrix}
\tag{95}
$$

Here I is the $N - 1$ by $N - 1$ identity matrix, and

$$c = \sqrt{\frac{\gamma p}{\rho}}, \qquad H = \frac{E + p}{\rho} \tag{96}$$

$$b_1 = \frac{\gamma - 1}{c^2}, \quad b_2 = 1 + b_1 u^2 - b_1 H, \quad b_3 = b_1 \sum_{i=1}^{N-1} Y_i z_i \tag{97}$$

$$z_i = h_i - h_N - c_p W \left(\frac{1}{W_i} - \frac{1}{W_N} \right) T \tag{98}$$

For more on the eigensystem, see (Fedkiw *et al.*, 1996).

ON SOME OUTSTANDING ISSUES IN CFD (1996)

RAINALD LÖHNER
GMU/CSI, The George Mason University
Fairfax, VA 22030-4444, USA

Abstract.
Against the general feeling that 'CFD is solved', I offer a few remarks from the trenches. When performing large-scale simulations of geometrically and physically complex flows, the notion of reliable, cost-effective, widely applicable and easy-to-use CFD still appears more as fiction than reality. The present monograph addresses some major areas where progress is urgently required. These areas are listed in the order a typical CFD run proceeds: model import/creation, grid generation, flow solvers, mesh adaptation and quality control. Finally, with a view towards the coming decade, code, data and project management are discussed.

1. Introduction

It has been a decade since a series of papers by Kutler (see (Kutler, 1985) for a representative example) set forth the notion that CFD is a solved problem. Following this line of thought with all its consequences, the CFD division at NASA Ames has disappeared. The central thesis of Kutler's thought was indeed correct: CFD understood as implicit, structured grid solvers for steady-state, Reynolds-Averaged Navier-Stokes equations with an algebraic turbulence model was a solved problem by the mid 1980's. The flaw in the thought process remains that such solvers can only address a small fraction of possible flows, namely those that are both geometrically and physically simple.

CFD comprises much more than this narrow class of problems, and it is here that major developments have taken place in the last decade. *Geometrical complexity* was addressed by introducing unstructured data representations, either on the macro-level (block-structured grids (Thompson, 1988), (Steinbrenner *et al.*, 1990), (Allwright, 1990), overset grids (Meakin

195

V. Venkatakrishnan et al (eds.), Barriers and Challenges in Computational Fluid Dynamics, 195-212.
© 1998 *Kluwer Academic Publishers.*

& Suhs, 1989), (Meakin, 1993)) or the micro-level (tetrahedral (Löhner *et al.*, 1987), (Peraire *et al.*, 1988), (Batina, 1990), (Peraire *et al.*, 1990), (Barth, 1991), (Mavriplis, 1991), (Peraire *et al.*, 1992), (Weatherill *et al.*, 1993), (Baum *et al.*, 1994), Cartesian (Gaffney *et al.*, 1987), (Berger & LeVeque, 1989), (Quirk, 1994), (Aftosmis *et al.*, 1995), (Bayyuk *et al.*, 1996) grids), in combination with fully automatic grid generation techniques (Baker, 1987), (Löhner, 1988),(Löhner & Parikh, 1988), (Peraire *et al.*, 1988), (Baker, 1989), (Peraire *et al.*, 1990), (Weatherill, 1992), (Weatherill *et al.*, 1993), (Löhner, 1996), (Löhner, 1996), (Löhner, 1996). These techniques proved so successful that the pre-processing stage (i.e. the acquisition of surface data, specification of boundary conditions and input of desired element size and shape in space) has become the main bottleneck of large-scale, geometrically complex runs. *Numerical accuracy* was improved via new flux vector splitting schemes (Liou & Steffen, 1993), (Coquel & Liou, 1993), (Jameson, 1993), (Liou, 1995), as well as Riemann solvers for complex equations of state [Colella]. The *reliability* limits and solution uniqueness of implicit schemes for time-accurate simulations were analyzed (Yee *et al.*, 1991), (Yee & Sweby, 1995). With the advent of computing power, large scale *direct simulations of turbulence* were carried out. *Mesh adaptation* became a major focus of research, and was used advantageously in some areas, achieving speed-ups and memory reductions in excess of 1:100 (Löhner & Baum, 1992), (Baum & Löhner, 1994). The improved flow solvers were not only used for analysis but also **design** (Jameson, 1995), (AIAA, 1995), and for *multidisciplinary applications*, e.g. fluid-structure interaction (Löhner *et al.*, 1995).

As one can see from this non-exhaustive list, CFD was rather far from being 'a solved problem' a decade ago. In the meantime, CFD has become an important prediction, analysis, design and discovery *tool* in many areas of science and engineering. As with any other technical product, a tool undergoes continuous improvements, leading constantly to new developments. Therefore, it is still hard to predict the demise of CFD at this point. We may predict a transformation, but not a demise.

In the following, developments required to take CFD from its present position to a reliable, cost-effective, widely applicable and easy-to-use tool for the prediction of geometrically and physically complex flows are considered. The areas are listed in the order a typical CFD run proceeds: model import/creation (Section 2), grid generation (Section 3), flow solvers (Section 4), mesh adaptation (Section 5), parallel processing (Section 6), management of large codes and quality control (Sections 7,8), and data management for large projects (Section 9). CFD as a tool to develop material models is considered in Section 10. Finally, a summary and conclusions are given in Section 11.

2. Model Import/Creation

Model import and/or creation, together with boundary condition specification, are currently the major bottlenecks for geometrically and physically complex flows. What follows in the simulation process - grid generation, flows solvers, data reduction - has been automated to a high degree. We may (and will) improve each of these subsequent stages of an analysis, but they are no longer as critical as before: given enough computer power, they will lead to acceptable results. However, if this first phase of any CFD run cannot be reduced to a matter of hours or at most days, competing ways of analysis and design will again play an increasing role. The argument that 'CFD is cheaper' or 'CFD yields more insight' does not hold for a company, where time-to-market is of paramount importance. It may be cheaper to design the product via CFD, but if the product is a year late because a team of engineers spent half a year creating and meshing a CFD model, the company may no longer be able to compete. For experiments, the time-to-measurable-model has been reduced drastically in recent years (e.g. via laser casting), allowing in some cases model construction and measurement in a matter of days. It is conceivable that this experimental process can be improved further, leading to CAD-to-experiment cycles that are only a few hours long. Design based on large data bases, in some cases in conjunction with neural net learning (Sellar & Batill, 1996) has also been improved steadily. Again, it is conceivable that very good evolutionary designs may be obtained by simply being able to intelligently extrapolate from our current data basis.

It is therefore imperative to reduce model import and creation to a matter of hours or at most days. As far as model import/creation are concerned, some unresolved issues are:

- Fast data input for complex geometries;
- Geometry repair and cleaning; and
- Link to Virtual Reality data sets.

In an attempt to cut through the maze of 'standards' and the sheer size and complexity of some of the data sets obtained from CAD packages, we have developed discrete data meshing (Löhner, 1996) and meshing from clouds of points (Löhner, 1996). Particularly the use of triangulations as a means to define surfaces has proven very useful, allowing the re-use of surface discretizations from other disciplines (e.g. structural dynamics), and the import of non-analytical, and hence discrete, data sets from medical and geological remote sensing (Löhner et al., 1995). At the same time, Virtual Reality (DataWare Inc., 1996) data sets can be used as a basis for model description. These data sets, which will play an ever increasing role in the next decade, are usually topologically inconsistent. Whenever a new surface is required, a triangular or trapezoidal element is introduced. As these elements are simply distributed in space, they do not form a topologically consistent subdivision of space. Figure 1 shows the surface triangulation of a rocket launcher that was obtained from a virtual reality set consisting of approximately 140,000 triangles. It took 6 man-weeks to generate, which is clearly unacceptable. The main difficulties were: management and local visualization of the large data set, reconstruction of topologically consistent surfaces, and surface intersection. The reconstruction of the proper 'wetted surface' for a CFD run from such a data set represents an interesting and challenging problem for the coming decade.

3. Grid Generation

Grid generation has been a major focus of research over the last decade. Particularly in the area of unstructured tetrahedral (Baker, 1987), (Löhner, 1988), (Löhner & Parikh, 1988), (Peraire et al., 1988), (Baker, 1989), (Peraire et al., 1990), (Weatherill, 1992), (Weatherill et al., 1993), (Löhner, 1993), (Löhner et al., 1995), and Cartesian (Berger & LeVeque, 1989), (Quirk, 1994), (Aftosmis et al., 1995), (Bayyuk et al., 1996) gridding, progress has been such that the generation of suitable grids for inviscid simulations of geometrically complex bodies (and this is not always a contradiction) may be considered as a solved problem.

As far as grid generation is concerned, some unresolved issues are:

- Minimal user-input meshing; and
- Proper gridding for Reynolds-Averaged Navier-Stokes, in particular for general geometries and wakes.

In order to address the first issue, we have developed adaptive background grids (Löhner, 1996). The idea here is to start from surface discretization tolerances. If required, these may be combined with user-specified mesh sizes at specific sites in space. The background grid is then adapted automatically to yield a smooth transition of element size and shape in space that satisfies all required tolerances and set element sizes. Figure 2, taken from (Baum et al., 1996), shows the surface discretization for a truck. An adaptive background grid with six levels of refinement was used to properly, and automatically, define the grid distribution in space.

4. Flow Solvers

The pervading themes here are applicability, reliability and speed. Many flows in engineering tend to be turbulent (model uncertainty), may have reacting species (stiff source terms, tens or hundreds of unknowns per grid point), exhibit boundary layers (meshing difficulties), and consist of multiple phases (cavitation, dusty flows). The current perception is that the simulation of these flows will only require evolutionary steps from present-day, mature Euler solvers. This perception may be seriously flawed. As a simple example, consider multiphase flows. When trying to apply high resolution schemes to shocks propagating through water/air interfaces, none of the commonly used approximate Riemann solvers worked (Luo *et al.*, 1996). The only schemes that gave reasonable results were first order Godunov (too diffusive for engineering applications) and FCT (mathematically incomplete). Among the outstanding unresolved issues for flow solvers we just mention a few:

- Higher order schemes;
- Fast time-accurate Navier-Stokes solvers;
- Real life physics (complex equations of state);
- Multiphase flows (cavitation, fluid/air interfaces);
- Chemically reacting flows; and
- Software maintenance for large-scale, multi-physics codes.

Each of these areas has been the center of considerable research efforts, but much more is required to arrive at the stated goal of reliable, cost-effective, widely applicable and easy-to-use CFD. Figure 3 shows a fragmentation calculation using a JWL equation of state for a high explosive, approximately 200 independently flying objects, and automatic local and global remeshing. Initially, the minimum distance between the fragments is of 0.5 *mm*, while the average size of each fragment is 30 *mm* × 20 *mm* × 15 *mm*. In contrast, the room size is several meters. Typical meshes for this simulation were of the order of 2.5 Mtets, and the simulations required of the order of 8 hours on the CRAY C-90. Figure 3a shows the fragments at a certain time, together with their projected impact points on the walls. The comparison to experiments is shown in Figure 3c. Although the correlation with experimental data is very good, the numerics required to do such runs with density differences of 1:1,200 require further attention.

5. Grid Adaptation

Over the last decade, grid adaptation techniques were developed with the following premises:

- An optimal grid leads to an optimal allocation of resources, and hence an optimal speed;
- If one regards grid adaptation as a form of grid generation, a further degree of automation, and hence rapid turnaround, is achievable;
- Error estimation, or at least error indication, becomes an integral part of any given solver, hence giving a bound on reliability.

These premises have not always proven correct, although in many cases spectacular savings in CPU and memory requirements, as well as solution enhancements were obtained. The question of error estimation has remained rather vague, and it is fair to say that unlike structural mechanics, 'hard' error bounds for Euler or high Reynolds-number flows have been difficult to achieve. The mandate still remains, that, as with experiments, CFD should give error margins on the solutions. Although this may seem impossible for many flows, grid adaptation and sequencing can at least take off some of the uncertainty, and in some cases yield proper error bounds. An unresolved issue here is:

- Grid Adaptation for transient, viscous, high Reynolds-number problems.

6. Parallel Processing

The very notion of 'large scale simulations' immediately conjures up the notion of speed as a critical item. Indeed, one may argue that the major portion of human endeavour in the computational sciences over the last 25 years has been devoted to speeding up the pertinent solvers. The last two decades have seen dramatic performance gains due to the following algorithmic improvements:

- Field solvers with limiting procedures which minimize numerical diffusion, allowing for crisp definition of discontinuities in advection-dominated problems (transport, fluids, wave propagation, etc.);
- Multi-level techniques, such as multigrid methods for grid-based field solvers, tree-codes for particle-based solvers, and fast multipole methods; and
- Automatic adaptive mesh refinement algorithms.

Each one of these techniques by itself can increase CPU performance by an order of magnitude, leading to a best-case scenario of 1:1000 for some problems. After such a spectacular gain, the immediate question is: what next ? The answer falls into three categories: better data structures, better use of single-processor hardware, and parallelization. The first two of these have been considered before (Löhner, 1993), (Löhner, 1994), and will not be discussed further. The third category deserves attention, as there can hardly be any doubt that the 90's are the decade of parallelism. Even though machines with several powerful vector-processors were installed at many places in the 80's, the operating system hardly allowed for the use of several processors during the same run. Moreover, in many instances the compilers were not mature, leading to meaningless gains in performance. With the advent of massively parallel machines, i.e. machines in excess of 500 nodes, the exploitation of parallelism in solvers has become a major

TABLE 1. Speed-Ups Obtain-
able (Amdahl's Law)

$\frac{R_p}{R_s}$	50%	90%	99%
10	1.81	5.26	9.17
100	1.98	9.90	50.25
1000	2.00	9.91	90.99

focus of attention. If we recall Amdahl's law, the speed-up s obtained by parallelizing a portion α of all operations required is given by

$$s = \frac{1}{\alpha \cdot \frac{R_s}{R_p} + (1 - \alpha)} \ , \tag{1}$$

where R_s, R_p denote the scalar and parallel processing rates (speeds) respectively. Table 1 shows the speed-ups obtained for different percentages of parallelization and different number of processors.

Note that even on a traditional shared-memory, multiprocessor vector machine, such as the CRAY C-90 with 16 processors, the maximum achievable speed-up between (badly written) scalar code and (well written) parallel vector code is a staggering $\frac{R_p}{R_s} = 240$. What is important to note is that as we migrate to higher numbers of processors, only the embarrassingly parallel codes will survive. Most of the applications ported successfully to parallel machines to date have followed the Single Program Multiple Data (SPMD) paradigm. For grid-based solvers, a spatial subdomain was stored and updated in each processor. For particle solvers, groups of particles were stored and updated in each processor. For obvious reasons, load balancing (Williams, 1990), (Simon, 1991), (Mehrotra *et al.*, 1992), (Vidwans *et al.*, 1993) has been a major focus of activity.

Despite the striking successes reported to date, a word of caution should be placed here. Only the simplest of all solvers: explicit timestepping or implicit iterative schemes, perhaps with multigrid added on, have been ported without major changes and/or problems to massively parallel machines with distributed memory. Many code options that are essential for realistic simulations will not be easy to parallelize. Among these, we mention local remeshing (Löhner, 1990), repeated h-refinement, such as required for transient problems (Löhner & Baum, 1992), contact detection and force evaluation (Haug *et al.*, 1991), some preconditioners, applications where particles, flow, and chemistry interact together, and applications with rapidly varying load imbalances. Even if 99% of these codes can be parallelized, the maximum achievable gain will be restricted to 1:100. Achieving full parallel performance and full code functionality for machines with more than 1,000 processors will require major algorithmic, software and hardware improvements in the coming decade. If we accept as fact that for most large-scale codes we may only be able to parallelize 99% of all operations, the shared memory paradigm, discarded for a while as non-scalable, will make a comeback. It is far easier to parallelize some of the more complex algorithms, as well as cases with large load imbalance, on a shared memory machine. And it is within present technological range to achieve a 100 processor, shared memory machine.

Immediate and more substantial gains in parallelism accrue from the fact that for vast areas where large-scale computing is used, the runs have to be carried out repeatedly. Consider for example the parametric study of wings for commercial aircraft. For each new design, the flowfields at different Mach-numbers and Reynolds-numbers have to be evaluated. This implies the equivalent of tens or hundreds of individual runs. These runs will have very similar CPU and memory requirements, so that if sufficient CPU and memory are available per processor, running them in (embarrassingly) parallel mode is the obvious choice. Besides parametric studies, design and optimization also offer inherent parallelism. For the most common design algorithms (see (AIAA, 1995) for a good cross-section), gradient estimations of the cost function are required. Every gradient calculation for n_d design variables requires n_d independent and very similar calculations. The same applies to genetic algorithms (Goldberg, 1989). Given that realistic designs have several hundred design variables, the idea is to solve on each processor an independent but similar (not only in physics, but also in CPU and memory cost) problem. As with parametric studies, this technique has been used with great success in those instances where the individual processors are sufficiently powerful, such as the Fujitsu VPP-128.

7. Management of Large Codes

Consider the typical life cycle of a large-scale computing code. The code starts as a basic research tool that is tried for a very limited number of cases. Let us denote this phase the proof-of-concept stage. If successful, and written with some generality in mind, the code will be expanded, adding new physical models, improving numerical techniques, adding practical engineering/physics options, etc. We will denote this phase as the expansion and acceptance stage. Again, if this phase is successful, a manual is written. This entails a number of consequences:

- the user base expands dramatically;
- the demands for expansion and inclusion of new options multiply;

- the code, and not the original application field, becomes the focus of attention of the developer;
- the code becomes a tool, with all the associated demands (quality assurance, bug control, training sessions, hot-line, etc.).

This is the reality for most of the successful large-scale computing codes currently in use. Given the uncertainty of getting past the proof-of-concept and expansion stages, no business venture will try to start a new code ab initio. The consequences of this are manifold and obvious:

- a) Maintenance: Most large-scale computing codes are the brainchild, and are maintained and updated, by one or at most two individuals. What typically happens is that while others may make changes or additions here and there in these codes, the faith of the community in these individuals is such that they have to 'bless' any changes and additions made by others, and re-incorporate them into the master-version. On the other hand, an individual can only code and debug a limited amount of additions per day without endangering the integrity of the code. Empirical evidence indicates that 1-2 debugged subroutines a day already represent some form of upper bound. Therefore, it takes a lot of duplication and delays to introduce changes into these codes.
- b) Style: The style in which a code is written will invariably reflect its originator. There are a lot of similarities between composing music and writing code. In the same way that we can differentiate between a composition by Bach or Mozart, so can we recognize a particular developer's code when reading it. Given that most of the code originators are not professional code writers, but scientists and engineers, it is not surprising that most codes used for large-scale computing are written using outdated styles, data structuring, and language options. One should also recognize that it is not the code developer's special interest to adher to the latest fad in computer science. Rather, it is to incorporate as quickly as possible improved physical models, better numerical techniques, or engineering options.
- c) Cottage Industry: Given that the code originator is preoccupied with new options and changes to the code, a cottage industry of knowledgeable users will form around it, in order to answer the more mundane questions of users, and to perform runs for those who only need a few results or do not want to invest in training for their personnel.

While we have progressed to a point where in the manufacturing industries we have a high degree of parallelization, the same does not apply to code writing. Object-oriented coding and CASE tools offer only a partial solutions here. After all, when all the pieces are put together, we have to solve the right ordinary or partial differential equations, with the right options and boundary conditions.

A recurring question has been the merit of can-do-all codes vs. specialized, one-task codes. Both exist, implying that both have their merit (clearly a Darwinian thought, not universally applicable). As far as limitations for CFD, the principal progress inhibitor is the way codes are written and maintained, and perhaps the human brain itself.

Empirical evidence indicates that good codes are written, maintained, and enhanced by one or two 'gurus'. The same applies in the opposite direction: a human being can not maintain and enhance more than a handful of codes. Given this situation, it is clear that progress within a single code is limited by the number of hours its author(s) can devote to it, and hence it is not scalable. This implies that the limitations of general purpose codes are due to time (and indeed general purpose codes typically lag behind the latest numerical developments), and those of highly specialized codes are due to cost (and indeed these codes are only maintained by professionals if a large user base exists (e.g. airfoil, wing codes), otherwise by students).

8. Quality Control

As portions of CFD transition from research to production tool, the danger of misuse by unknowing users is a major concern. Efforts are underway at some of the larger agencies and laboratories to 'certify' CFD codes. Anyone who has ever written a sufficiently complex code can but wonder at such efforts. Any code is only error-free until the next error is found. Sometimes, errors are hidden for years. The blind use of certified codes poses a grave danger, as it takes away the healthy critical distance we should have with anything made by humans. One can already envision scientific papers with sentences of the form 'the problem was run with code X'. No mention - and hence no thought - of:

- Physical accuracy of the model (Euler/RANS/LES/DNS, turbulence model and suitability, similar studies, etc.);
- Geometrical accuracy of the model (level of geometrical detail omitted and its effect on the simulation);
- Mesh suitability and convergence studies (minimum spacing, stretching factors, boundary layer or other physical length scale resolution, mesh refinement studies, etc.);
- Numerical schemes used (amount of artificial dissipation, suitability, order of accuracy, etc.);
- Comparison to any experimental, analytical or numerical results available.

'The problem was run with code X' will be sufficient to give the reader the comfort that what follows is proper, validated, serious, meaningful - in short correct. When CFD has reached this state - and judging from similar experience in structural dynamics, it will happen some time in the next decade - the intellectual death of CFD as a vibrant, dynamic new tool, will be at hand. What will follow is the incorporation of CFD into large software packages with all the constraints on openness, honesty and originality the marketplace imposes.

As a final thought on quality control: Who will train and certify the user ? The same software company that produced the code ! The danger is obvious.

9. Data Management for Large Projects

While streamlining individual code development has not yielded measurable gains, the opposite is true for multi-code and data management. The current trend is towards integrated analysis toolkits combined with integrated data base management. The idea is to steer and document properly all the stages required for large-scale computing. When a new project is started, data bases are created automatically. The toolkit allows for different levels of accuracy. For fluids, one may envision lifting line codes, linear and nonlinear potential, Euler, and Navier-Stokes codes, all available from a single mouse-driven menu. The codes would automatically be assigned to machines suitable for their execution, and all steps recorded in a database as the projects progresses through its stages. Visualization tools that are suitable to each one of the individual codes would allow the user to assess immediately the results, and alert him automatically to physically relevant phenomena and/or possible numerical problems. Although this may sound like fiction, several commercial software providers already have primitive versions of such an environment in place. The potential benefits of performing large-scale computing in this way are enormous: the complete project history is available, different codes may be used and integrated simultaneously allowing for cross-checking, and empirical data may be integrated and compared instantaneously, leading to shortened design cycles and/or improved understanding.

10. CFD as a Tool to aid in the Development of Material Models

Beyond the admittedly very large range of familiar Newtonian flows, such as air and water, lies the whole range of non-Newtonian flows. Flows that fall under this category include many biological fluids (blood, serum, etc.) and polymers (plastics, melts, etc.). The stress-strainrate relationships for non-Newtonian flows may be complex, non-linear, and multivariate (temperature, material history, etc.). Material models for some of these flows are still under development, and it is here that CFD can be used to aid experiments in the determination of the physical parameters. A similar tendency can be observed in structural dynamics, where large, commercial 'standard codes' offer the possibility to incorporate new material models. Indeed, a major portion of structural dynamics research is devoted to improved material models. A similar shift may be observed for present-day CFD as far as turbulence models are concerned. CFD-assisted development of material models for non-Newtonian flows is already occurring, and is expected to expand in the coming decade.

11. Summary

CFD, as well as all the other disciplines that comprise Computational Mechanics, have seen major advances over the last decade. New ideas and approaches are still blossoming. The present book gives ample evidence of this. Given the enormous number of unsolved or suboptimally solved problems, it would be foolish to predict an end to developments in CFD. The present monograph has highlighted some of the areas where progress is urgently needed in order to arrive at a reliable, cost-effective, widely applicable and easy-to-use tool for the prediction of geometrically and physically complex flows. At the same time, non-traditional aspects of CFD, such as management of large codes, parallelism at the project level, and data preprocessing were discussed.

Despite some worrisome developments, looking from the last decade of CFD towards the next decade, we can but only hope that the best in CFD is yet to come.

12. Acknowledgements

I would like to acknowledge the many fruitful and stimulating discussions during the course of the years with my colleagues and students at the Institute for Computational Sciences and Informatics of the George Mason University, Science Applications International Corp., the Naval Research Laboratory, NASA LaRC, NASA GSFC, as well as other universities throughout the world.

References

AIAA, *Proc. 12th AIAA CFD Conf.*, San Diego, CA, June 1995.
Aftosmis, M.J., Berger, M.J. and Melton, J.E., "Adaptation and Surface Modeling for Cartesian Mesh Methods," AIAA-95-1725-CP, 1995.
Allwright, S., "Multiblock Topology Specification and Grid Generation for Complete Aircraft Configurations," AGARD-CP-464, 11, 1990.

Baker, T.J., "Three-Dimensional Mesh Generation by Triangulation of Arbitrary point Sets," AIAA-CP-87-1124, 8th CFD Conference, Hawaii, 1987.

Baker, T.J., "Developments and Trends in Three-Dimensional Mesh Generation," *Appl. Num. Math.*, Vol. 5, 1989, pp. 275-304.

Barth, T., "A 3-D Upwind Euler Solver for Unstructured Meshes," AIAA-91-1548-CP, 1991.

Batina, J.T., "Unsteady Euler Airfoil Solutions Using Unstructured Dynamic Meshes," *AIAA Journal*, Vol. 28, No. 8, 1990, pp. 1381-1388.

Baum, J.D., Luo, H. and Löhner, R., "A New ALE Adaptive Unstructured Methodology for the Simulation of Moving Bodies," AIAA-94-0414, 1994.

Baum, J.D. and Löhner, R., "Numerical Simulation of Shock-Box Interaction Using an Adaptive Finite Element Scheme," *AIAA Journal*, Vol. 32, No. 4, 1994, pp. 682-692.

Baum, J.D., Luo, H., Löhner, R., Yang, C., Pelessone, D. and Charman, C., "A Coupled Fluid/Structure Modeling of Shock Interaction with a Truck," AIAA-96-0795, 1996.

Bayyuk, S., Powell, K. and vanLeer, B., "An Algorithm for Simulation of Flows with Moving Boundaries and Fluid-Structure Interactions," *First AFOSR Conf. on Dynamic Motion CFD*, Rutgers, June 3-5, 1996.

Berger, M.J. and LeVeque, R., "An Adaptive Cartesian Mesh Algorithm for the Euler Equations in Arbitrary Geometries," AIAA-89-1930, 1989.

Coquel, F. and Liou, M.-S., "Field by Field Hybrid Upwind Splitting Methods," AIAA-93-3302-CP, 1993.

DataWare Inc. Objects Manual, 1996.

Gaffney, R., Hassan, H. and Salas, M., "Euler Calculations for Wings Using Cartesian Grids," AIAA-87-0356, 1987.

Goldberg, D.E., *Genetic Algorithms in Search, Optimization and Machine Learning*, Addison-Wesley, 1989.

Haug, E., Charlier, H., Clinckemaillie, J., DiPasquale, E., Fort, O., Lasry, D., Milcent, G., Ni, X., Pickett, A.K. and Hoffmann, R., "Recent Trends and Developments of Crashworthiness Simulation Methodologies and their Integration into the Industrial Vehicle Design Cycle; *Proc. Third European Cars/Trucks Simulation Symposium (ASIMUTH)*, Oct. 28-30, 1991.

Jameson, A., "Artificial Diffusion, Upwind Biasing, Limiters and Their Effect on Accuracy and Multigrid Convergence in Transonic and Hypersonic Flows," AIAA-93-3359, 1993.

Jameson, A., "Optimal Aerodynamic Design Using CFD and Control Theory," AIAA-95-1729, 1995.

Kutler, P., "A Perspective of Theoretical and Applied Computational Fluid Dynamics," *AIAA Journal*, Vol. 23, No. 3, 1985, pp. 328-341.

Liou, M.-S. and Steffen, C.J., "A New Flux Splitting Scheme," *J. Comp. Phys.*, Vol. 107, No. 23, 1993.

Liou, M.-S., "Progress Towards an Improved CFD Method: AUSM$^+$," AIAA-95-1701-CP, 1995.

Löhner, R., Morgan, K., Peraire, J. and Vahdati, M., "Finite Element Flux-Corrected Transport (FEM-FCT) for the Euler and Navier-Stokes Equations," *Int. J. Num. Meth. Fluids*, Vol. 7, 1987, pp. 1093-1109.

Löhner, R., "Some Useful Data Structures for the Generation of Unstructured Grids," *Comm. Appl. Num. Meth.*, Vol. 4, 1988, pp. 123-135.

Löhner, R. and Parikh, P., "Three-Dimensional Grid Generation by the Advancing Front Method," *Int. J. Num. Meth. Fluids*, Vol. 8, 1988, pp. 1135-1149.

Löhner, R., "Three-Dimensional Fluid-Structure Interaction Using a Finite Element Solver and Adaptive Remeshing," *Computer Systems in Engineering*, Vol. 1, Nos. 2-4, 1990, pp. 257-272.

Löhner, R. and Baum, J.D., "Adaptive H-Refinement on 3-D Unstructured Grids for Transient Problems," *Int. J. Num. Meth. Fluids*, Vol. 14, 1992, pp. 1407-1419.

Löhner, R., "Some Useful Renumbering Strategies for Unstructured Grids," *Int. J. Num.*

Meth. Eng., Vol. 36, 1993, pp. 3259-3270.

Löhner, R., "Matching Semi-Structured and Unstructured Grids for Navier-Stokes Calculations," AIAA-93-3348-CP, 1993.

Löhner, R., "Edges, Stars, Superedges and Chains," *Comp. Meth. Appl. Mech. Eng.*, Vol. 111, 1994, pp. 255-263.

Löhner, R., Yang, C., Cebral, J., Baum, J.D., Luo, H., Pelessone, D. and Charman, C., "Fluid-Structure Interaction Using a Loose Coupling Algorithm and Adaptive Unstructured Grids," AIAA-95-2259 [Invited], 1995.

Löhner, R., "Extending the Range of Applicability and Automation of the Advancing Front Grid Generation Technique," AIAA-96-0033, 1996.

Löhner, R., "Surface Reconstruction from Clouds of Points," *Publication CIMNE* 88, Universidad Politécnica de Catalunya, Barcelona, Spain, March 1996.

Löhner, R., "Re-Gridding Surface Triangulations," *J. Comp. Phys.*, Vol. 126, 1996, pp. 1-10.

Luo, H., Baum, J.D. and Löhner, R., "Edge-Based Finite Element Scheme for the Euler Equations," *AIAA Journal*, Vol. 32, No. 6, 1994, pp. 1183-1190.

Luo, H., Baum, J.D. and Löhner, R., "A Finite Volume Scheme for Hydrodynamic Free Boundary Problems on Unstructured Grids," AIAA-95-0668, 1995.

Luo, H., Baum, J.D. and Löhner, R., "A Hybrid Interface Capturing Method for Compressible Multi-Fluid Flows on Unstructured Grids," AIAA-96-0416, 1996.

Mavriplis, D., "Three-Dimensional Unstructured Multigrid for the Euler Equations," AIAA-91-1549-CP, 1991.

Meakin, R.L. and Suhs, N., "Unsteady Aerodynamic Simulations of Multiple Bodies in Relative Motion," AIAA-89-1996, 1989.

Meakin, R.L., "Moving Body Overset Grid Methods for Complete Aircraft Tiltrotor Simulations," AIAA-93-3350-CP, 1993.

Mehrota, P., Saltz, J. and Voigt, R., (eds.), *Unstructured Scientific Computation on Scalable Multiprocessors*, MIT Press, 1992.

Peraire, J., Peiro, J., Formaggia, L., Morgan, K. and Zienkiewicz, O.C., "Finite Element Euler Calculations in Three Dimensions," *Int. J. Num. Meth. Eng.*, Vol. 26, 1988, pp. 2135-2159.

Peraire, J., Morgan, K. and Peiro, J., "Unstructured Finite Element Mesh Generation and Adaptive Procedures for CFD," AGARD-CP-464, 18, 1990.

Peraire, J., Peiro, J. and Morgan, K., "A Three-Dimensional Finite Element Multigrid Solver for the Euler Equations," AIAA-92-0449, 1992.

Quirk, J.J., "An Alternative to Unstructured Grids for Computing Gas Dynamics Flows Around Arbitrarily Complex Two-Dimensional Bodies," *Comp. Fluids*, Vol. 32, 1994, pp. 125-142.

Sellar, R. and Batill, St., "A Neural Network-Based, Concurrent Subspace Optimization Approach to MDO," AIAA-96-0714, 1996.

Simon, H., "Partitioning of Unstructured Problems for Parallel Processing," *NASA Ames Tech. Rep.*, RNR-91-008, 1991.

Steinbrenner, J.P., Chawner, J.R. and Fouts, C.L., "A Structured Approach to Interactive Multiple Block Grid Generation," AGARD-CP-464, 8, 1990.

Thompson, J.F., "A Composite Grid Generation Code for General 3D Regions - The EAGLE Code," *AIAA Journal*, Vol. 26, No. 3, 1988, p. 271ff.

Vidwans, A., Kallinderis, Y. and Venkatakrishnan, V., "A Parallel Load Balancing Algorithm for 3-D Adaptive Unstructured Grids," AIAA-93-3313-CP, 1993.

Weatherill, N.P., "Delaunay Triangulation in Computational Fluid Dynamics," *Comp. Math. Appl.*, Vol. 24, Nos. 5/6, 1992, pp. 129-150.

Weatherill, N.P., Hassan, O. and Marcum, D.L., "Calculation of Steady Compressible Flowfields with the Finite Element Method,"f AIAA-93-0341, 1993.

Williams, D., "Performance of Dynamic Load Balancing Algorithms for Unstructured Grid Calculations," CalTech Rep. C3P913, 1990.

Yee, H.C., Sweby, P.K. and Griffiths, D.F., "Dynamical Approach Study of Spurious

Steady-State Numerical Solutions of Nonlinear Differential Equations. I. The Dynamics of Time Discretization and its Implications for Algorithm Development in Computational Fluid Dynamics," *J. Comp. Phys.*, Vol. 97, 1991, pp. 249-310.

Yee, H.C. and Sweby, P.K., "Dynamical Approach Study of Spurious Steady-State Numerical Solutions of Nonlinear Differential Equations. II. Global Asymptotic Behaviour of Time Discretizations," *Comp. Fluid Dyn.*, Vol. 4, 1995, pp. 219-283.

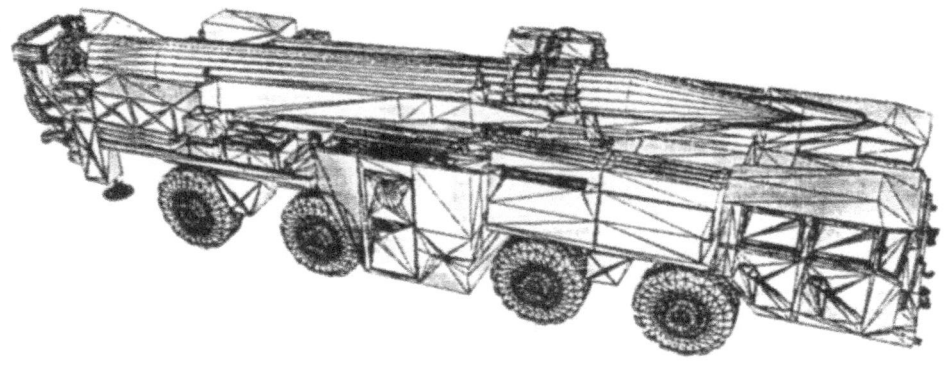

Figure 1. Surface Triangulation (CAD Definition from VR Data Set. a) VR Data. b) Topologically Consistent Surface Triangulation.

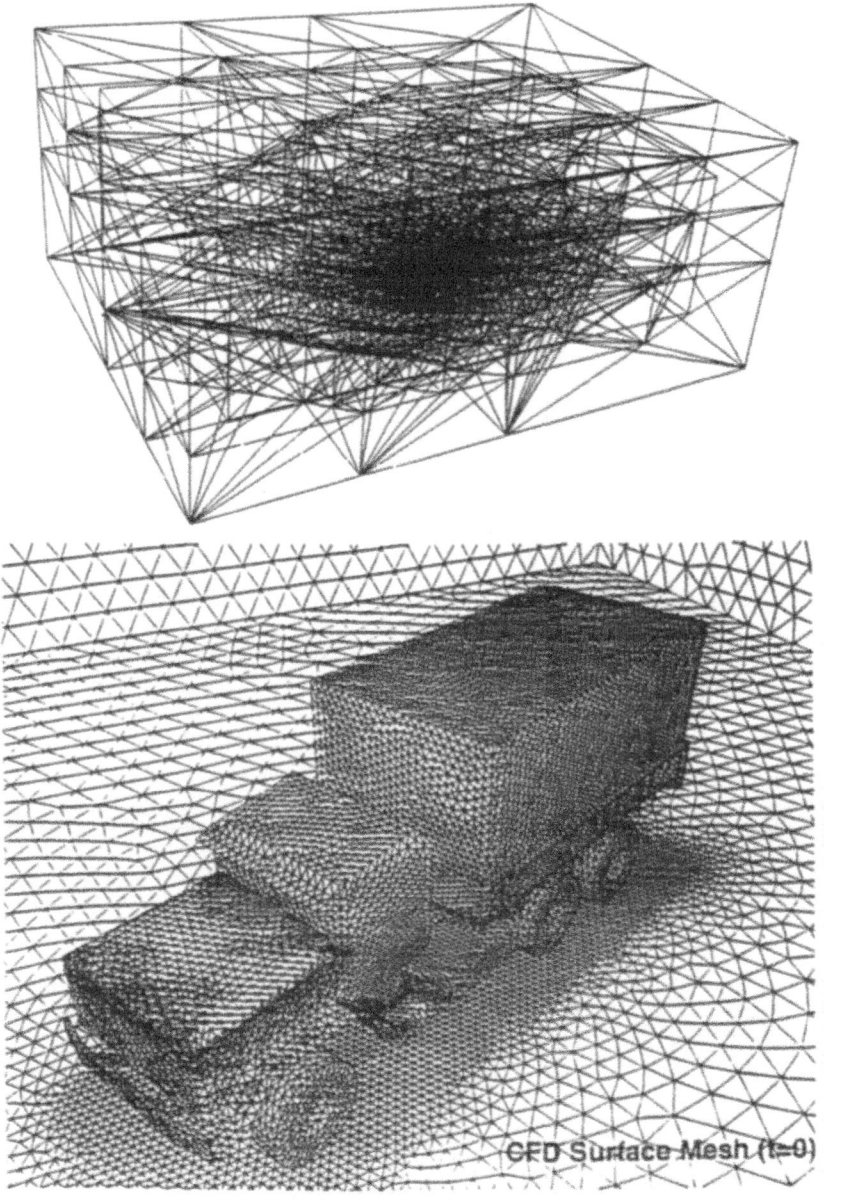

Figure 2. Surface Triangulation (Adaptive Background Grids. a) Adaptive Background Grid. b) Surface Triangulation.

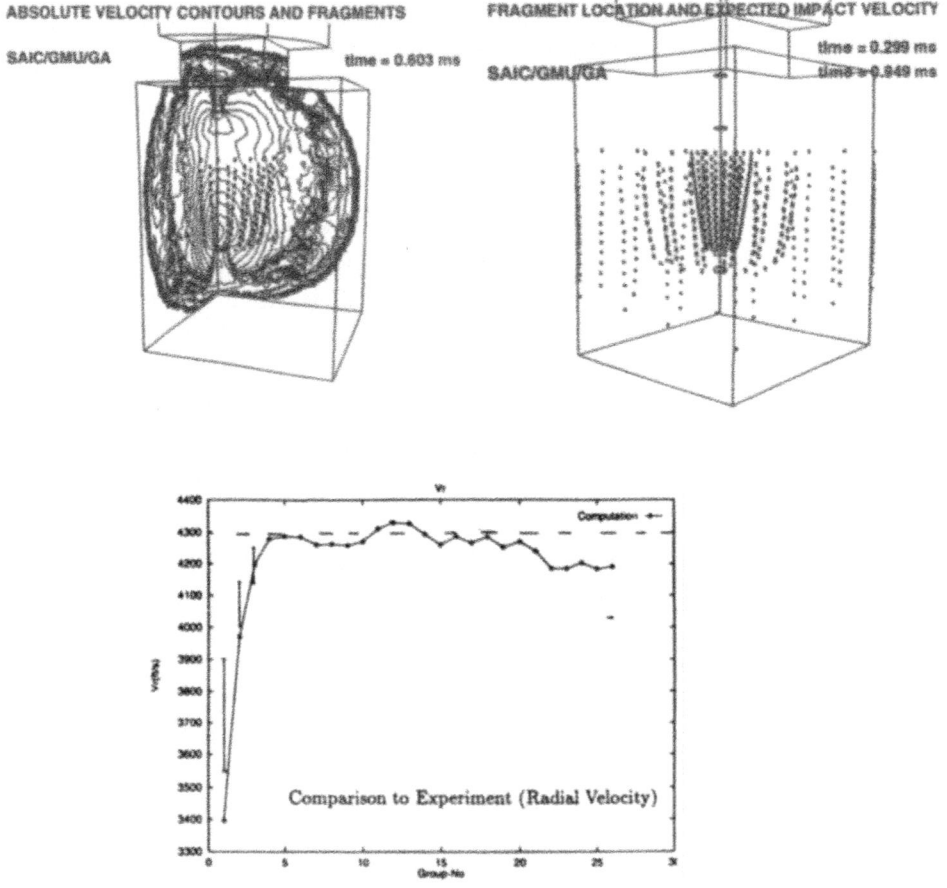

Figure 3. Fragmenting Object (JWL EOS, ALE, 200 Objects, Moving Meshes, Automatic Regridding)

ACCURATE AND ROBUST METHODS FOR VARIABLE DENSITY INCOMPRESSIBLE FLOWS WITH DISCONTINUITIES

W. J. RIDER AND D. B. KOTHE
Los Alamos National Laboratory
Los Alamos, NM 87545

AND

E. G. PUCKETT AND I. D. ALEINOV
University of California, Davis
Mathematics Department
Davis, CA 95616

Abstract.

We are interested in the solution of incompressible flows possessing densities that vary both discontinuously and smoothly. Smooth density variations might be caused by temperature effects, whereas abrupt variations are present at immiscible fluid interfaces. If interfaces are present, we wish to model their gross topological changes. The design of an incompressible flow algorithm that maintains solution accuracy and simulation robustness in the presence of density variations presents challenges, especially if the variations are discontinuous and topologically complex.

We discuss the construction of robust, high-fidelity fractional-step projection methods for computing such flows. We focus on algorithms for the projection, the numerical linear algebra, and interfacial physics such as surface tension. Also discussed are algorithms for multi-dimensional advection and volume tracking. Numerical examples are presented to illustrate our current capabilities.

1. Introduction

The flow of incompressible fluids with discontinuous density variations (interfaces) occurs in widespread applications. Water/air free surface flow is

213

V. Venkatakrishnan et al (eds.), Barriers and Challenges in Computational Fluid Dynamics, 213-230.
© 1998 *Kluwer Academic Publishers.*

a classical example, e.g., a water drop falling into a pool of water. Another important example is the filling of a cast metal mold with a molten metal alloy. Yet another is the production and transport of micron-sized ink drops during inkjet printer operation. Reliable simulation of these types of flows demands a numerical model with accuracy, fidelity, and robustness.

Accuracy is defined as the quality of deviating slightly from fact. For our purposes, this definition is refined as the measured error for a given solution. There is also a distinction between order of accuracy and numerical accuracy. For reasonable grid resolution, methods with a higher order of accuracy can be accompanied by significantly larger numerical error than the lower order method. This naturally leads to our next definition.

Fidelity is defined as exact correspondence with fact. A solution that possesses fidelity is one that is physically meaningful. A method is considered to be high-fidelity when it produces solutions that are accurate relative to the computational resources (the mesh size) applied to them. For example, interface tracking mechanisms can increase solution fidelity by maintaining interface discontinuities as the interface is advected and/or undergoing topological change.

Robustness is the property of being powerfully built or sturdy. A robust method will not fail in a catastrophic manner, but rather "degrade gracefully." Robustness implies that the algorithm can be used with confidence on a difficult problem. The degree to which the degradation is graceful is subject to interpretation. A robust method should produce physically reasonable results beyond the point where accuracy is expected or achieved.

In the next several sections we will focus on the key elements in our incompressible flow solvers. In Section 2 we introduce our projection algorithms for discretizing the incompressibility constraint in a robust manner. Next, in Section 3, we discuss issues related to solving the pressure equation effectively. The methods for computing interface and flow kinematics (advection) are discussed in Section 4. Section 5 follows with an introduction and discussion of our surface tension model. Next we present a set of sample applications to amplify our arguments. Finally, in Section 7, we conclude with a summary of various outstanding numerical issues.

2. Projection Methods

Here we introduce the principal aspects of a projection method. Our basic goal with projection methods is to advance a velocity field, $\mathbf{u} = (u, v)^T$ without regard for the solenoidal nature of $\mathbf{u}$, then recover the proper solenoidal velocity field, $\mathbf{u}^d$ ($\nabla \cdot \mathbf{u}^d = 0$). The means to this end is a projection operator, $\mathcal{P}$, which projects $\mathbf{u}^d$ out of $\mathbf{u}$:

$$\mathbf{u}^d = \mathcal{P}(\mathbf{u}).$$

The velocity $\mathbf{u}$ can be decomposed into a solenoidal vector, $\mathbf{u}^d$, and a curl-free vector, expressed as the gradient of a potential, $\boldsymbol{\nabla}\varphi$. This decomposition is written

$$\rho\mathbf{u} = \rho\mathbf{u}^d + \boldsymbol{\nabla}\varphi, \tag{1}$$

or

$$\mathbf{u} = \mathbf{u}^d + \sigma\boldsymbol{\nabla}\varphi, \tag{2}$$

where $\sigma = 1/\rho$. Taking the divergence of (2) yields an elliptic equation for φ:

$$\boldsymbol{\nabla}\cdot\mathbf{u} = \boldsymbol{\nabla}\cdot\sigma\boldsymbol{\nabla}\varphi. \tag{3}$$

Given the solution φ to equation (3), $\mathbf{u}^d$ results from the correction,

$$\mathbf{u}^d = \mathbf{u} - \sigma\boldsymbol{\nabla}\varphi.$$

2.1. PROJECTIONS: THE BASIC IDEA

Our fractional-step algorithm consists of a predictor step, in which the solenoidal nature of the velocity field is ignored, and a corrector step, in which a projection recovers the solenoidal velocity field. In the predictor step, the time n velocity in cell (i,j), $\mathbf{u}_{i,j}^n$, is advanced with the convection-diffusion equation:

$$
\begin{aligned}
\mathbf{u}_{i,j}^{*,n+1} = {} & \mathbf{u}_{i,j}^n - \Delta t\left[(\mathbf{u}\cdot\boldsymbol{\nabla}\mathbf{u})_{i,j}^{n+\frac{1}{2}} + \sigma_{i,j}^{n+\frac{1}{2}}\mathbf{G}_{i,j}\phi^{n-\frac{1}{2}}\right.\\
& \left. - \frac{\nu\sigma_{i,j}^{n+\frac{1}{2}}}{2}L_{i,j}\left(\mathbf{u}^n + \mathbf{u}^{*,n+1}\right) - \mathbf{f}_{i,j}^{n+\frac{1}{2}}\right],
\end{aligned}
\tag{4}
$$

where $\mathbf{G}_{i,j}$ is the discrete gradient and $L_{i,j}$ is the discrete Laplacian. This provides a nominally second-order discretization. The advection term is discretised with an unsplit high-order Godunov method (Bell $et\ al.$, 1989; Colella, 1990). For variable density flows, this method is described in (Bell & Marcus, 1992; Puckett $et\ al.$, 1997).

Several variations of the projection implied by (4) are possible. By removing the gradient of pressure from (4), φ in equation (3) is actually a pressure rather than an increment in pressure. The form of $\mathbf{u}$ in the discrete divergence on the LHS of (3) can be chosen several ways, e.g., the advanced-time predictor velocity, $\mathbf{u}^{*,n+1}$, or the predicted change, $(\mathbf{u}^{*,n+1} - \mathbf{u}^n)$. One might assume these differences to be higher order effects, but experience has shown otherwise for both exact and approximate projections, as discussed later.

An exact projection results when the Laplacian operator (L) on the RHS of equation (3) is derived from the discrete discrete divergence (D) and gradient $(\mathbf{G})$ operators: $L = D\sigma\mathbf{G}$. The discrete velocity divergence in an exact projection is zero to within the convergence tolerance of the solution to equation (3).

The exact discrete projections given above provide a good foundation in the numerical implementation of projection methods, but have some practical difficulties. These problems are discussed in (Almgren et al., 1996; Lai et al., 1993). The pressure/velocity decoupling can interact poorly with localized source terms (e.g., chemical reactions), leading to instabilities. Additionally, the local decoupling renders multigrid techniques cumbersome (Howell, 1993), and hampers the implementation of adaptive grid techniques (Almgren et al., 1993; Howell, 1993).

To address these problems, new types of projection algorithms have been developed. In "approximate" projections, introduced in (Almgren et al., 1996), the operator L is derived from a discretization of the continuous projection operator. The discrete velocity divergence in an approximate projection is not zero, but is rather a function of the truncation error. The operators D and $\mathbf{G}$ have the same form as the exact projection, but the Laplacian is modified.

2.2. ROBUST PROJECTION METHODS

In approximate projections, the velocity divergence is not constrained to be zero (to some small tolerance), hence robust algorithms can be elusive. We will demonstrate this with a single test problem, then describe improvements.

The principal problem with approximate projections is the presence and growth of null spaces in the discrete operators, which is manifested as high-frequency noise. This noise can be controlled by identifying and filtering unphysical velocity modes (Lai, 1993; Rider, 1994) or by carefully formulating the form of the approximate projection (Rider, 1994). Without these steps, approximate projection methods are prone to failure on more difficult problems.

We currently damp these spurious modes in two ways: the explicit addition of a high-order viscosity, or the use of an iterated projection derived from a discrete stencil that differs from the approximate projection stencil. These methods are most effective when used in concert.

The formulation of the projection directly affects the time evolution of the discrete divergence. If the divergence on the LHS of (3) is the difference between the predicted and old time velocity,

$$\nabla \cdot \left(\mathbf{u}^{*,n+1} - \mathbf{u}^n\right), \tag{5}$$

then the discrete divergence errors *accumulate* in time. On the other hand, if a predictor velocity is used,

$$\nabla \cdot \mathbf{u}^{*,n+1}, \tag{6}$$

than the discrete divergence errors *do not accumulate* in time, which is preferable. It should be noted that equation (5) is the form standard in the literature.

Consider the following test problem, which will illuminate these subtle issues. A circular drop with radius 0.15 is placed at $(0.5, 0.75)$ in a unit square computational domain that is partitioned with a 64×64 grid. Gravity is unity (downward) and all boundaries are frictionless (free-slip). The drop fluid is 1000 times more dense than the background fluid (having unit density). The flow is integrated forward in time to $t = 1$ using the Euler equations. A high-order Godunov method and an unsplit piecewise linear volume tracking algorithm (discussed in Section 4) is used to advect the flow. The CFL number is $\frac{1}{2}$ unless otherwise stated. The unsteady flow is computed with variations of both the exact and approximate projection methods. Each method demonstrates second-order convergence (in space and time) on sufficiently smooth problems.

Solutions obtained with the standard exact and approximate projection methods (i.e., without filters) are shown in Figure 1[1]. Both solutions exhibit spurious features in the velocity field in the flow above the drop. The exact projection solution (Figure 1a) displays some velocity field decoupling and slight asymmetries. Despite the use of a smaller time step in integrating the flow (CFL=1/4), the approximate projection solution in Figure 1b is unacceptable. As discussed later, this solution is compromised in part because projection in equation (3) is the projecting equation (5) and solving for a pressure increment rather than a total pressure.

When the predictor velocity $(\mathbf{u}^{*,n+1})$ is projected (rather than the velocity difference) and the total pressure (rather than the pressure increment) is solved for in equation (3), the drop solutions improve significantly, as shown in Figure 2. The exact projection solution in Figure 2a now exhibits symmetry and a coupled velocity field. The approximate projection solution (Figure 2b) additionally requires velocity filters before its solution quality matches and surpasses that of the exact projection. The decoupled velocity field and asymmetries evident in Figure 1b are effectively suppressed.

---

[1] The standard formulation solves for an increment in pressure and projects $\nabla \cdot \left(\mathbf{u}^{*,n+1} - \mathbf{u} \right)$

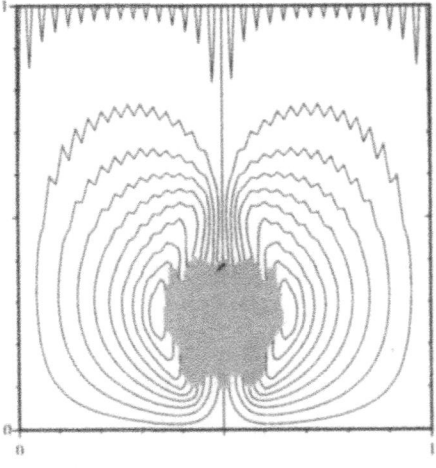 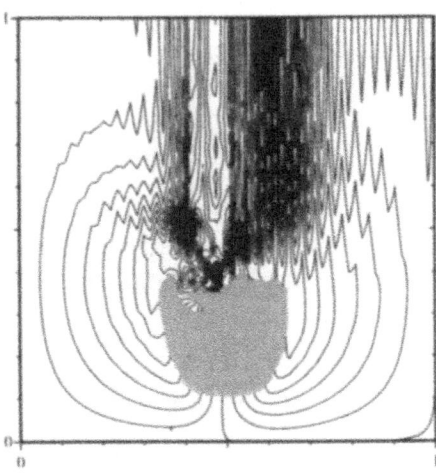

(a) Exact pressure increment projection of the velocity difference.

(b) Approximate pressure increment projection of the velocity difference.

Figure 1. Drop solutions for our standard exact and approximate projections. Both solutions use a grid where all variables are cell-centered. The droplet outline and the velocity streamlines are shown. In both cases the projection equation LHS is given by equation (5).

3. Linear Algebra

The cost of incompressible flow solutions based on projection methods is typically dominated by the effort required to find solutions to the elliptic pressure equation (3). Designing an efficient and scalable method for solving this system of linear equations is therefore of paramount importance. We have used three methods in our study: a preconditioned conjugate gradient (CG) method, a multigrid (MG) method, and a multigrid preconditioned conjugate gradient (MGCG) method. Our results indicate that the MGCG method is the most effective.

3.1. CONJUGATE GRADIENT METHODS

Because the exact and approximate projections produce symmetric semi-positive-definite and positive-definite systems of linear equations, we can use CG methods (Golub & Van Loan, 1989). We can also employ preconditioning to improve convergence, which is especially important when density ratios across interfaces are large. Typically we use an incomplete Cholesky

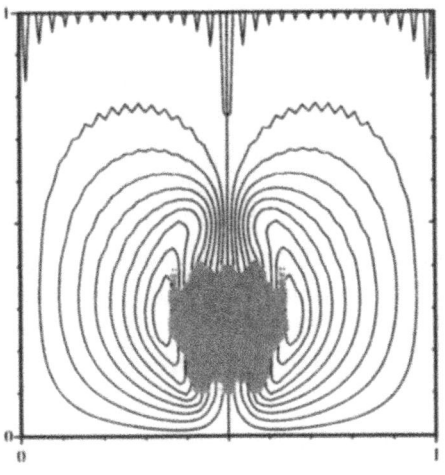 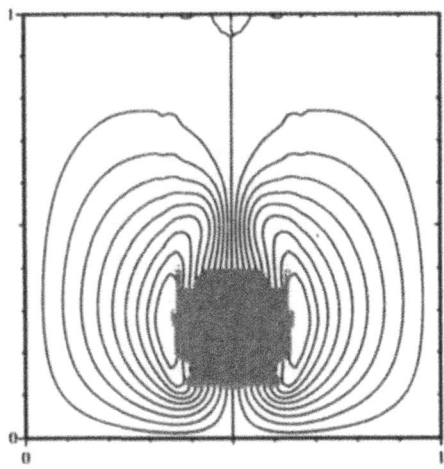

(a) Exact pressure projection of the pre-   (b) Filtered approximate pressure pro-
dictor velocity.   jection of the predictor velocity.

Figure 2.   The end product of the changes in the projection formulation and the filtering
of approximate projections. The projection equation LHS is given by equation (6).

method or some other iterative procedure (SSOR, Jacobi) in finding approximate solutions to the preconditioning equation.

Several properties of preconditioned CG method's ability to solve equation (3) are worth noting. First, it is extremely robust (having never failed in our experience), and is also general, being applicable to any of our exact or approximate projections. On the other hand, the amount of work required by the preconditioned CG method grows as $N^{3/2}$ (N is the total number of unknowns). Therefore the CG method demands an ever-increasing proportion of the CPU time as the grid is refined.

3.2. MULTIGRID METHODS

Solutions to the linear systems arising from approximate projections can also be obtained with a MG algorithm (Briggs, 1987). Use of a MG algorithm is desirable because of its attractive scaling. The operation count for classical direct linear algebra solution techniques (e.g., Cholesy) scale like N^3. This scaling improves to N^2 for banded solvers that take advantage of the structure of the linear system. MG scales linearly with N. Thus, MG (where it works) will eventually provide the fastest route to a solution as the grid is refined.

One of the most important tasks in formulating a MG algorithm is the approximation of the coarse-grid linear equations. One approach is using intergrid transfer functions to define variational or Galerkin coarse grid operators. The complication and expense of this task, however, has motivated a simpler approach based on suggestions in (Liu *et al.*, 1992; Lai, 1993). The operator remains the same, as do the boundary conditions on the course grids, but σ must be defined at each level. The basic idea is to construct coarse grid approximations that produce an average cell value for $\sigma \nabla \varphi$ that is identical to the fine-level value. In using the cell-centered MG framework, this control volume derivation of the equations comes quite naturally.

We find that our MG algorithm converges quickly to a solution in most cases, but fails on occasion with flows having interfaces possessing large density variations and complex topology (e.g., a drop splashing into a pool).

3.3. MULTIGRID PRECONDITIONED CONJUGATE GRADIENT

Our current solution to the MG robustness problem is to employ the symmetric MG algorithm to precondition a standard CG method. The idea was first proposed and demonstrated by Tatebe (Tatebe, 1993). Experimentation has proven the utility of combining these two methods. This combined MGCG method usually scales like a MG algorithm, but occasionally CG-like scaling is exhibited. Nevertheless, we have found it to be robust. Ultimately we wish to design a robust method that consistently exhibits MG-like scaling.

4. Advection and Interface Tracking

4.1. HIGH-ORDER GODUNOV METHODS

We now discuss the basic principles of our advection method, which is based on the framework established in (van Leer, 1984; Bell *et al.*, 1988; Colella, 1990). Our method has many similarities with Colella's formulation, as modified for incompressible flow algorithms based on projection methods (Bell *et al.*, 1989). These methods are "unsplit", i.e., a full multi-dimensional solution is updated in a single time step. Single-step, multi-dimensional integration is important because the incompressibility constraint is inherently multi-dimensional. The flow solver should therefore reflect the intrinsic coupling of the velocity field.

Multi-dimensional advection algorithms are constructed via the time-centered approximation of the dependent variables at cell-edges. Time-centering is accomplished with the variable's full PDE (with all terms included). While the details of the methods are given in the above-stated references, important contributions in the use of these methods have recently been made. Brown and Minion discuss the nature of these solutions when resolution is not adequate (Brown & Minion, 1995), and Minion has suggested stability-enhancing improvements (Minion, 1996).

4.2. VOLUME TRACKING OF INTERFACES

The essential features of volume tracking methods are as follows. First, an initial (known) fluid interface geometry is used to generate initial fluid volume fractions in each computational cell. This task requires computing the volume truncated by the fluid interface in each cell containing an interface. Exact interface information is then discarded in favor of the discrete volume fraction data.

Interfaces are subsequently "tracked" by evolving fluid volumes in time with the solution of a standard advection equation. At any time in the solution, an exact interface location is not known, i.e., a given distribution of volume fraction data does not guarantee a unique interface topology. Interface geometry is instead inferred (based on assumptions of the particular algorithm) and its location is "reconstructed" from local volume fraction data. Interface locations are then used to compute the volume fluxes necessary for the advective term in the volume evolution equation. Volume fluxes are therefore approximated geometrically rather than algebraically. Typical implementations of these algorithms are one-dimensional, with multi-dimensionality built up through operator splitting. The assumed interface geometry, interface reconstruction, and volume flux calculation typically comprise the unique features of a given volume tracking method.

Our piecewise linear volume tracking algorithm, as implemented, is straightforward, simple, and extensible. This is accomplished by drawing upon the extensive literature available in the field of of computational geometry (O'Rourke, 1993). The algorithm is robust, second-order accurate in time and space, and is constructed from a set of simple geometric functions. A detailed account of our volume tracking algorithm, including comparison with other methods, is given in (Rider & Kothe, 1996). Pilliod and Puckett (Pilliod & Puckett, 1997) also review and analyze volume tracking methods, as well as introducing their own unsplit time integration scheme.

5. Surface Tension

Our current models for interfacial surface tension begin with methodology established in the continuum surface force (CSF) method (Brackbill et al., 1992). The basic premise of the CSF method to model physical processes specific to and localized at fluid interfaces (e.g, surface tension) by applying the process to fluid elements everywhere within interface transition regions. Surface processes are replaced with volume processes whose integral effect properly reproduces the desired interface physics. This approach falls under the general class of immersed interface methods (Leveque & Li, 1994) whose origin dates back to the pioneering work of Peskin (Peskin, 1977). The CSF method lifts all topological restrictions without sacrificing accuracy, robustness, or reliability. It has been verified extensively in 2-D flows through its implementation in a classical algorithm for free surface flows (Kothe *et al.*, 1991; Kothe & Mjolsness, 1992), where complex interface phenomena such as breakup and coalescence have been modeled.

In the CSF model, surface tension is reformulated as a volumetric force given by

$$\mathbf{F_s} = \mathbf{f_s}\delta_s. \tag{7}$$

Here δ_s is a surface delta function and $\mathbf{f_s}$ is the surface tension force per unit interfacial area (Brackbill *et al.*, 1992):

$$\mathbf{f_s} = \tau\kappa\hat{\mathbf{n}} + \nabla_s\tau, \tag{8}$$

where τ is the surface tension coefficient, ∇_s is the surface gradient, $\hat{\mathbf{n}}$ is the interface unit normal, and κ is the mean interfacial curvature:

$$\kappa = -(\nabla \cdot \hat{\mathbf{n}}). \tag{9}$$

The first term in (8) is a force acting normal to the interface, proportional to the curvature κ. The second term is a force acting along the interface (tangentially) toward regions with higher surface tension coefficient values. The normal force tends to smooth and propagate regions of high curvature, whereas the tangential force tends to force fluid along the interface toward regions of higher τ.

The surface delta function was proposed in the original CSF model to be (Brackbill *et al.*, 1992)

$$\delta_s = \frac{|\nabla c|}{[c]} = |\mathbf{n}| \tag{10}$$

where c is the characteristic (color) function uniquely identifying each fluid in the problem and $[c]$ is the jump in the color function across the interface in question, which is unity since volume fractions serve as the color function in this work. If a wide stencil is used for $\hat{\mathbf{n}}$ in (10), then δ_s will be nonzero in cells that are in close proximity to the interface. We currently force δ_s to be zero in these cells, which causes the CSF to be non-zero only within the interface transition region. A proper δ_s insures that the CSF is normalized to recover the conventional description of surface tension as the local product $\kappa h \to 0$.

Despite the success of the CSF model and related immersed interface methods, outstanding issues remain (Kothe *et al.*, 1996). If these issues can be resolved adequately, a wider range of surface tension-driven flows will be modeled reliably. For example, improved forms for δ_s, displaying better convergence and smoothness properties, are needed. Our current numerical results are very sensitive to the form used for δ_s, indicating that the quality of CSF model relies heavily on the quality of the form used for δ_s (Kothe *et al.*, 1996). Recent results by Aleinov and Puckett (Aleinov & Puckett, 1995) motivate the use of other kernels, such as the Peskin kernel (Peskin, 1977) or higher-order Nordmark (Nordmark, 1991) kernel.

Perhaps the most stringent test for a surface tension model is a test of the ability to maintain an equilibrium (minimal energy) configuration. A 2-D or 3-D static drop is such an example (Kothe *et al.*, 1996). Here a perfectly spherical drop is placed in a lighter-density background fluid, and all forces are ignored except the drop interfacial surface tension. The drop should remain stationary, as the net surface tension force is zero. An incompressible flow solution for this system, however, generates false flow dynamics (dubbed "parasitic currents") that can grow with time (sometimes unbounded) (Kothe *et al.*, 1996). The source of these currents originates in part with the surface tension model, as the computed pressure gradient at the drop interface does not exactly cancel the surface tension force. New developments in surface tension models must address this inability to maintain an equilibrium configuration.

6. Applications

6.1. A MOLD-FILLING PROBLEM

As an example of our current 3-D capabilities, consider the following sample "mold-filling" problem. A rectangular box, spanning $0 \leq x, y \leq 24$ and $0 \leq z \leq 30$, is partitioned with $24 \times 24 \times 30$ unit cubical cells. The box is initially filled with a quiescent background fluid having a density ten times less than the filling fluid. At time zero, filling fluid is injected with velocity $(0.0, 0.0, -88.6)$ through a hole in the box at $0 \leq x, y \leq 5$ and $z = 30$ (the top corner). The background fluid is allowed to escape through a vent at $19 \leq x, y \leq 24$ and $z = 30$ (the opposite top corner). Gravity is $(0.0, 0.0, -980.6)$, and both the background and filling fluid are assumed to be incompressible and inviscid. Surface tension at interfaces between the background and filling fluid is neglected. The filling dynamics (as inferred from the interface topology) are followed up to a time of 3.0.

As is evident from Figure 3, which depicts the filling fluid interface topology at four different times, this idealized calculation presents an energetic and rigorous test of our ability to model complex topology free surface flows encountered in the mold-filling process. Solutions here are obtained with a cell-centered approximate pressure projection method without filters. Equation (6) is used for the LHS of the projection. Interfaces are tracked with our piecewise-planar volume tracking method (Kothe et al., 1996). Additional results of this simulation (including animations) can be found in (Kothe et al., 1995).

Next we consider another difficult interfacial flow application, namely the production and transport of ink drops in an inkjet printing process.

6.2. INKJET PRINTER NOZZLE

Surface Tension The presence of the surface delta function in the expression for the continuum surface tension force is often a source for instabilities and poor convergence of projection methods. If we suppose that the surface tension coefficient τ is constant, then the singularity can be eliminated. Upon substitution of (8) and (10) into equation (7) for the surface tension force, and using the relation $\mathbf{n}|\nabla c| = \nabla c$, we can write

$$\mathbf{F_s} = \tau\kappa\nabla c = \nabla(\tau\kappa c) - \tau c\nabla\kappa. \tag{11}$$

Of course, this equation is valid only if κ is defined in the entire domain and is smooth. This is not the case, since κ is defined only on the interface, but if the interface is smooth then κ can be spread smoothly over the entire domain by some simple averaging procedure. In this case the first term in the RHS of (11), which contains the main singularity, can be included into a definition of the pressure and the remaining surface tension force expression will be less singular (it will contain a Heaviside function instead of a delta function). In other words, we can introduce new definitions for the pressure and surface tension force

$$\tilde{p} = p - \sigma\kappa c \tag{12}$$

$$\tilde{\mathbf{F}}_\mathbf{s} = -\sigma c\nabla\kappa. \tag{13}$$

for which the Navier-Stokes equations will have the same form, except the volumetric force will not have a delta function. This method was used by Sussman (Sussman et al., 1996) with a level set method. We have found this method to be more robust than the standard CSF method. It enabled us to perform computations for which the standard CSF method failed (e.g., the computation presented here).

Geometry If the computational domain has a complex geometry, special techniques are required to impose the boundary conditions. We use a Cartesian grid, in which a regular rectangular grid is cut by external boundaries. Such a grid remains rectangular away from the boundary but has irregular "incomplete" cells on the fluid-body interface. The algorithms described above need to be reformulated for the irregular boundary cells.

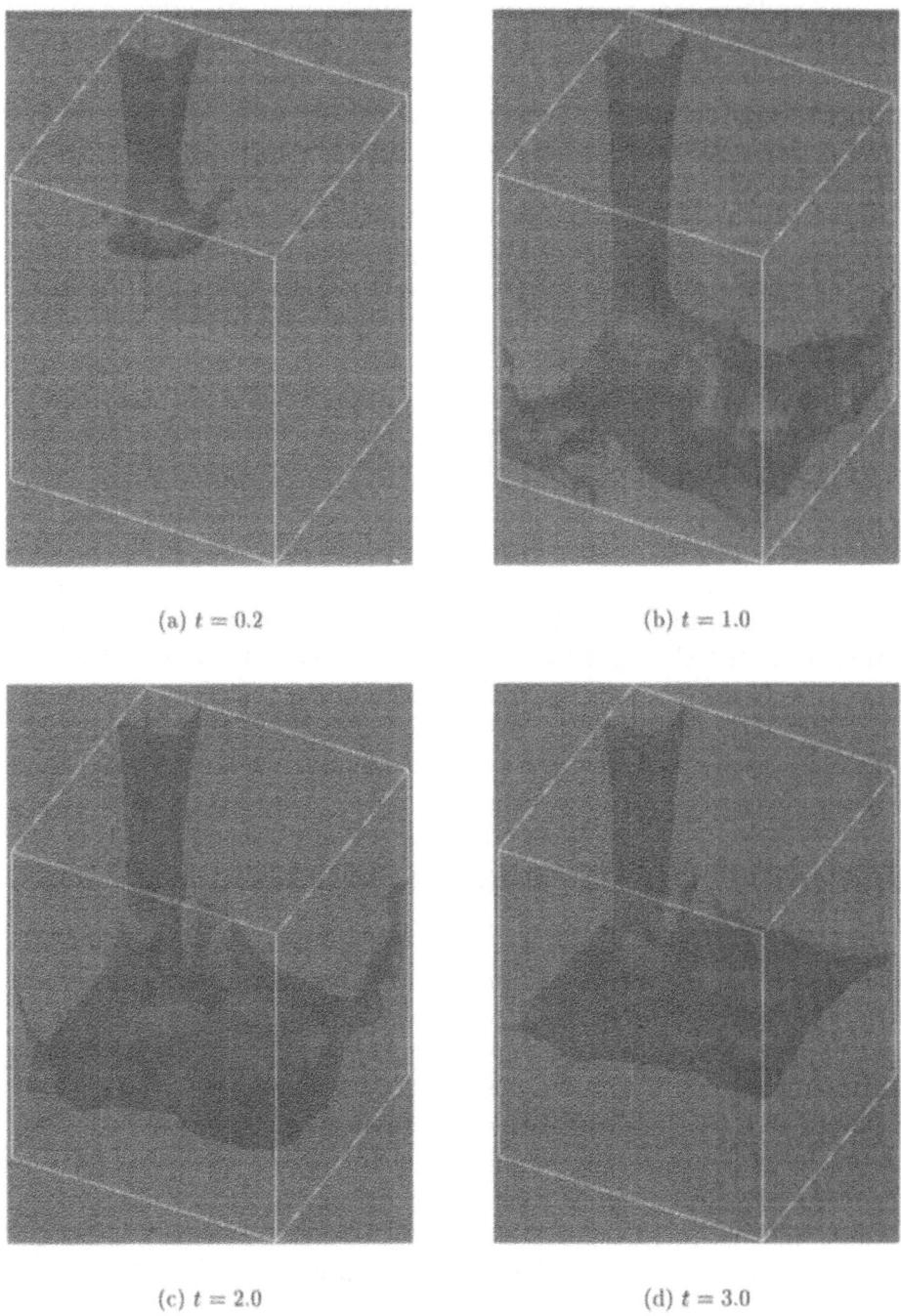

(a) $t = 0.2$ (b) $t = 1.0$

(c) $t = 2.0$ (d) $t = 3.0$

Figure 3.   Three-dimensional simulation of a box being filled with a heavier fluid from the top corner. Shown is the filling fluid interface topology (the 0.5 volume fraction isosurface) at four different times.

The problem that arises in the velocity advection algorithm is due to the CFL constraint that requires a time step size to be proportional to the cell size. Since we want to model an arbitrary geometry, we may have very small cells on the fluid-body boundary. This places a severe restriction on the time step size, slowing down the computations and causing other difficulties. We solve this problem using the flux redistribution algorithm, described in (Almgren *et al.*, 1997), which allows us to use a time step size computed according to the regular cell CFL constraint. The general idea is to first advect velocities using a conservative transport algorithm while ignoring the fluid-body boundary. Next, in each boundary cell we apply a correction computed according to the stability requirement. Extra fluxes resulting from this correction are then distributed over the neighboring cells. Though stability of this algorithm has not been proven, its behavior is robust in our experience.

A general approach to elliptic solvers is to use a standard stencil for elliptic operator away from the boundary, but modify it for irregular cells. We use a finite element approach to the approximate projection in which pressure is located at the vertices of a regular grid. This algorithm is transferred to a Cartesian grid by changing the domain of integration in the weak form of the evolution equations. Integrals over the entire cells are replaced by integrals over those parts of the cells which are inside the fluid. The viscous solver can also be reformulated for a Cartesian grid, though it requires more work. At the moment we use a "stair-step" approximation for the viscous solver in which boundary cells are treated as if they are entirely in the fluid and no-slip boundary conditions are imposed at edges entirely in the wall.

Volume tracking methods are more complicated on Cartesian grids, since they need an interface to be reconstructed and transported within non-rectangular cells. Currently we employ a "stair-step" approximation for interface reconstruction in those cells. Another issue in the presence of multiple fluids and boundary cells are contact angles between fluid-fluid interfaces and the fluid-body boundary. At the moment we allow only two contact angles for wettable and non-wettable surfaces.

Computations Figures 4 and 5 depict computations made for ejection of water into air from the nozzle of an ink jet printer. The surface inside the nozzle is assumed to be wettable and the surface outside the nozzle to be non-wettable. Although our numerical method is purely two-dimensional, the surface tension force is computed using cylindrical symmetry. This is necessary, because curvature in the third dimension is significant in such problems, therefore neglecting it gives unrealistic results. Uniform Dirichlet boundary conditions on velocity are used on the lower boundary. Other boundary conditions for velocity are free flow on the upper boundary and

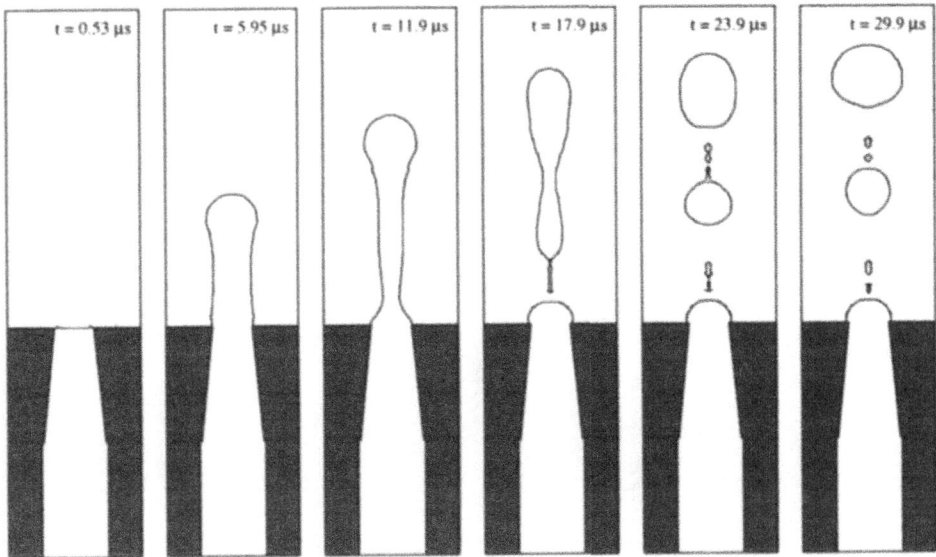

Figure 4.   Ejection of water into air from a nozzle: 1/2 cosine wave impulse. The dimensions of the domain are 60 x 240 microns, with the duration of the impulse being 10 μs. The grid is 32 × 128 cells. The fluid densities are 1000 kg/m^3 and 1 kg/m^3 with viscosities of 0.001137 kg/m s and 0.00001776 kg/m s.

no-slip on the left and right boundaries. The inflow velocity at the bottom boundary is given by an impulse model using 1/2 and 1/4 of a cosine wave to $\approx$ 30 μs, followed by no further inflow. In both cases the duration of the impulse and the amount of the injected fluid are the same, except the impulse shapes are slightly different. We are interested in following the formation of the droplet and its separation from the nozzle. In both cases we observe the formation of a satellite, with its size depending on the shape of the impulse. The slightly unphysical meniscus shape at 0.53 microseconds is due to our initial conditions (flat surface) being incompatible with the specified contact angle.

7. Next Steps

We have made significant progress in achieving the goal of constructing accurate, robust solvers for variable density incompressible flows with discontinuities. Despite significant progress, much work remains. Application of our methodology to a wider variety of flows is focusing our attention on current weaknesses. Chief among these is the extension of our method to nonorthogonal, three-dimensional grids, which is currently being pursued in a new casting simulation tool (Kothe *et al.*, 1995). Additional research effort is targeting improved surface tension models, as their robustness and

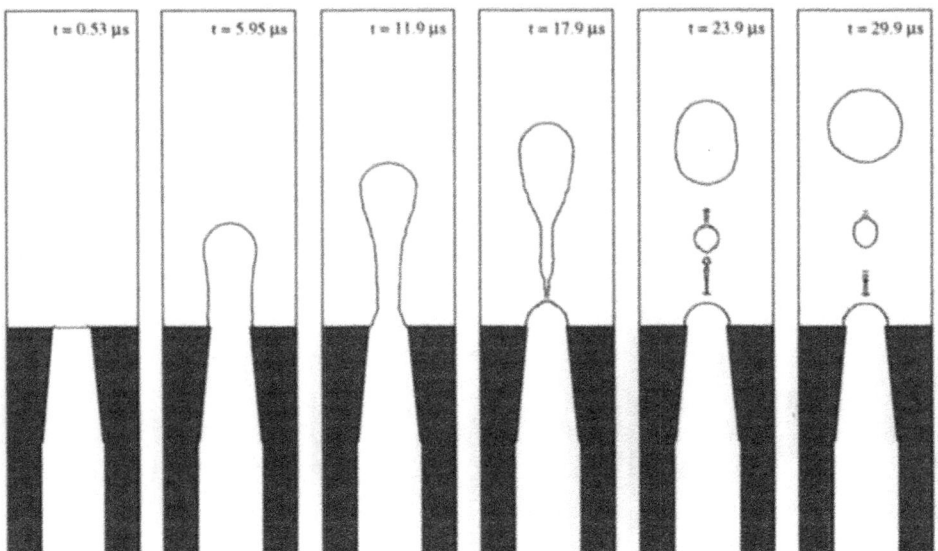

Figure 5.   Ejection of water into air from a nozzle: 1/4 cosine wave impulse.

accuracy is currently less than satisfactory. Efficient linear algebra on 3-D complex grids is needed, hence will also be the focus of future efforts.

References

Aleinov, I. and Puckett, E. G. Puckett, "Computing Surface Tension with High-Order Kernels," *6th International Symposium on Computational Fluid Dyanamics*, 1995.

Almgren, A. S., Bell, J. B., Colella, P., and Howell, L. H., "An adaptive projection method for the incompressible Euler equations," *Proceedings of the AIAA Eleventh Computational Fluid Dynamics Conference*, J.L. Thomas, ed., AIAA Paper 93-3345, pp. 530-539, 1993.

Almgren, A. S., Bell, J. B., Colella, P., and Marthaler, T., "A Cartesian grid projection method for the incompressible Euler equations in complex geometries," *SIAM J. Sci. Comput.*, Vol. 18, No. 1, September 1997.

Almgren, A. S., Bell, J. B., and Szymczak, W. G., "A numerical method for the incompressible Navier-Stokes equations based on an approximate projection," *SIAM Journal of Scientific Computing*, Vol. 17, 1996, pp. 358-369.

Bell, J. B., Colella, P., and Glaz, H. M., "A second-order projection method of the incompressible Navier-Stokes equations," *Journal of Computational Physics*, Vol. 85, 1989, pp. 257-283.

Bell, J. B., Dawson, C. N., and Shubin, G. R., "An unsplit, higher order Godunov method for scalar conservation laws in multiple dimensions," *Journal of Computational Physics*, Vol. 74, 1988, pp. 1-24

Bell, J. B. and Marcus, D. L., "A second-order projection method variable-density flows," *Journal of Computational Physics*, Vol. 101, 1992, pp. 334-348.

Brackbill, J. U., Kothe, D. B., and Zemach, C., "A continuum method for modeling surface tension," *Journal of Computational Physics*, Vol. 100, 1992, pp. 335-354.

Briggs, W. L., "A Multigrid Tutorial," *SIAM*, 1987.

Brown, D. L. and Minion, M. L., "Perfromance of underresolved two-dimensional incompressible flow simulations," *Journal of Computational Physics*, Vol. 122, 1995, pp. 165-183.

Colella, P., "Multidimensional upwind methods for hyperbolic conservation laws," *Journal of Computational Physics*, Vol. 87, 1990, pp. 171-200.

Golub, G. H. and Van Loan, C. F., *Matrix Computations*, Johns Hopkins University Press, 1989.

Howell, L. H., "A multilevel adaptive projection method for unsteady incompressible flow," *Proceedings of the Sixth Copper Mountain Conference on Multigrid Methods*, N. D. Melson, T. A. Manteuffel, and S. F. McCormick, eds., 1993, pp. 243-257.

Kothe, D. B. et al., "Computer simulation of metal casting processes: A new approach," Technical Report LALP-95-197, Los Alamos National Laboratory, 1995. Available on World Wide Web at http://gnarly.lanl.gov/Telluride/Telluride.html.

Kothe, D. B. and Mjolsness, R. C., "Ripple: A new model for incompressible flows with free surfaces," *AIAA Journal*, Vol. 30, 1992, pp. 2694-2700.

Kothe, D. B., Mjolsness, R. C., and Torrey, M. D., "Ripple: A computer program for incompressible flows with free surfaces," Technical Report LA-12007-MS, Los Alamos National Laboratory, 1991.

Kothe, D. B., Rider, W. J., Mosso, S. J., Brock, J. S., and Hochstein, J. I., "Volume tracking of interfaces having surface tension in two and three dimensions," Technical Report AIAA 96-0859, AIAA, 1996. Presented at the 34rd Aerospace Sciences Meeting and Exhibit.

Lai, M. F., "A Projection Method for Reacting Flow in the Zero Mach Number Limit," Ph.D. thesis, University of California at Berkeley, 1993.

Lai, M., Bell, J. B., and Colella, P., "A projection method for combustion in the zero Mach number limit," *Proceedings of the AIAA Eleventh Computational Fluid Dynamics Conference*, J.L. Thomas, ed., 1993, pp. 776-783; AIAA Paper 93-3369.

Leveque, R. J. and Li, Z., "The immersed interface method for elliptic equations with discontinuous coefficients and singular sources," *SIAM Journal on Numerical Analysis*, Vol. 31, 1994, pp. 1019-1044.

Liu, C., Liu, Z., and McCormick, S., "An efficient multigrid scheme for elliptic equations with discontinuous coefficients," *Communications in Applied Numerical Methods*, Vol. 8, 1992, pp. 621-631.

Minion, M. L., "On the stability of Godunov projection methods for incompressible flow," *Journal of Computational Physics*, Vol. 123, 1996, pp. 435-449.

Nordmark, H. O., "Rezoning for higher order vortex methods," *Journal of Computational Physics*, Vol. 97, 1991, pp. 366-397.

O'Rourke, J., *Computational Geometry in C*, Cambridge, 1993.

Peskin, C. S., "Numerical analysis of blood flow in the heart," *Journal of Computational Physics*, Vol. 25, 1977, pp. 220-252.

Pilliod, J. E., Jr. and Puckett, E. G., "Second-Order Volume-of-Fluid Algorithms for Tracking Material Interfaces," in preparation.

Puckett, E. G., Almgren, A. S., Bell, J. B., Marcus, D. L., and Rider, W. J., "A High-Order Projection Method for Tracking Fluid Interfaces in Variable Density Incompressible Flows," *Journal of Computational Physics*, Vol. 130, 1997, pp. 269-282.

Rider, W. J., "Filtering nonsolenoidal modes in numerical solutions of incompressible flows," Technical Report LA-UR-94-3014, Los Alamos National Laboratory, 1994. Available on World Wide Web at http://www-xdiv.lanl.gov/XHM/personnel/wjr-/Web_papers/pubs.html.

Rider, W. J., "The robust formulation of approximate projection methods for incompressible flows," Technical Report LA-UR-94-3015, Los Alamos National Laboratory, 1994. Available on World Wide Web at http://www-xdiv.lanl.gov/XHM/personnel-/wjr/Web_papers/pubs.html.

Rider, W. J. and Kothe, D. B., "Reconstructing volume tracking," Technical Report LA-UR-96-2375, Los Alamos National Laboratory, 1996. Available on

World Wide Web at http://www-xdiv.lanl.gov/XHM/personnel/wjr/Web_papers-/pubs.html, also submitted to the *Journal of Computational Physics*.

Sussman, M., Almgren, A., Bell, J., Colella, P., Howell, L., and Welcome, M., "An adaptive level set approach for incompressible two-phase flows," *Proceedings of the ASME Fluids Engineering Summer Meeting*, July 1996.

Tatebe, O., "The multigrid preconditioned conjugate gradient method," *Proceedings of the Sixth Copper Mountain Conference on Multigrid Methods*, N. D. Melson, T. A. Manteuffel, and S. F. McCormick, eds., 1993, pp. 621-634.

van Leer, B., "Multidimensional explicit difference schemes for hyperbolic conservation laws," *Computing Methods in Applied Sciences and Engineering VI*, In R. Glowinski and J.-L. Lions, eds., 1984, pp. 493-497.

A VARIATIONAL APPROACH TO DERIVING SMEARED-INTERFACE SURFACE TENSION MODELS

DAVID JACQMIN
NASA Lewis Research Center
Cleveland, Ohio

Abstract.
Smeared-interface surface tension models can be derived variationally. The advantage of the variational approach is that the resulting models, both continuum and discrete, conserve total energy. This paper explains the variational approach and shows how to apply it to derive energetically consistent CSF (Brackbill et al., 1992) and distributed force (Unverdi & Tryggvason, 1992) surface tension models.

1. Introduction

Smeared interface models have recently become popular for conducting numerical simulations of multi-phase flow. These models use fixed grids and model the interface as having finite width that is resolvable on the grid. The fractional presence of a phase can then be represented by a field variable that varies continuously through the interface. Fluid properties are keyed to this variable and surface tension forces are distributed through the width of the model interface. These models have slow convergence but they allow the approximate solution of problems that would otherwise be intractable.

This paper gives a brief introduction to the use of a variational method for generating such models. The method starts with a model of the thermodynamics, or, if the system is isothermal, just the free energy density, of the model two-phase fluid. The surface tension forces are derived from the fact that advection can change fluid energy, by changing shape, but doesn't change total energy. An advective process that changes fluid energy must therefore be subject to forces that change kinetic energy so as to keep the total energy constant. Calculus of variations can be used to investigate arbitrary advective perturbations to the system and to deduce the required

231

V. Venkatakrishnan et al (eds.), Barriers and Challenges in Computational Fluid Dynamics, 231-240.
© 1998 *Kluwer Academic Publishers.*

offsetting forces. The method is really the same as the method of virtual work commonly used in elasticity and to analyze mechanical systems. The variational approach is applicable to both discrete and continuum energy models. A great advantage of the method is that its resulting surface tension models are always energy conserving.

The two main smeared surface tension techniques are the distributed force method of Unverdi & Tryggvason (1992) and the continuum surface tension method of Brackbill, Kothe & Zemach (1992). A newer technique is the phase-field method (Anderson & McFadden, 1996; Antonovskii, 1995; Chella & Viñals, 1996; Jacqmin, 1995b, 1996; Jasnow & Viñals, 1996; Nadiga & Zaleski, 1996). This method is the source of the ideas to be discussed and applied here.

The method of Unverdi & Tryggvason tracks interfaces by following the advection of control points. These points mark the smeared interface's center. The interfaces are further defined by connecting the control points by curves or line segments (in 2-D) or triangular surfaces (in 3-D). Surface tension forces are calculated from the control point positions and distributed to the fixed grids. Changes in fluid properties across the interface are smoothed so as to occur over several grid cells.

In the continuum-surface-tension method of Brackbill et al. the surface tension forcing is set equal to the interface gradient times its curvature. The total forcing on the fluid through an interface is thus proportional to the interface's gradient-averaged curvature. The method has been applied using volume-of-fluid (Lafaurie et al. 1994; Rider et al., 1995; Kothe et al., 1996), TVD (Jacqmin,1995a), and level-set (Sussman et al., 1994; Sussman et al., 1996) methods. VOF methods actually use what amounts to two functions to define an interface. The transition from one phase to another is represented by a "color function" which varies from 0 to 1. VOF advection operating on this color function generates a very steep interface, less than two cells wide. This is too steep to allow smooth curvature calculations so surface tension forcing and material property variations are computed using a locally averaged (mollified) form of the color function. The level-set method makes use of a smooth "approximate-distance" function. The zero value of this function marks the smeared-interface center. Convection methods appropriate for smooth functions can be used. The function is initialized as a distance function and thereafter maintained against excessive advective distortion by relaxation, downgradient to the (evolving) distance function, and/or by occasional reinitializations. Surface tension forces are found by calculating the curvature of the function in the vicinity of its zero level.

The above methods are based on models of surface tension *forces*. Phase field methods are based on models of fluid *energy*. The simplest model of energy density that gives two phase flow is

$$e = \frac{1}{2}\alpha|\vec{\nabla}C|^2 + \beta\Psi(C) \tag{1}$$

a formulation that goes back to van der Waals (1893). The first term is gradient energy, the second bulk energy. Two phases are possible if Ψ has two minima. The phase field method uses smooth advection routines with interface profiles maintained against distortion by energy-downgradient antidiffusion.

This paper will show how the variational approach works and how the distributed force method and the VOF version of the CSF method can be derived variationally . A motive for this work has been to find out how the mollified color function can be made consistent with energy conservation. I show how this can be done and also briefly indicate some possibly useful additional smoothing techniques.

2. Variational Derivation of Surface Tension Forces for an Incompressible, Isothermal Fluid System

The surface energy of an isothermal fluid system is equal to its surface area times its surface tension σ. To keep the analysis simple, consider a 2D system with surface height h a single valued function of x. Then the energy of the system is

$$\mathcal{E} = \sigma\int\sqrt{1 + \left(\frac{\partial h}{\partial x}\right)^2}\,dx + \int\int\frac{1}{2}\rho(u^2 + v^2)\,dx\,dy = \mathcal{F} + \mathcal{K} \tag{2}$$

$\mathcal{F}$ is the free energy of the system, $\mathcal{K}$ the kinetic energy. The variation of the free energy with time is

$$\frac{d\mathcal{F}}{dt} = \sigma\int\frac{\partial h/\partial x}{\sqrt{1 + (\partial h/\partial x)^2}}\frac{\partial^2 h}{\partial t\partial x}\,dx \tag{3}$$

Upon integrating by parts this becomes

$$\frac{d\mathcal{F}}{dt} = -\sigma\int\frac{\partial}{\partial x}\left(\frac{\partial h/\partial x}{\sqrt{1 + (\partial h/\partial x)^2}}\right)\frac{\partial h}{\partial t}\,dx = -\sigma\int\kappa(x)\frac{\partial h}{\partial t}\,dx \tag{4}$$

where $\kappa(x)$ is the interface curvature. The integration by parts requires boundary conditions for h. Either h fixed or contact angle fixed can be used.

The evolution of h is given by the kinematic equation

$$\frac{\partial h}{\partial t} = v - u\frac{\partial h}{\partial x} \tag{5}$$

and the evolution of $\mathcal{K}$ is given by

$$\frac{d\mathcal{K}}{dt} = \int\int (F_x u + F_y v)\,dx\,dy \tag{6}$$

The effective part of $\vec{F}$, since the pressure in an incompressible fluid doesn't change total kinetic energy, is due only to surface tension forcing. The total energy $\mathcal{E}$ remains constant. This gives

$$\int\int \left(\left(F_x + \sigma\delta(y-h)\kappa(x)\frac{\partial h}{\partial x}\right)u + \left(F_y - \sigma\delta(y-h)\kappa(x)\right)v \right)dx\,dy = 0 \tag{7}$$

This is true for arbitrary incompressible $\vec{u}$ and therefore $\vec{F}$ must, to within the gradient of an arbitrary scalar field, be

$$F_x = -\sigma\delta(y-h)\kappa(x)\frac{\partial h}{\partial x} \quad ; \quad F_y = \sigma\delta(y-h)\kappa(x) \tag{8}$$

$\vec{F}$ is a vector delta function with magnitude equal to $\sigma\sqrt{1+(\partial h/\partial x)^2}\,\kappa$, the surface tension times the curvature times interface length per dx.

In terms of u and v the evolution of free energy is given by

$$\frac{d\mathcal{F}}{dt} = \sigma\int\int \delta(y-h)\kappa(x)\left(\frac{\partial h}{\partial x}u - v\right)dx\,dy \tag{9}$$

Comparing to (8), we see that

$$F_x = -\frac{\delta}{\delta u}\frac{d\mathcal{F}}{dt} \quad ; \quad F_y = -\frac{\delta}{\delta v}\frac{d\mathcal{F}}{dt} \tag{10}$$

This relationship is generally applicable to smeared interface models, both continuum and discretized.

3. Variational Derivation of Surface Tension Forces for a Discretized Interface

The simplest case is given by an interface with straight line segments connecting the control points. For this the free energy is

$$\mathcal{F} = \sigma\sum_i \sqrt{(x_{i+1} - x_i)^2 + (y_{i+1} - y_i)^2} \tag{11}$$

The evolution of free energy is given by

$$\frac{d\mathcal{F}}{dt} = \sum_i \frac{\partial\mathcal{F}}{\partial x_i}\frac{dx_i}{dt} + \sum_i \frac{\partial\mathcal{F}}{\partial y_i}\frac{dy_i}{dt} \tag{12}$$

where

$$\frac{\partial \mathcal{F}}{\partial x_i} = \sigma \left(\frac{x_i - x_{i-1}}{S_{i-1/2}} + \frac{x_i - x_{i+1}}{S_{i+1/2}} \right) \tag{13}$$

and similarly for $\partial \mathcal{F}/\partial y_i$. $S_{m+1/2}$ denotes $\sqrt{(x_{m+1} - x_m)^2 + (y_{m+1} - y_m)^2}$.

If the discretized interface is modeled as being in a continuum fluid then $d\vec{x}_i/dt = \vec{u}(x_i, y_i)$. Application of (10) yields

$$\vec{F} = -\delta(x - x_i, y - y_i)\sigma \left(\frac{\vec{x}_i - \vec{x}_{i-1}}{S_{i-1/2}} + \frac{\vec{x}_i - \vec{x}_{i+1}}{S_{i+1/2}} \right) \tag{14}$$

delta-function forces located at the control points. If the fluid velocity field is discretized on a fixed grid then movement of a control point must be determined by an interpolation from nearby velocity grid points. In general, $\frac{dx_i}{dt} = \sum_j a_{ij} u_j$, $\frac{dy_i}{dt} = \sum_j b_{ij} v_j$. Then

$$\frac{d\mathcal{F}}{dt} = \sum_j \left(\sum_i a_{ij} \frac{\partial \mathcal{F}}{\partial x_i} \right) u_j + \sum_j \left(\sum_i b_{ij} \frac{\partial \mathcal{F}}{\partial y_i} \right) v_j \tag{15}$$

giving

$$F_{x_j} = -\sum_i a_{ij} \frac{\partial \mathcal{F}}{\partial x_i} \quad ; \quad F_{y_j} = -\sum_i b_{ij} \frac{\partial \mathcal{F}}{\partial y_i} \tag{16}$$

We see that the interpolation stencils for calculating grid velocities set the distribution of surface tension forces to the grid. Interpolation of grid velocities to the control points and distribution of forces to the grid cannot be done independently while maintaining energy conservation. If, for smoothness, a wide distribution of surface tension forces to the grid is desired, then a wide stencil should be used for the interpolation of the velocities.

Another device could be used to help ensure smoothness, the use of diffusion or relaxation to prevent small scale jaggedness on the interface. The simplest relaxation process that is downgradient in free energy, and therefore guaranteed smoothing because it moves in the direction of reducing surface area, is

$$\frac{d\vec{x}_i}{dt} = -D \frac{\partial \mathcal{F}}{\partial \vec{x}_i} \tag{17}$$

This, however, works equally on all wavelengths. A process that works mainly on small scales is

$$\frac{d\vec{x}_i}{dt} = D \left(\frac{\partial \mathcal{F}}{\partial \vec{x}_{i-1}} - 2 \frac{\partial \mathcal{F}}{\partial \vec{x}_i} + \frac{\partial \mathcal{F}}{\partial \vec{x}_{i+1}} \right) \tag{18}$$

This can be further modified, while remaining proveably downgradient, to meet the constraint of no net change of phase volume.

All the above can be done in three dimensions, though some of the geometric calculations are then much lengthier. The simplest general case is that of control points defining a triangulated surface. Rates of change of triangular areas must be calculated in order to find $d\mathcal{F}/dt$ and the forces.

4. Variational Derivation of Surface Tension Forces for the Continuum Surface Tension Model

The CSF model of Brackbill et al. assumes smeared interfaces. The fluid phases are represented by a field variable that will be taken here to be the color function C. This is assumed to change in a continuous fashion through an interface from 0 to 1. The forces in the fluid are modeled as proportional to the curvature of the color function field times its gradient:

$$\vec{F} = -\sigma \left(\vec{\nabla} \cdot \left(\vec{\nabla} C / |\vec{\nabla} C| \right) \right) \vec{\nabla} C = -\sigma \kappa \vec{\nabla} C \tag{19}$$

The free energy density corresponding to this is $\sigma |\vec{\nabla} C|$. This can be shown using the same approach as in the previous sections. We have

$$\mathcal{F} = \int \sigma |\vec{\nabla} C| \, dV \tag{20}$$

from which

$$\frac{d\mathcal{F}}{dt} = \sigma \int \frac{\vec{\nabla} C}{|\vec{\nabla} C|} \cdot \vec{\nabla} \left(\frac{\partial C}{\partial t} \right) dV \tag{21}$$

After integration by parts (21) becomes

$$\frac{d\mathcal{F}}{dt} = -\sigma \int \kappa \frac{\partial C}{\partial t} dV \tag{22}$$

For an incompressible fluid

$$\frac{\partial C}{\partial t} = -u \frac{\partial C}{\partial x} - v \frac{\partial C}{\partial y} = -\frac{\partial u C}{\partial x} - \frac{\partial v C}{\partial y} \tag{23}$$

The middle of (23) inserted into (22) immediately leads to the CSF forcing given by (19). Insertion of the right hand side followed by an integration by parts leads to

$$\vec{F} = \sigma C \vec{\nabla} \kappa \tag{24}$$

(24) differs from (19) by the gradient of $\sigma \kappa C$ but this difference is negated by the fluid's incompressibility. For (19) the scalar potential that maintains incompressibility is the true pressure, for (24) it is the pressure plus $\sigma \kappa C$.

There are many possible discrete versions of (19). Since VOF CSF uses a mollified C for computing curvatures I consider here a free energy expressed in terms of M_i, the discrete mollified color function. M_i is formed by a local averaging of C_i; $M_i = \sum_j a_{ij} C_j$, where $\sum_j a_{ij} = 1$. The discrete free energy densities formed from the M_i have the general form

$$f_i = \sigma \sqrt{\sum_j \sum_k \gamma_{ijk} (M_k - M_j)^2} \tag{25}$$

A simple discretization on a square grid is

$$f_{k,l} = \frac{\sigma}{\sqrt{2}h} \sqrt{\sum_{r=\pm 1} (M_{k,l} - M_{k+r,l})^2 + \sum_{s=\pm 1} (M_{k,l} - M_{k,l+s})^2} \tag{26}$$

where $f_{k,l}$ is the free energy density at grid point (x_k, y_l) and h is the grid spacing.

The free energy functional is the sum of the free energy densities times area factors, $\mathcal{F} = \sum_i A_i f_i$. Its variation is

$$\frac{d\mathcal{F}}{dt} = \sum_i A_i \frac{df_i}{dt} = \sum_j \left(\sum_i A_i \frac{\partial f_i}{\partial M_j} \right) \frac{dM_j}{dt} = \sigma \sum_j A_j \kappa_j \frac{dM_j}{dt} \tag{27}$$

$\sum_i A_i \dfrac{\partial f_i}{\partial M_j} / \sigma A_j$ is the discrete curvature κ_j. As an example, the curvature corresponding to discretization (26) is

$$\kappa_{k,l} = \frac{1}{\sqrt{2}h} \sum_{r=\pm 1} \left(\frac{1}{f_{k+r,l}} + \frac{1}{f_{k,l}} \right) \left(M_{k,l} - M_{k+r,l} \right)$$

$$+ \frac{1}{\sqrt{2}h} \sum_{s=\pm 1} \left(\frac{1}{f_{k,l+s}} + \frac{1}{f_{k,l}} \right) \left(M_{k,l} - M_{k,l+s} \right) \tag{28}$$

From the relationship between M and C

$$dM_j/dt = \sum_k a_{jk} dC_k/dt \tag{29}$$

The evolution of C_k is

$$\frac{dC_k}{dt} = \sum_m \left(\alpha_{km} \overline{C}_m u_m + \beta_{km} \overline{C}_m v_m \right) \tag{30}$$

(30) is applicable to a general unstructured grid in which the u_m and v_m are located at cell interfaces and the C_k are located at cell centers. The $\overline{C}_m$

are located at the cell interfaces. In the VOF method the $\overline{C}_m$ are calculated via nonlinear interpolations between the C in nearby cells. The summation in m includes just those interfaces m that border the cell k. The α_{km} and β_{km} are zero except for the two ks, say $k_-(m)$ and $k_+(m)$, that are divided by the interface m. The values of $\alpha_{k_+,m}$ and $\beta_{k_-,m}$ depend on interface orientation, interface length, and cell area. Also, if the cells k_- and k_+ have the same area, $\{\alpha_{k_-,m},\beta_{k_-,m}\} = -\{\alpha_{k_+,m},\beta_{k_+,m}\}$.

Substituting (30) into (29) and then the result into (27), we obtain

$$\frac{d\mathcal{F}}{dt} = \sigma \sum_m \overline{C}_m \left(\sum_{k=k_-,k_+} \alpha_{km} \sum_j a_{jk} A_j \kappa_j \right) u_m$$

$$+ \sigma \sum_m \overline{C}_m \left(\sum_{k=k_-,k_+} \beta_{km} \sum_j a_{jk} A_j \kappa_j \right) v_m \quad (31)$$

from which

$$F_x = -\sigma \overline{C}_m \sum_{k=k_-,k_+} \alpha_{km} \sum_j a_{jk} A_j \kappa_j = -\sigma \overline{C}_m \sum_{k=k_-,k_+} \alpha_{km} A_k \overline{\kappa}_k \quad (32)$$

$$F_y = -\sigma \overline{C}_m \sum_{k=k_-,k_+} \beta_{km} \sum_j a_{jk} A_j \kappa_j = -\sigma \overline{C}_m \sum_{k=k_-,k_+} \beta_{km} A_k \overline{\kappa}_k \quad (33)$$

at (x_m, y_m). The $\overline{\kappa}_k$ are locally averaged curvatures, $\overline{\kappa}_k = \sum_j a_{jk} A_j \kappa_j / A_k$. (32-33) shows that mollifying C results in a secondary smoothing: u and v are then forced by mollified curvatures. *If a mollified C is used to compute curvature then this curvature must be further mollified in order to conserve energy. This mollified curvature field is what is used to force $\vec{u}$.*

An additional smoothing technique is available, the use of smeared or averaged velocities to convect the C. Instead of (30) one can use

$$\frac{dC_k}{dt} = \sum_m \left(\alpha_{km} \overline{C}_m \overline{u}_m + \beta_{km} \overline{C}_m \overline{v}_m \right) \quad (34)$$

where $\{\overline{u}_m, \overline{v}_m\} = \{\sum_n b_{mn} u_n, \sum_n b_{mn} v_n\}$, $\sum_n b_{mn} = 1$. $\overline{u}$ and $\overline{v}$ must satisfy the discretized continuity equation. A diffusive process that maintains continuity constraints can be used to derive and define the smoothed velocities. The surface tension forcing that results is

$$F_x = -\sigma \sum_m b_{mn} \left(\overline{C}_m \sum_{k=k_-,k_+} \alpha_{km} A_k \overline{\kappa}_k \right) \quad (35)$$

$$F_y = -\sigma \sum_m b_{mn} \left(\overline{C}_m \sum_{k=k_-,k_+} \beta_{km} A_k \overline{\kappa}_k \right) \quad (36)$$

at (x_n, y_n). This constitutes a local averaging of the forces given in (32-33). As with the distributed force model, this averaging mirrors the averaging given the velocities.

Before closing, one minor difficulty with the CSF method needs discussing, the indefiniteness of the curvature in locations where gradients are zero. Where this occurs curvature calculations must be made using some limiting procedure. In discretized systems a zero gradient means the discretized free energy is zero. This results in divide-by-zero problems when calculating curvatures. Numerators are also zero so the curvatures are undefined.

In the continuum case the problem is resolved by using the momentum forcing given by equation (19). (19) gives either a zero or a smoothly varying non-zero forcing when the gradient is zero. In the latter case, which holds at maxima and minima with non-zero second derivatives, the local forcing can be calculated by interpolation from nearby points.

The form of the surface tension forcing (32-33) for the VOF CSF model is equivalent to that of (24). It is difficult to transform it to the form of (19) because of the nonlinear calculation of the $\overline{C}_m$. There are, however, at least two alternative ways to resolve the curvature issue.

One way is to use free energy densities for the curvature calculation that are averaged over each time step. An averaged free energy density will be zero only if it is zero initially and remains unchanged throughout the time step. In this case, however, it can be dropped from the calculation - to be specific, from the first summation in (27) - because it makes no contribution to free energy change during the time step and can thus make none to changes in momentum.

Another, perhaps easier, way to resolve the problem is to use a modified free energy density. One possibility is

$$f = \sigma\left(\sqrt{|\vec{\nabla} M|^2 + \epsilon^2} - \epsilon\right) \tag{37}$$

The modified curvature corresponding to this is zero whenever the gradient is zero for any positive ϵ. A non-zero ϵ also gives smoother surface tension calculations. It introduces an $O(\epsilon)$ error into the calculation of the surface energy but this can be made small compared to the error already introduced by interface smearing.

References

Anderson, D. M., and McFadden, G. B., "A Diffuse-Interface Description of Fluid Systems, National Institute of Standards and Technology," Gaithersburg, MD, Report No. NISTER 5887, 1996.

Antanovskii, L. K., "A Phase Field Model of Capillarity," *Phys. Fluids*, Vol. 7, 1995, pp. 747-753.

Brackbill, J., Kothe, D. B., and Zemach, C., "A Continuum Method for Modeling Surface Tension," *J. Comp. Phys.*, Vol. 100, 1992, pp. 747-753.

Chella, R. and Viñals, J., "Mixing of a Two-Phase Fluid by Cavity Flow," *Phys. Rev. E.*, Vol. 53, 1996, pp. 3832-3840.

Jacqmin, D., "Three-Dimensional Computations of Droplet Collisions, Coalescence, and Droplet/Wall Interactions using a Continuum Surface Tension Method," AIAA 95-0833, presented at the 33rd Aerospace Sciences Meeting, Reno, NV, 1995.

Jacqmin, D., "An Energy Approach to the Continuum Surface Tension Method: Applications to Droplet Coalescences and Droplet/Wall Interactions," Proceeedings of the 1995 ASME IMECE, San Francisco, CA, 1995.

Jacqmin, D., "An Energy Approach to the Continuum Surface Tension Method," AIAA 96-0858, presented at the 34th Aerospace Sciences Meeting, Reno, NV, 1996.

Jasnow, D. and Viñals, J., "Coarse-Grained Description of Thermo-Capillary Flow," *Phys. Fluids*, Vol. 8, 1996, pp. 660-669.

Kothe, D. B., Rider, W. J., Mosso, S. J., and Brock, J. S., "Volume Tracking of Interfaces Having Surface Tension in Two and Three Dimensions," AIAA 96-0859, presented at the 34th Aerospace Sciences Meeting, Reno, NV, 1996.

Nadiga, B. T. and Zaleski, S., "Investigations of a Two-Phase Fluid Model," to appear in *Eur. J. Mech., B/Fluids*, Vol. 15, 1996.

Lafaurie, B., Nardone, C., Scardovelli, R., Zaleski, S., and Zanetti, G., "Modeling Merging and Fragmentation in Multiphase Flows with SURFER," *J. Comp. Phys.*, Vol. 113, 1994, pp. 134-147.

Rider, W. J., Kothe, D. B., Moddo, S. J., and Cerutti, J. H., "Accurate Solution Algorithms for Incompressible Multiphase Flows," AIAA 95-0699, presented at the 33rd Aerospace Sciences Meeting, Reno, NV, 1995.

Sussman, M., Fatemi, E., Smeraka, P., and Osher, S. J., "An Improved Level Set Method for Incompressible Two-Phase Flow," to appear in *J. Comp. and Fluids*, 1996.

Sussman, M., Smeraka, P., and Osher, S. J., "A Level Set Approach for Computing Solutions to Incompressible Two-Phase Flow," *J. Comp. Phys.*, Vol. 114, 1994, pp. 146-159.

Unverdi, S. 0. and Tryggvason, G., "A Front-Tracking Method for Viscous, Incompressible, Multi-Fluid Flows," *J. Comp. Phys.*, Vol. 100, 1992, pp. 25-37.

van der Waals, J. D., "The Thermodynamic Theory of Capillarity Flow under the Hypothesis of a Continuous Variation of Density," *Verhandel/Konink. Akad. Weten.*, Vol. 1, 1893.

COMPOUNDED OF MANY SIMPLES

Reflections on the Role of Model Problems in CFD

PHILIP ROE
W.M. Keck Foundation Laboratory
for Computational Fluid Dynamics
Department of Aerospace Engineering
University of Michigan
Ann Arbor, Michigan 48109-2118, USA

1. Introduction

JACQUES: *But it is a melancholy of my own, compounded of many simples, extracted from many objects, and indeed the sundry contemplation of my travels (in which my often rumination wraps me in a most humorous sadness).*
WILLIAM SHAKESPEARE, *As You Like It*, Act IV, Scene 1.

In the above extract from Shakespeare's comedy, Jacques (a character whose principal role is that of an ironic commentator on the action) is replying to a young woman who says she has heard that he has a somewhat melancholy temperament. He concedes this, but will not admit it to be a weakness, because it arises as the distillation of a life's experience. Scientific researchers also know and value the melancholy that derives from not knowing the solution to a problem; it is often the driving force for greater efforts, and has an uniquely personal quality deriving from ones past attempts to make sense of the evidence. That evidence is usually, in a quite literal sense, 'compounded of many simples' because of the natural belief that a difficulty should always be studied in its simplest manifestation.

A beginning student in CFD will therefore study the linear advection equation to learn that the timestep in a hyperbolic marching problem is limited by the CFL rule. She will see from the heat equation that explicit methods are very inefficient for parabolic problems, and from Laplace's

[0]This research was partly supported by the National Aeronautics and Space Administration under NASA Contract No. NAS1-19480 while the author was in residence at the Institute for Computer Applications in Science and Engineering (ICASE), NASA Langley Research Center, Hampton, Va 23681-0001.

V. Venkatakrishnan et al (eds.), Barriers and Challenges in Computational Fluid Dynamics, 241-258.
© 1998 *Kluwer Academic Publishers.*

equation that Gauss-Seidel converges much quicker than point-Jacobi. With time, these lessons become part of the instinctive mental apparatus of the researcher or developer. The melancholy fact is that these lessons are not always adequate to solve our practical questions. In this talk I want to examine why this is, and what new 'simples' might be of value.

My personal, biased view of current CFD is that the most rapid progress is being made in new application areas, and I would cite MHD and Maxwell's equations as examples where techniques developed in other contexts are reaping rapid harvests. On the other hand, codes for industrial aerodynamics (high Reynolds number compressible flow) seem to have reached a state of relative stagnation; they usually run, give reasonable answers, and don't converge as slowly as they used to. There have been big advances in generating grids to handle complex geometry, but improvements to the actual flow algorithms are coming rather slowly and painfully.

There are several possibilities to explain this state of affairs. One is just that the problems really are that hard. Everything reasonable has already been done to make the schemes as good as they can be; if we are unhappy with the results we can only buy newer computers. Another possibility is that there still are new tricks to be played with the old knowledge. I am sure this is true, but doubt that there are any major breakthroughs to come from this source. Thirdly, we may hope to be more precise in identifying the simple behaviour hidden within complex systems. Fourthly, we may seek new perspectives from which the previously complex becomes simple. I will try to point out some possibilities in these last two directions.

As regards the recognition of simplicity, I will identify in Section 2 a precise machinery for identifying the simplest possible model problems contained in a given linear differential system. The general first-order homogeneous system in two-dimensions can be completely solved, and is equivalent to a certain number of linear advection equations, and a certain number of Cauchy-Riemann systems. All subsystems are completely decoupled, and this represents an optimal decomposition of the problem into its simplest elements. Of course it is not trivial to find the discretizations that take best advantage of the situation, but if this were the level of our most complex problems I feel confident that this path would solve them. For certain systems that are more complicated, such as the three-dimensional steady Euler equations, the simplest model problems can be identified by symmetry arguments; they turn out to present canonical problems that are simple but unstudied. Much progress may result from their investigation.

For problems that are more complex, the question posed in Section 2 is still precise. It amounts to seeking the invariant (in a certain sense) subspaces of several matrices, but there is apparently no systematic way to arrive at the answers. However, because physically-motivated problems tend to have some structure, *ad hoc* methods can be used to tackle individual problems. It then becomes clear that some problems contain no subproblems. They are their own simplest models, and the diagonalization approach breaks down.

In speculating about what should be done then, I return to a point that has bothered me for a long time. There is a huge abundance of analytical work on the fluid flow equations. Aside from the conservation form, there is the (usually much simpler) primitive form, there are characteristic and bicharacteristic forms, Crocco's law and a host of vorticity theorems, stream function and velocity potential formulations, and so on. The usual view of algorithm design is that one should pick one of these forms, discretise it, and hope that the others will follow 'within truncation error'. Conservation form is, of course, the overwhelmingly popular candidate; the other forms are usually employed only as accuracy checks. For example, the constancy of entropy along an inviscid streamline is seldom explicitly enforced in an Euler code, but commonly used to demonstrate that a clean solution has been achieved. The only example I can think of where independent formulations of the fluid equations have been combined is the combination of characteristic form and conservation form in Godunov-type codes. This particular combination of course has been outstandingly successful.

Some effort has been made to satisfy several forms of the equations simultaneously; notably by devising grids and storage strategies such that discrete versions of the vector identities used to prove many of the theorems are also discrete identities (Nicolaides, 1992; Nicolaides, 1993). This has had some success for the simpler forms of governing equations, but nothing has yet been devised that is universal. A by-product of this work is often to prove a precise well-posedness result for the discrete system; there may turn out to be exactly as many degrees of freedom as there are equations and boundary conditions to be satisfied (Ta'asan, 1993). This goal too proves elusive in full generality.

A possible way out of the dilemma, discussed in Section 3, is to abandon the notion of actually *solving* any of the specific forms of equation, but to approximate several of them. One deliberately creates an overspecified problem and then mimimises the error. A natural anxiety is that including more equations will greatly increase the work to be done. The trick will be to add precisely the right amount of redundant information, and it must still be verified that good choices exist for important practical problems.

An exciting opportunity is that by including the nodal coordinates among the variables with respect to which the objective is to be minimized, we create a moving-grid algorithm at very little additional cost. Properties of the grid motion turn out to reflect very accurately the physics of the governing equations. For hyperbolic problems the grid points tend to a steady state in which they form a characteristic mesh, whereas for elliptic problems they tend to group themselves smoothly in a way that recognises local anisotropy in the form of the equations. 'Diagonal-swapping' can be used as an independent way to decrease the objective, and failure to achieve small local residuals can be used to motivate point insertion. In either case, the grid that is created is not just a good general-purpose (e.g. Delaunay) grid, but one that is highly specific to the particular governing equations and the data of the particular problem. Grids that respond to features of the data have begun to be investigated in the context of approximation theory only comparatively recently (Dyn *et al.*, 1990; d'Azevedo & Simpson, 1991; Rippa, 1992; Baines, 1994; Tourigny & Hülsemann, 1997), and can have a dramatic impact on the solution quality (see (Castro Diaz *et al.*, 1995) and the contribution to these procedings by Habashi). The approach discussed here has the potential to go even further by responding to properties of the governing equations also.

I take some risk in choosing to write the latter part of this article on topics with which I have only just begun to experiment, and for which few solid results, analytical or experimental, are yet available. However, the theme of this workshop is to identify broad strategic issues and speculate about their resolution. I believe that is what I am doing. Moreover, the methods are not without precedent. They have much in common with First-Order System Least-Square (FOSLS) finite-element methods (Jiang & Povinelli, 1990; Jiang, 1992; Aziz *et al.*, 1985; Cai *et al.*, 1994; Zeitoun it et al., 1995). See especially in (Cai *et al.*, 1994) the stress placed on choosing an objective function. The moving grid interpretation links into Moving Finite Elements (MFE) of the kind described by Baines (Baines, 1994). There might well be a sense in which this apparent complexity is actually simplicity, because we have created additional degrees of freedom with which to combat the problem.

2. Recognising Simplicity

A physical theory should be as simple as possible, but no simpler.
ALBERT EINSTEIN

Consider a set of linear partial differential equations in d space dimensions and m unknowns, written as

$$\sum_{i=1}^{i=d} A_i \partial_{x_i} \mathbf{u} = 0. \tag{1}$$

We will define simplicity in terms of the smallest number (n) of scalar parameters required to define a solution, thus

$$\mathbf{u}(\mathbf{x}) = \mathbf{u}_0 + \sum_{j=1}^{j=n} p_j(\mathbf{x})\,\mathbf{r}_j. \tag{2}$$

Note that all gradients in this solution lie in span$\{\mathbf{r}_j\}$. We will call this the *gradient subspace*. The $\{\mathbf{r}_j\}$ are vectors defining certain solution modes and the scalars $\{p_j\}$ are their amplitudes. With $n = 1$, this is a classical 'simple wave'. For $1 < n < m$ we might describe it as a n-simple solution. Inserting this as a trial solution into the governing equations yields

$$\sum_{i=1}^{i=d} \sum_{j=1}^{j=n} A_i \partial_{x_i} p_j\,\mathbf{r}_j = 0. \tag{3}$$

There are a total of nd vectors $\{A_i\mathbf{u}_j\}$ in this equation. Since some linear combination of them vanishes they cannot be independent but must span a subspace of dimension $k \leq m$, which we will call the *residual subspace*. The interesting case is when $k = n < m$. Then we can choose some set of vectors $\{\mathbf{b}_j\}$ as a basis for the residual subspace, and expand all terms in (3) in that basis. Equating the coefficient of each basis vector to zero will give n relationships among the scalar quantities $\partial_{x_i} p_j$; in effect, a set of n partial differential equations for the n parameters. If these are satisfied, (2) will be a solution of (1). In contrast with simple waves, the modes $\{\mathbf{r}\}$ cannot individually satisfy (1), but can do so in the right combination.

Formally, the problem can be stated as follows: Given a set of $m \times m$ real matrices $\{A_i\}, i = 1 \ldots d$, find a set of real vectors $\{\mathbf{u}_j\}, j = 1 \ldots n$, such that $\dim\{A_i\mathbf{u}_j\} = n < m$.

There appears to be no standard algorithm for solving this problem, which is related to, although weaker than, the problem of simultaneous block diagonalization of $\{A_i\}$.[1] However, it can be solved completely in the two-dimensional homogeneous case (where we just have a pair of matrices) by solving the generalised eigenvalue problem for A_1, A_2,

$$\det(A_1 - \lambda A_2)\mathbf{r} = 0 \tag{4}$$

If λ is always real we have a complete decomposition into scalar problems. If there are a pair of complex roots $\lambda_{1,2} = \lambda_R \pm i\lambda_I$ then there are a pair of eigenvectors $\mathbf{r}_{1,2} = \mathbf{r}_R \pm i\mathbf{r}_I$, satisfying

$$(A_1 - \lambda_R A_2 - i\lambda_I A_2)(\mathbf{r}_R + i\mathbf{r}_I) = 0 \tag{5}$$

---

[1]Block diagonalization occurs when it is required that span$\{A_i\mathbf{u}_j\} \equiv$ span$\{\mathbf{u}_j\}$. This will be the case if the set of matrices $\{A_i\}$ is augmented by the identity matrix, and is the requirement for a decomposition of the unsteady flow. A preconditioning matrix is one that can be added to the set $\{A_i\}$ without changing the invariant subspaces.

From the real and imaginary parts of this equation it is easy to see that

$$\dim\{A_1 \mathbf{r}_R, A_1 \mathbf{r}_I, A_2 \mathbf{r}_R, A_2 \mathbf{r}_I\} = 2.$$

Moreover it can be shown that if the complex left eigenvalues are $(\ell_R \pm i\ell_I)$ then the scalar parameters are $p_1 = \ell_R \cdot \mathbf{u}$, $p_2 = \ell_I \cdot \mathbf{u}$ (see (Roe & Turkel, 1996)). The 2-simple elliptic solutions to these equations reside in the subspace spanned by $\{\mathbf{r}_R, \mathbf{r}_I\}$. For the system of linearized Euler equations for (ρ, u, v, p) made dimensionless by $(\rho_0, a_0, a_0, \rho_0 a_0^2)$ we obtain the elliptic system

$$(1 - M^2)p_x - v_y = 0 \tag{6}$$
$$v_x + p_y = 0 \tag{7}$$

Solutions of (6),(7) are completed by solutions to the hyperbolic problems $S_x = H_x = 0$ (where S, H are entropy and enthalpy) residing in the complementary subspace (Roe & Mesaros, 1996).

Anticipating a little the developments of the next section, suppose that a residual for the Euler equations $A\mathbf{u}_x + B\mathbf{u}_y \neq 0$ has been found in some computational cell during an iteration toward steady state. Projecting this vector into the subspaces inhabited by different solutions gives the decomposition of the residual into components due to different physical effects.

$$\Pi_A = \begin{bmatrix} 0 & \frac{-1}{M} & 0 & 1 \\ 0 & 0 & 0 & 0 \\ 0 & 0 & 1 & 0 \\ 0 & \frac{-1}{M} & 0 & 1 \end{bmatrix}, \Pi_H = \begin{bmatrix} 0 & \frac{1}{M} & 0 & 0 \\ 0 & 1 & 0 & 0 \\ 0 & 0 & 0 & 0 \\ 0 & \frac{1}{M} & 0 & 0 \end{bmatrix}, \Pi_S = \begin{bmatrix} 1 & 0 & 0 & -1 \\ 0 & 0 & 0 & 0 \\ 0 & 0 & 0 & 0 \\ 0 & 0 & 0 & 0 \end{bmatrix}. \tag{8}$$

If the residual is $\mathbf{R}$ then $\Pi_A \mathbf{R}, \Pi_H \mathbf{R}, \Pi_S \mathbf{R}$ are its components due to acoustic (or potential) behaviour, enthalpy variation and entropy variation respectively. It may be verified that these projections are orthonormal, i.e. $\Pi_i \Pi_j = \delta_{ij} \Pi_i$ for any pair; therefore the problem is reduced to completely independent subproblems. This happens of course because they were sought in separate subspaces.

At least in the linear case, numerical schemes can be derived that apply completely different methodology to the different components of the residual. For example, the hyperbolic parts can be solved by marching in space and the elliptic parts by relaxation, as our beginning student would know. Since the problems are completely decoupled, the whole system can be solved in an optimal manner (Roe & Mesaros, 1996).

Consider now the linearised Euler equations in three dimensions written, in the same variables as above, as $A\mathbf{u}_x + B\mathbf{u}_y + C\mathbf{u}_z = 0$, where

$$A = \begin{bmatrix} M & 1 & 0 & 0 & 0 \\ 1 & M & 0 & 0 & 0 \\ 0 & 0 & M & 0 & 0 \\ 0 & 0 & 0 & M & 0 \\ 0 & 0 & 0 & 0 & M \end{bmatrix}, B = \begin{bmatrix} 0 & 0 & 0 & 0 & 0 \\ 0 & 0 & 1 & 0 & 0 \\ 0 & 0 & 0 & 0 & 0 \\ 1 & 0 & 0 & 0 & 0 \\ 0 & 0 & 0 & 0 & 0 \end{bmatrix}, C = \begin{bmatrix} 0 & 0 & 0 & 1 & 0 \\ 0 & 0 & 0 & 0 & 0 \\ 0 & 0 & 0 & 0 & 0 \\ 1 & 0 & 0 & 0 & 0 \\ 0 & 0 & 0 & 0 & 0 \end{bmatrix}.$$

(9)

It turns out that there are two invariant subspaces of dimension one, containing, as before, the effects of entropy and shear waves, and one of dimension three, for which the governing equations are

$$(1 - M^2)p_x - v_y - w_z = 0 \tag{10}$$
$$v_x + p_y = 0 \tag{11}$$
$$w_x + p_z = 0 \tag{12}$$

As noted in (Roe & Turkel, 1996), these imply a purely elliptic equation for the pressure,

$$(1 - M^2)p_{xx} + p_{yy} + p_{zz} = 0$$

but the transverse velocities obey a third-order equation, and are also linked to each other by a hyperbolic equation

$$(w_y - v_z)_x = 0$$

showing constancy of streamwise vorticity along a streamline. So in three dimensions the distinction between elliptic and hyperbolic problems begins to blur a little. There seems to be no minimal set of equations from which all of the relevant physics can be directly read off. The numerical method must therefore be adapted to handling information in more than one form.

3. Coping with Complexity

What I tell you three times is true.
LEWIS CARROLL, *The Hunting of the Snark*

We begin by dividing the domain of interest into a set $\{T\}$ (see Figure 1) of triangular or tetrahedral elements[2]. All quantities are thought of as stored at the vertices and vary linearly over the element, implying interelement continuity. We consider first-order differential operators that are locally linearised within each element; such operators acting on the locally constant gradients give rise to residuals that are constant within elements. We call these the element fluctuations Φ_T and note that the number of

---

[2] Such a grid is, of course, compounded of many simplices.

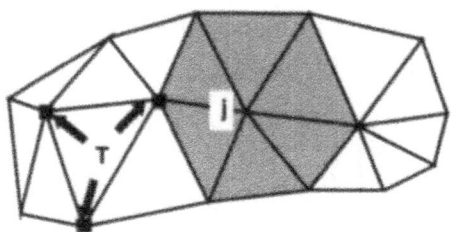

Figure 1. Part of an unstructured grid. A typical point is denoted by j and $\{T_j\}$ is the set of triangles (shown shaded) that have it for a vertex. A typical triangle is denoted by T and $\{j_T\}$ is the set of points (marked with squares) forming its vertices.

fluctuations typically exceeds the number of vertex values. Therefore the fluctuations cannot be made to vanish individually, but some weighted sum can be made to vanish at each vertex. The strategies for picking the weights have been extensively debated (Mesaros & Roe, 1995; Paill'ere *et al.*, 1995; Sidilkover & Roe, 1995).

Here we attempt to minimise some functional

$$F = \frac{1}{2} \sum_{T \in \{T\}} \Phi_T^t Q_T \Phi_T \tag{13}$$

where Q_T is a positive definite symmetric matrix that serves two purposes. Within the element it assigns relative weight to the different equations that are being approximated, and through a scaling factor it also weights the errors in each triangle relative to the other triangles. It is through the choice of Q_T that the scheme may be endowed with physical insight. Such a scheme can be coded with remarkable elegance. The derivative of F with respect to a particular nodal quantity (solution value or coordinate) is the sum of terms arising from each of the simplices that share that vertex. Thus

$$\frac{\partial F}{\partial \mathbf{u}_j} = \sum_{T \in \{T_j\}} \Phi_T^t \left(Q_T \frac{\partial \Phi_T}{\partial \mathbf{u}_j} + \frac{1}{2} \frac{\partial Q_T}{\partial \mathbf{u}_j} \Phi_T \right). \tag{14}$$

The vanishing of this sum has a dual significance. Clearly it enforces a minimum of F, but also it signifies local equilibrium of a numerical scheme of distributive type. Suppose that we compute Φ_T within each triangle and then make adjustments to the vertex values, and to the nodal coordinates,

$$\delta \mathbf{u}_j = -\Phi_T^t \left(Q_T \frac{\partial \Phi_T}{\partial \mathbf{u}_j} + \frac{1}{2} \frac{\partial Q_T}{\partial \mathbf{u}_j} \Phi_T \right). \tag{15}$$

Such a scheme reaches equilibrium everywhere only at a local minimum [3] of F. For homogeneous problems there is, as there should be, a conservation property. Since $\Phi_T = 0$ when all vertex values are equal, and Φ_T is linear in the vertex values, we must have

$$\sum_{j \in jT} \frac{\partial \Phi_T}{\partial \mathbf{u}_j} = 0. \tag{16}$$

which implies that

$$\sum_{j \in jT} \delta \mathbf{u}_j = 0. \tag{17}$$

Therefore, any interior simplex merely redistributes nodal quantities. This applies to nodal coordinates also, the centroid of the element does not move. Naturally, integrated quantities can change due to events at a boundary.

3.1. SCALAR ADVECTION

The prototype problem is two-dimensional advection on a triangular grid

$$u_t + (\vec{a} \cdot \nabla)u = u_t + au_x + bu_y = 0. \tag{18}$$

The advection speed $\vec{a}$ need not be a constant, but must be assigned a suitable average value within each cell. The first step is to integrate u_t over the triangle. This works out as

$$\phi_T = -\int \int (\vec{a} \cdot \nabla) u \, dx \, dy \tag{19}$$

$$= \sum_{j \in \{jT\}} k_j u_j, \tag{20}$$

where u_j is the value of u at vertex j, and k_{jT} is given by

$$k_{jT} = -\frac{1}{2}\vec{a} \cdot \vec{n}_{jT}$$

with $\vec{n}_{jT}$ the scaled inward normal to the side of T opposite vertex j. Note that ϕ_T is an area-weighted residual; it seems natural to minimise

$$F = \frac{1}{2} \sum_{T \in \{T\}} \frac{\phi_T^2}{S_T}, \tag{21}$$

---

[3]The minus sign is inserted in (15) to ensure that we go *down* the gradient. In fact the scheme naturally follows a path of steepest descent. In the linear case it will inevitably find a global minimum, although probably not in the most efficient way.

i.e. $Q_T = 1/S_T$ The area S is of course given by

$$S = \frac{1}{2}\left(x_1(y_2 - y_3) + x_2(y_3 - y_1) + x_3(y_1 - y_2)\right).$$

The changes made at vertex 1, for example, are then

$$\delta \begin{pmatrix} u_1 \\ x_1 \\ y_1 \end{pmatrix} = - \begin{pmatrix} \frac{\phi_T}{S_T}\frac{\partial \phi_T}{\partial u_1} - \frac{\phi_T^2}{2S_T^2}\frac{\partial S_T}{\partial u_1} \\ \frac{\phi_T}{S_T}\frac{\partial \phi_T}{\partial x_1} - \frac{\phi_T^2}{2S_T^2}\frac{\partial S_T}{\partial x_1} \\ \frac{\phi_T}{S_T}\frac{\partial \phi_T}{\partial y_1} - \frac{\phi_T^2}{2S_T^2}\frac{\partial S_T}{\partial y_1} \end{pmatrix} \tag{22}$$

$$= - \begin{pmatrix} \frac{\phi_T}{2S_T}k_1 \\ \frac{b\phi_T}{2S_T}(u_2 - u_3) - \frac{\phi_T^2}{4S_T^2}(y_2 - y_3) \\ -\frac{a\phi_T}{2S_T}(u_2 - u_3) + \frac{\phi_T^2}{4S_T^2}(x_2 - x_3) \end{pmatrix}, \tag{23}$$

and the other changes follow by cyclic permutation. It is clear that this can be written as a very simple loop over the triangles.

Without allowing nodal movement, this is an interesting but not a very recommendable scheme. It has a centrally-symmetric stencil if the grid is symmetric, and although it is third-order accurate at the steady state on such grids, convergence is extremely slow; the iteration corresponds to solving $u_t = u_{ss}$ where s is distance along the streamline. Of course the converged solution is not monotone.

When grid motion is allowed something rather dramatic happens; the method finds an exact solution by aligning grid edges with characteristics. The principal mechanism for this is the first term in $\delta x, \delta y$, which produces motion normal to the characteristic direction. Such motion ceases when the fluctuation vanishes. This happens when every triangle has one edge aligned with the characteristic, and equal nodal values at the ends of that edge. Therefore there are many nodal configurations for which $F = 0$ is possible, and a method that follows steepest descent will find one of them. I am indebted to Prof. M. J. Baines for the very interesting observation that both types of motion implied by (23) are norm-reducing, and that the first term by itself can reduce the residuals to zero. The second term is the mechanism by which cells compete for area.

To demonstrate the technique numerically on a problem with non-constant coefficients, consider convection in a circle, such that

$$yu_x - xu_y = 0. \tag{24}$$

The fluctuation, on a general triangle, is given by

$$\phi = \oint (uy\,dy + ux\,dx) = \oint u\,d\left(\frac{x^2 + y^2}{2}\right) \simeq \frac{1}{4}\sum_{j\in\{j_T\}} u_j \Delta_{j_T}(x^2 + y^2) \tag{25}$$

where Δ_{jT} is a difference taken counterclockwise along the edge of triangle T that is opposite vertex j. From this it is easy to calculate the derivatives required for (22). The distribution formulae to vertex j from triangle T are

$$\delta u_j = \frac{1}{2}\left(\frac{\phi_T}{S_T}\right)\Delta_{jT}(x^2 + y^2), \tag{26}$$

$$\delta\vec{x}_j = \left(\frac{\phi_T}{S_T}\right)\vec{x}_j\Delta_{jT}u - \frac{1}{4}\left(\frac{\phi_T}{S_T}\right)^2\vec{n}_{jT}. \tag{27}$$

These changes are accumulated at a node with a relaxation factor small enough for stability. For faster convergence Newton iteration suggests itself and sequential (Gauss-Seidel-like) methods have been proposed in related contexts (Baines, 1994; Tourigny & Baines, 1997). An experiment was performed using the coarse 21×11 grid shown in Figure 2(left). The initial solution was taken to be zero everywhere, and on the inflow boundaries zero was also specified, except at just two grid points ($x = -0.5, -0.6$). With grid adaptation supressed the results in Figure 3(left) were obtained. They are not monotone due to the fact that an unconstrained least-squares method is employed.

Then the nodes were allowed to move[4], except at the inflow boundaries, where they were held fixed, and at the outflow boundaries, where they were permitted to move only within the boundary. As predicted the code finds the exact solution! The grid is continuously deformed so that certain cell edges become aligned with the circular characteristics, as seen in Figure 2(right). This only happens in regions affected by the transient solution. Note that some of the edges that converge into the characteristic paths were initially vertical, others horizontal or diagonal.

3.2. 2×2 SYSTEMS

The model problem here will be taken to be inviscid, irrotational flow arising from a small perturbation of a uniform stream and satisfying

$$\delta = (1 - M^2)u_x + v_y = 0, \tag{28}$$

$$\omega = v_x - u_y = 0, \tag{29}$$

We expect to find a distinction betwen the hyperbolic case $M^2 > 1$ and the elliptic case $M^2 < 1$. On a triangle T, the discrete versions are

$$D_T = \frac{1}{2}\sum_{j\in\{jT\}}(\beta^2 u_j\Delta_{jT}y - v_j\Delta_{jT}x), \tag{30}$$

---

[4]For the sake of simplicity in these exploratory calcalations, no diagonal-swapping was employed, but the diagonals were chosen in the initial grid in a favorable sense.

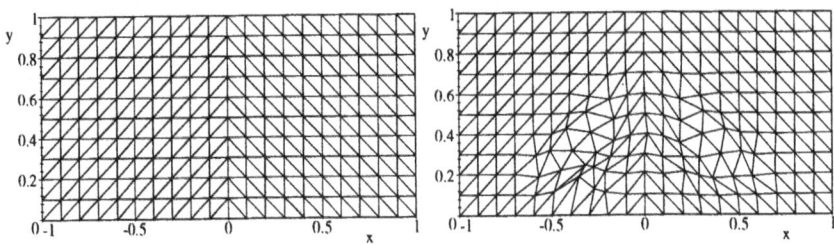

Figure 2. The initial grid (left) for the circular advection problem, and the final grid (right).

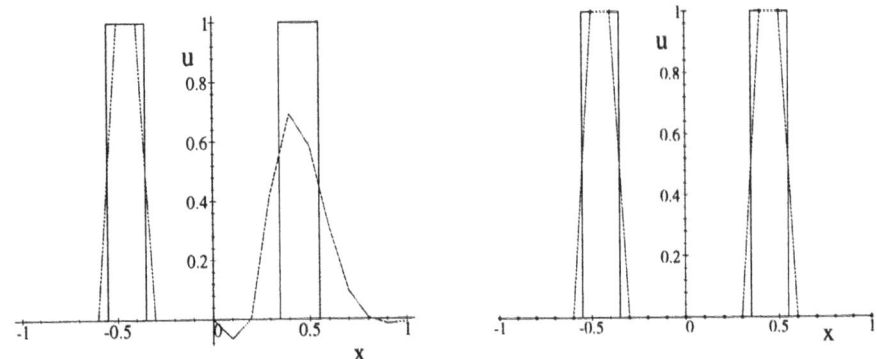

Figure 3. The circular advection problem, solved on a fixed grid (left) and an adaptive grid (right).

$$\Omega_T = \frac{1}{2} \sum_{j\in\{j_T\}} (v_j \Delta_{jT} y + u_j \Delta_{jT} x). \tag{31}$$

with $\beta^2 = 1 - M^2$. We will seek $\{u\}, \{v\}, \{x\}, \{y\}$ to minimise

$$F = \sum_T \frac{1}{2S_T} (D_T^2 + |M^2 - 1|\Omega_T^2), \tag{32}$$

where the choice of norm will be justified below.

Applying the minimization procedure leads straightforwardly to

$$\begin{pmatrix} \delta u_j \\ \delta v_j \\ \delta x_j \\ \delta y_j \end{pmatrix} \propto \frac{D_T}{S_T} \begin{pmatrix} (M^2 - 1)\Delta_{jT} y \\ \Delta_{jT} x \\ -\Delta_{jT} v \\ (1 - M^2)\Delta_{jT} u \end{pmatrix} + \frac{|M^2 - 1|\Omega_T}{S_T} \begin{pmatrix} -\Delta_{jT} x \\ -\Delta_{jT} y \\ \Delta_{jT} u \\ \Delta_{jT} v \end{pmatrix}$$

$$+\frac{D_T^2 + |M^2 - 1|\Omega_T^2}{2S_T^2} \begin{pmatrix} 0 \\ 0 \\ \Delta_{jT} y \\ -\Delta_{jT} x \end{pmatrix} \tag{33}$$

In the supersonic case, we can write $\beta = \sqrt{M^2 - 1}$ and rearrange to get

$$\begin{pmatrix} \delta(u_j \pm \beta v_j) \\ \delta(y_j \mp \beta x_j) \end{pmatrix} \propto \frac{\beta^2 \Omega_T \mp \beta D_T}{S_T} \begin{pmatrix} -\Delta_{jT}(x \pm \beta y) \\ \Delta_{jT}(v \mp \beta u) \end{pmatrix} - \frac{D_T^2 + \beta^2 \Omega_T^2}{2S_T^2} \begin{pmatrix} 0 \\ \Delta_{jT}(x \pm \beta y) \end{pmatrix} \tag{34}$$

This represents a discrete diagonalization. The RHS is driven by differences of characteristic variables, either in physical or velocity (hodograph) space, The changes produced by these on the left are normal to the characteristic directions. Only in the norm (32) does this occur. Mesh adjustment will cease when each triangle has one side aligned with each characteristic, and the corresponding characteristic equation is satisfied along it. When this happens F will vanish and we have a minimum configuration. Therefore the scheme will automatically generate a characteristic mesh.

Subsonically, diagonalization occurs in a different sense. If we introduce Prandtl-Glauert coordinates $X = x, Y = \beta y$, it can be shown that when the nodal values u, v are converged, both of these variables individually satisfy

$$\sum_{i \in \{i_{T_j}\}} (u_j - u_i)(\cot \theta_{ij} + \cot \theta_{ji}) = 0 \tag{35}$$

where θ_{ij}, θ_{ji} are the angles opposite to the edge ij in the triangles either side of the edge. Only the choice (32) produces this discrete decoupling, that exactly imitates the analytical decoupling into the pair of elliptic problems $(1 - M^2)(u, v)_{xx} + (u, v)_{yy} = 0$. This discretisation is positive (h-elliptic) provided

$$\cot \theta_{ij} + \cot \theta_{ji} > 0$$

for all edges, implying $\theta_{ij} + \theta_{ji} \leq \pi$. This is a property of Delaunay triangulations. We conclude that the grid should have the Delaunay property in the Prandtl-Glauert plane (Figure 4), implying that in the physical plane it should be elongated normal to the flow direction.

Grid movement does not necessarily cease when the nodal values reach equilibrium. There is a nice duality between grid and solution. The grid reaches equilibrium when $X = x, Y = \sqrt{1 - M^2}y$ satisfy a discrete Laplacian in the plane $U = \sqrt{1 - M^2}u, V = v$. In effect, the hodograph equations are solved simultaneously with the flow equations.

3.2.1. *Diagonal Swapping*

Allowing the grid to move does not by itself guarantee a good grid; the minimization assumes a fixed grid connectivity, and the initial connectivity

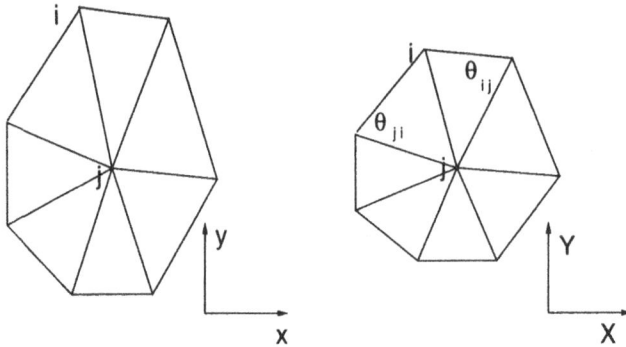

Figure 4. The triangles neighbouring a point in physical space (left) and Prandtl-Glauert space (right).

might be unsatisfactory. A well-known trick to achieve Delaunay triangulations is to consider all pairs of adjacent triangles as quadrilaterals and to replace the existing diagonal whenever the other one would be shorter. With the proviso that length should be measured in the Prandtl-Glauert plane, this procedure can be shown always to reduce F. It has the side-effect of improving the condition number of the discretization also. For a scalar problem the diagonal is chosen that has the smallest projected length normal to the characteristic direction.

3.2.2. *Boundary Conditions*
At a typical node k on the boundary, one might wish to impose a tangency condition

$$(u, v) \cdot (n_{x,k}, n_{y,k}) = 0.$$

This can easily be incorporated into the minimisation procedure by adding to F a term involving a Lagrange multiplier. Thus we seek the minimum of

$$F = \sum_{\{T\}} \frac{1}{2S_T}(\Delta_T^2 + \beta^2 \Omega_T^2) + \sum_{\{j\}} \lambda_j(u_j, v_j) \cdot (n_{x,j}, n_{y,j}) = 0. \qquad (36)$$

We now have additional terms in the derivative formulae:

$$\frac{\partial F}{\partial u_j} = \sum_{T \in \{T_j\}} \frac{1}{S_T}(\Delta y_j \Delta_T + \Delta x_j \Omega_T) + \lambda_j n_{x,j}, \qquad (37)$$

$$\frac{\partial F}{\partial v_j} = \sum_{T \in \{T_j\}} \frac{1}{S_T}(-\frac{1}{\beta^2}\Delta x_j \Delta_T + \Delta y_j \Omega_T) + \lambda_j n_{y,j} \qquad (38)$$

Interpreting these as distribution formulae, we see that they are altered only by the presence of additional changes due to the boundary condition,

$$\delta u_j^b = \lambda_j n_{x,j}, \tag{39}$$

$$\delta v_j^b = \lambda_j n_{y,j}. \tag{40}$$

Although the Lagrange multiplier λ_j is not known, we know that it enforces tangency. Therefore the correct boundary procedure is to add a normal velocity sufficient to do this. In other words, we should merely neglect the normal velocity component predicted by the regular procedure. This is common practice, often justified heuristically. Here it emerges with rigorous justification as a projection of the gradient path into the constraint surface.

3.3. STOKES FLOW

In two dimensions, Stokes flow (the flow of a viscous fluid at such low Reynolds number that inertia forces can be neglected) is governed by

$$\partial_x p = \mu(\partial_{xx} u + \partial_{yy} u), \tag{41}$$

$$\partial_y p = \mu(\partial_{xx} v + \partial_{yy} v), \tag{42}$$

$$\partial_x u + \partial_y v = 0. \tag{43}$$

These equations contain both first and second-order derivatives. If they are written in the form $A\mathbf{u}_x + B\mathbf{u}_y + C\nabla^2\mathbf{u} = 0$ then applying the methods of Section 2 we find that the matrices A, B, C have no invariant subspaces of dimension less than three, so the lowest-order problem cannot be simplified. Diagonalization can be achieved by going to a higher-order formulation; it is well-known that the pressure and the vorticity $\omega = \partial_x v - \partial_y u$, both obey Laplace's equation, and the velocities obey the biharmonic equation.

Reduction to fully first-order form comes by introducing the divergence $\delta = \partial_x u + \partial_y v$ and the vorticity, so that

$$\begin{array}{rcccl} m_x &=& \partial_x p + \mu\partial_y\omega &=& 0 \\ m_y &=& \partial_y p - \mu\partial_x\omega &=& 0 \\ \delta &=& \partial_x u + \partial_y v &=& 0 \\ \xi &=& \partial_x v - \partial_y u - \omega &=& 0 \end{array} \tag{44}$$

We now have four equations rather than the original three. At first glance, the first pair of equations, being a Cauchy-Riemann system for the pressure and vorticity, could be solved by themselves, and then the second pair solved to obtain the velocity. However, since the surface boundary conditions $u = v = 0$ are contained in the second pair it would be better to solve all the equations simultaneously. We are led to consider minimisation of the sum over all elements of

$$F_T = M_x^2 + M_y^2 + k^2(\Delta^2 + \Xi^2) \tag{45}$$

where upper-case symbols denote the natural elementwise discretizations.

Within an element, it seems natural to weight the first pair equally with each other, and probably the second pair also, but purely on dimensional grounds there is a need to distinguish between the pairs. The quantities that might reasonably enter into k are the viscosity μ and the local cell area S. Simple dimensional analysis reveals that

$$k^2 = C'\frac{\mu^2}{S}. \tag{46}$$

The pressure can then be shown to satisfy the same discrete scalar Laplace equation that turned up in the previous problem. Vorticity is linked to velocity, and the discrete velocity equations are linked to the vorticity but not to each other.

The full compressible Navier-Stokes equations can also be reduced to a first-order system by invoking the divergence and vorticity as new variables, but the best way to formulate a norm may depend (in three dimensions) on clearing up the three-dimensional Euler equations. [5]

4. Conclusions

Only he who continually attempts the absurd is capable of achieving the impossible.
MIGUEL DE CERVANTES, *Don Quixote*

I believe that we must build upon simplicity; the human mind has no other option. The barrier I identify is that old simplicities no longer serve. The challenge is to devise new simplicities.

Historically, simplicity has been sought in the form of model problems rich enough to encapsulate some essential difficulty, and it is amazing how much has actually been learned in this way. The approach could take us to our final goals, however, only if our real problems were simply the sum of the model problems. A diagonalization into independent subproblems would allow difficulties, and their solutions, just to be superposed. Considering the three-dimensional Euler equations, however, and a fortiori the Navier-Stokes equations, it becomes clear that no simple subproblems are to be

---

[5] Since the writing of the first draft, my student Hiroaki Nishikawa has obtained results for the driven cavity problem for the 2D incompressible Navier-Stokes equations. The grid adapts in a very appropriate way.

found. The essence of the problem becomes the interaction between different kinds of process. The resolution of conflict is the province of optimization theory. It makes sense to use all of the available degrees of freedom, so it is natural to find the grid playing a vital role.

However, we cannot suppose that all CFD problems will become trivial when reformulated as minimization problems. The folk-lore of many cultures tells cautionary tales whose point is to beware of what you wish for, because getting it may not bring happiness. In this context, the wish that must be carefully expressed is the norm in which the error is to be minimized. Although the example of linearized compressible flow shows that a correct norm can certainly induce remarkable properties, there is no guarantee that every problem posesses such a distinguished norm. Indeed, the choice may turn out to be context-dependent, even for a given equation set. One application might call for clean handling of shockwaves, but another for the accurate convection of vorticity. To deal with such diversity, any known theorem relating to the system at hand can be incorporated into the objective function, with greater or lesser weight. If it happens to be a consequence, at the discrete level, of laws already employed the solutions will not change. Otherwise, the new theorem will now be more faithfully obeyed, perhaps at the expense of some other property.

The work described above represents only the first preliminary steps in a lengthy program, but it is hard not to be impressed by a simple concept that is independent of dimension, addresses both solution error and grid optimization, spontaneously adapts to the physics of diverse problems, and automatically implements a hodograph transformation.

References

Aziz, A. K., Kellogg, R. B., and Stephens, A. B., "Least-square methods for elliptic systems," *Math Comp.* Vol. 44, 1985, pp. 53-75.

Baines, M. J., "Algorithms for optimal piecewise linear and constant L_2 fits to continuous functions with adjustable nodes in one and two dimensions," *Math. Comp.*, Vol. 62, 1994, pp. 645-669.

Baines, M. J., *Moving Finite Elements*, Oxford University Press, 1994.

Cai, Z., Lazarov, R. D., Manteuffel, T., and McCormick, S., "First-order system least-squares for partial differential equations; Part 1," *SINUM*, Vol. 31, 1994, pp. 1785-1799.

Castro Diaz, M. J., Hecht, F., and Mohammedi, B., "New progress on anisotropic grid adaptation for inviscid and viscous flow simulations," *INRIA Report RR 2671*, Roquencourt, 1995.

d'Azevedo, E. F., and Simpson, R. B., "On optimal triangular meshes for minimizing the gradient error," *Numerische Mathematik*, Vol. 59, 1991, pp. 321-348.

Dyn, N., Levin, D., and Rippa, S., "Data-dependent triangulations for piecewise linear interpolation," *IMA J. Num. Anal.*, Vol. 10, 1990, pp. 137-154.

Jiang, B. N., "A least-squares finite-element method for incompressible Navier-Stokes problems," *Int. J. Num. Meth. Fluids*, Vol. 14, 1992, pp. 843-859.

Jiang, B. N. and Povinelli, L. A., "Least-squares finite-element method for fluid dynamics," *Comput. Methods Appl. Mech. and Eng.*, Vol. 81, 1990, pp. 13-37.

Mesaros, L. and Roe, P. L., "Multidimensional fluctuation-splitting schemes based on decomposition methods," *AIAA 12th CFD Conference*, 1995, pp. 582-591.

Nicolaides, R. A., "Direct discretization of planar div-curl problems," *SINUM*, Vol. 29, No. 1, 1992, pp. 32-56.

Nicolaides, R. A., "Three-dimensional covolume algorithms for viscous flow," in *Algorithmic Trends in Computational Fluid Dynamics*, Hussaini, Kumar, and Salas, eds., Springer, 1993.

Paillère, H., Deconinck, H., and Roe, P. L., "Conservative upwind residual-distribution schemes based on the steady characteristics of the Euler equations," *AIAA 12th CFD Conference*, 1995, pp. 592-605.

Rippa, S., "Long and thin triangles can be good for linear interpolation," *SINUM*, Vol. 29, No. 1, 1992, pp. 257-270.

Roe, P. L. and Mesaros, L., "Solving steady mixed conservation laws by elliptic/hyperbolic splitting," 15th Int. Conf. Num. Meth. Fluid Dyn. *Lecture Notes in Physics*, Springer, to appear, 1996.

Roe, P. L. and Turkel, E., "The quest for diagonalisation of differential systems," *in this proceedings*, 1996.

Sidilkover, D. and Roe, P. L., "Unification of some advection schemes in two dimensions," *ICASE Report No. 95-10*, 1995.

Ta'asan, S., "Canonical forms of multidimensional steady inviscid flows," *ICASE Report No. 93-34*, 1993.

Tourigny, Y. and Baines, M. J., "Analysis of an algorithm for generating locally optimal meshes for L_2 approximation by discontinuous piecewise polynomials," *Math. Comp.*, to appear, 1997.

Tourigny, Y. and Hülsemann, F., "A new moving mesh algorithm for the finite element solution of variational problems," *Math. Comp.*, submitted 1997.

Zeitoun, D. G., Laible, J. P., and Pinder, G. F., "A weighted least-squares method for first-order hyperbolic systems," *Int. J. Num. Meth. Fluids*, Vol. 20, 1995, pp. 191-212.

A UNIFIED CFD-BASED APPROACH TO A VARIETY OF PROBLEMS IN COMPUTATIONAL PHYSICS

RAMESH K. AGARWAL
Aerospace Engineering
Wichita State University
Wichita, Kansas 67260

Abstract.

It is shown that many of the equations of mathematical physics describing a diverse range of physical phenomena, for example the Euler and Navier-Stokes equations of fluid dynamics, Maxwell equations, Schroedinger equation, semiconductor device simulation equations, relativistic gas dynamics equations, equations governing the elastic deformation of solids, equations of general relativity etc. can be written as a set of first-order partial differential equations in conservation law form. As a consequence, the well-developed CFD grid-generation techniques and solution algorithms can be easily applied for the numerical solution of these equations. However the gridding and the numerical algorithm employed must capture the physics of the problem and these requirements may differ substantially depending upon the problem to be solved. For example, the gridding requirements for scattering problems in computational electromagnetics are substantially different from those needed for flowfield simulations in computational aerodynamics. Examples from several disciplines are considered to describe the unified approach and to highlight the application specific differences which must be considered in gridding and in selecting/developing an efficient numerical algorithm.

1. Governing Equations

In two-dimensions, a variety of equations in mathematical physics can be expressed in conservation law form as follows:

$$\frac{\partial Q}{\partial t} + \frac{\partial F}{\partial x} + \frac{\partial G}{\partial y} = S, \tag{1}$$

259

V. Venkatakrishnan et al (eds.), Barriers and Challenges in Computational Fluid Dynamics, 259-282.
© 1998 *Kluwer Academic Publishers.*

where Q, F, G and S are defined below.

1.1. EULER/NAVIER-STOKES EQUATIONS

Let $\rho, p, (u, v), e, T, \mu$, and κ denote the density, pressure, velocity components, energy/unit mass, temperature, viscosity, and thermal conductivity in a fluid. Then the compressible unsteady Navier-Stokes equations are given by (Anderson $et\ al.$, 1984),

$$Q = \begin{bmatrix} \rho \\ \rho u \\ \rho v \\ \rho e \end{bmatrix}, F = \begin{bmatrix} \rho u \\ \rho u^2 + p - \tau_{xx} \\ \rho u v - \tau_{xy} \\ (\rho e + p)u - u\tau_{xx} - v\tau_{xy} + q_x \end{bmatrix}, G = \begin{bmatrix} \rho v \\ \rho u v - \tau_{xy} \\ \rho v^2 + p - \tau_{yy} \\ (\rho e + p)v - u\tau_{xy} - v\tau_{yy} + q_y \end{bmatrix}, S = 0,$$

$$(2)$$

where $p = (\gamma - 1)\left\{\rho e - \frac{1}{2}(u^2 + v^2)\right\}, \tau_{xx} = \frac{2\mu}{3}\left(2\frac{\partial u}{\partial x} - \frac{\partial v}{\partial y}\right).$
$\tau_{yy} = \frac{2\mu}{3}\left(2\frac{\partial v}{\partial y} - \frac{\partial u}{\partial x}\right), \tau_{xy} = \mu\left(\frac{\partial u}{\partial y} + \frac{\partial v}{\partial x}\right), q_x = -\kappa\frac{\partial T}{\partial x}$, and $q_y = -\kappa\frac{\partial T}{\partial y}.$
Euler equations are obtained by setting $\mu = 0$ in Equations (2).

1.2. FLUID EQUATIONS WITH CHEMICAL KINETICS

For the calculation of flowfields involving finite-rate chemistry effects, for example hydrogen/air combustion or dissociated air at high altitude, Equations (2) are solved in conjunction with transport equations for conservation of species created or annihilated during the combustion or any other chemically reacting process. The transport equation for each species "i" can be expressed in conservation law form as:

$$Q = [\rho Y_i], F = \left[\rho u Y_i - \rho D\frac{\partial Y_i}{\partial x}\right], G = \left[\rho v Y_i - \rho D\frac{\partial Y_i}{\partial y}\right], S = [\dot{m}_i], \quad (3)$$

where
Y_i = mass fraction of species "i",
D = Diffusion constant, and
$\dot{m}_i$ = amount of production or depletion of species "i".

For chemically reacting flows, the expressions for internal energy, heat flux, and the specific heat of the mixture change are as follows:

$$e = \frac{p}{\rho(\gamma - 1)} + \frac{1}{2}(u^2 + v^2) + \sum_{i=1}^{\#spec} Y_i h_i^0$$

$$q_x = -\kappa\frac{\partial T}{\partial x} - \rho D\sum_{i=1}^{\#spec} h_i\frac{\partial Y_i}{\partial x}$$

$$(C_p)_{mixture} = \sum_{i=1}^{\#spec} Y_i C_p^i \qquad (4)$$

In Equations (3) and (4), h_i^0 denotes the enthalpy of formation of the species "i" and h_i denotes the specific enthalpy. $\dot{m}_i$ is determined by the reaction rate modeling of the combustion process. Using the reaction rates and the law of mass action, $\dot{m}_i$ can be determined. For details, the reader is referred to (Gielda et $al.$, 1994).

1.3. ACOUSTICS EQUATIONS

Let (u_0, v_0) denote the steady mean flow velocity components and $\tilde{\rho}, \tilde{p}, (\tilde{u}, \tilde{v})$ denote the acoustic perturbation density, pressure and velocity components, respectively. Then for the linearized compressible Euler equations (acoustic equations) we can write (Agarwal & Huh, 1995b),

$$Q = \begin{pmatrix} \overline{\rho} \\ \overline{\rho u} \\ \overline{\rho v} \\ \overline{\rho e} \end{pmatrix}, \quad F = \begin{pmatrix} \overline{\rho u} \\ u_0\,[2\overline{\rho u} - u_0\tilde{\rho}] + \tilde{p} \\ v_0\,[\overline{\rho u} - u_0\tilde{\rho}] + \overline{\rho v} u_0 \\ u_0\,[\overline{\rho h} - h_0\tilde{\rho}] + \overline{\rho u} h_0 \end{pmatrix}, \quad G = \begin{pmatrix} \overline{\rho v} \\ u_0\,[\overline{\rho v} - v_0\tilde{\rho}] + \overline{\rho u} v_0 \\ v_0\,[2\overline{\rho v} - v_0\tilde{\rho}] + \tilde{p} \\ v_0\,[\overline{\rho h} - h_0\tilde{\rho}] + \overline{\rho v} h_0 \end{pmatrix}, S = 0. \qquad (5)$$

A state equation is required for closure and is of the form,

$$\tilde{p} = (\gamma - 1)\left\{ \overline{\rho e} - \frac{1}{2}[u_0(2\overline{\rho u} - u_0\tilde{\rho}) + v_0(2\overline{\rho v} - v_0\tilde{\rho})] \right\}.$$

The variables $\overline{\rho u}, \overline{\rho v}, \overline{\rho e}$ and $\overline{\rho h}$ are defined as,

$$\overline{\rho u} = \rho_0\tilde{u} + \tilde{\rho}u_0, \overline{\rho v} = \rho_0\tilde{v} + \tilde{\rho}v_0, \overline{\rho h} = \overline{\rho e} + \tilde{p}$$

and

$$\overline{\rho e} = \frac{\tilde{p}}{\gamma - 1} + \rho_0 u_0 \tilde{u} + \rho_0 v_0 \tilde{v} + \frac{1}{2}\tilde{\rho}(u_0^2 + v_0^2).$$

1.4. MAXWELL EQUATIONS

Let D denote the electric field displacement and B the magnetic field induction. For transverse electric (TE) polarization we can write (Shu & Agarwal, 1992),

$$Q = \begin{pmatrix} B_z \\ D_x \\ D_y \end{pmatrix}, \quad F = \begin{pmatrix} D_y / \varepsilon \\ 0 \\ B_z / \mu \end{pmatrix}, \quad G = \begin{pmatrix} -D_x / \varepsilon \\ -B_z / \mu \\ 0 \end{pmatrix}, \quad S = 0, \tag{6}$$

where μ and ε represent the permeability and permittivity of the homogeneous medium, respectively. Similarly, $Q, F, G,$ and S can be written for transverse magnetic (TM) polarization as,

$$Q = \begin{bmatrix} D_z \\ B_x \\ B_y \end{bmatrix}, \quad F = \begin{bmatrix} -B_y / \mu \\ 0 \\ -D_z / \varepsilon \end{bmatrix}, \quad G = \begin{bmatrix} B_x / \mu \\ D_z / \varepsilon \\ 0 \end{bmatrix}. \tag{7}$$

1.5. SCHROEDINGER EQUATION

For the wave function Ψ, the Schroedinger equation can be written as (Landau & Lifshitz, 1959), $i\hbar \frac{\partial \Psi}{\partial t} = \hat{H}\Psi$, where $\hat{H}$ = non-relativistic Hamiltonian in Coulomb-variant form

$$\hat{H} = \frac{\hat{p}^2}{2m} - \frac{e}{m}(A \cdot \hat{p}) + \frac{e^2}{2m}A^2 + U(x,y). \tag{8}$$

Here e = electron charge, $\hat{p}$ = momentum operator $= (\hat{h}/i)\nabla, m$ = electron mass, A = vector potential ($\nabla \cdot A = 0$, homogeneous field), $\hat{h} = h/2\pi, U$ = scalar potential, and h = Plank's constant. Letting $\Psi = \alpha exp(i\beta)$, it can be shown after considerable algebraic manipulation that (Agarwal et al., 1992),

$$Q = \begin{pmatrix} \rho \\ \rho u \\ \rho v \end{pmatrix}, \quad F = \begin{pmatrix} \rho \\ \rho u^2 \\ \rho v u \end{pmatrix}, \quad G = \begin{pmatrix} \rho v \\ \rho u v \\ \rho v^2 \end{pmatrix}, \quad S = -\frac{\rho}{m} \begin{pmatrix} -\frac{e}{\rho}A \cdot \nabla \rho \\ \hat{U}_x \\ \hat{U}_y \end{pmatrix}, \tag{9}$$

where $\rho = \alpha^2$ = probability density, $V = (u,v) = (\hat{h}/m)\nabla\beta$, and $\hat{U} = U - \frac{\hbar^2}{4m\rho}\left[\nabla^2\rho - \frac{(\nabla\rho)^2}{2\rho}\right]$.

1.6. SEMICONDUCTOR DEVICE SIMULATION EQUATIONS

The hydrodynamic form of the semiconductor device simulation equations can be written as (Fatemi et $al.$, 1991):

$$
Q = \begin{bmatrix} n \\ a \\ b \\ W \end{bmatrix}, \; F = \begin{bmatrix} a/m \\ (2W - p^2/3\alpha + a^2/\alpha)/3 \\ ab/\alpha \\ a(5W - p^2)/3\alpha \end{bmatrix}, \; G = \begin{bmatrix} a/m \\ ba/\alpha \\ (2W - p^2/3\alpha + b^2/\alpha)/3 \\ b(5W - p^2)/3\alpha \end{bmatrix},
$$

$$
S = \begin{bmatrix} 0 \\ \left(\dfrac{\partial a}{\partial t}\right)_c - enE_1 \\ \left(\dfrac{\partial b}{\partial t}\right)_c - enE_2 \\ \left(\dfrac{\partial W}{\partial t}\right)_c = \dfrac{e}{m}(aE_1 + bE_2) - \nabla\cdot(\kappa\nabla T) \end{bmatrix},
$$

(10)

$$
p = a^2 + b^2, \text{ and } \alpha = mn.
$$

These equations are solved in conjunction with an equation for the electric potential ϕ,

$$
\nabla \cdot (\varepsilon\nabla\phi) = -e(N_D - N_A - n), \vec{E} = -\nabla\phi = E_1\hat{i} + E_2\hat{j}, \quad (11)
$$

where m = effective electron mass, n = electron density, e = electron charge, $\vec{E}$ = electric field, T = temperature, N_D = density of donors, N_A = density of receptors, ε = dielectric constant, ϕ = electric potential, κ = thermal conductivity, $p = mn\vec{v}$ = momentum density, a = x-component of momentum density, b = y-component of momentum density, W = energy density = $\frac{3}{2}nT + \frac{1}{2}mnv^2$, and $\left(\frac{\partial a}{\partial t}\right)_c$, $\left(\frac{\partial b}{\partial t}\right)_c$, and $\left(\frac{\partial W}{\partial t}\right)_c$ denote the collision terms.

1.7. RELATIVISTIC GAS DYNAMICS EQUATIONS

For the relativistic gas dynamics equations we can write (Schneider et $al.$, 1993),

$$Q = \begin{bmatrix} R \\ M \\ N \\ E \end{bmatrix}, \quad F = \begin{bmatrix} Ru \\ Mu+p \\ Nu \\ (E+p)u \end{bmatrix}, \quad G = \begin{bmatrix} Rv \\ Mv \\ Nv+p \\ (E+p)v \end{bmatrix}, \quad S = 0, \qquad (12)$$

where

$$R = \gamma n, E = \gamma^2(e+p) - p, p = (\Gamma - 1)(e - n), M = \gamma^2(e+p), \text{ and}$$

$$\gamma = \frac{1}{1 - (u^2 + v^2)}. \qquad (13)$$

In Equations (12), R, M, E, are the inertial mass density, momentum density, and energy density of the fluid. These inertial quantities are related to the rest frame energy density e, mass density n, and the fluid velocity v through the nonlinear transformations expressed by Equations (13). Γ is an adiabatic constant.

1.8. EINSTEIN'S EQUATIONS OF GENERAL RELATIVITY

In compact form, these equations are written as (Arnowitt *et al.*, 1962),

$$G_{\mu v} = 8\pi T_{\mu v}, \qquad (14)$$

where $G_{\mu v}$ = Einstein's tensor and $T_{\mu v}$ = Stress energy tensor.

In Equation (14), space and time are intertwined. The resulting second-order partial differential equations (PDEs) are not suitable for performing computer evolution. The space and time are split by the Arnowitt-Deser-Misner (ADM) decomposition. Einstein's equations are thus split into two kinds of equations: (a) constraint equations and (b) evolution equations. Constraint equations contain no time derivatives and relate field variables at a given instant of time. Evolution equations are first-order PDEs that propagate the initial data to next instant of time. If the constraints equations are satisfied initially, the evolution equations guarantee that they will be satisfied at all subsequent times. There are a total of thirty-four evolution equations which can be expressed in hyperbolic conservation law form. The details are given in (Seidel & Suen, 1994).

1.9. EQUATIONS OF ELASTICITY

For an isotropic Hookean solid, equations governing the elastic deformation can be written as (Fung, 1977),

$$
Q = \begin{bmatrix} \alpha \\ \beta \\ \rho \\ \rho u \\ \rho v \end{bmatrix}, \quad
F = \begin{bmatrix} 0 \\ 0 \\ \rho u \\ \rho u^2 - \tau_{xx} \\ \rho u v - \tau_{xy} \end{bmatrix}, \quad
G = \begin{bmatrix} 0 \\ 0 \\ \rho v \\ \rho u v - \tau_{xy} \\ \rho v^2 - \tau_{yy} \end{bmatrix}, \quad
S = \begin{bmatrix} u \\ v \\ 0 \\ f_x \\ f_y \end{bmatrix}, \tag{15}
$$

where $\tau_{xx} = (\lambda + G)(\frac{\partial u}{\partial x}) + G\frac{\partial v}{\partial y}$, $\tau_{yy} = \lambda\frac{\partial u}{\partial x} + (\lambda + G)\frac{\partial v}{\partial y}$, $\tau_{xy} = G\left(\frac{\partial u}{\partial y} + \frac{\partial v}{\partial x}\right)$, ρ = density of the material, (α, β) = particle displacement at (x, y), (u, v) = particle velocity at (x, y), (f_x, f_y) = external body forces, λ, G = Lame constants, G = shear modulus, and v = Poisson's Ratio = $\frac{\lambda}{2(\lambda + G)}$.

2. Numerical Method

The governing Equations (1) expressed in conservation law form can be solved by a variety of numerical algorithms described in the literature (Anderson *et al.*, 1984). The choice of a particular algorithm is determined by the physics of the problem. For some problems steady state solution is required, in other instances the solution has a wave-like behavior or the determination of the time history of the solution is important. The order of accuracy of the algorithm, its stability characteristics, the numerical implementation of the boundary conditions, and the mesh all play an important role in obtaining an accurate and efficient solution to a given physical problem. For many problems in physics described here, the geometry modeling, grid generation and flow solver technology developed for the solution of Euler and Navier-Stokes equations in Computational Fluid Dynamics (CFD) can be efficiently employed with minor modifications.

It should be noted that Equations (1) are solved in time-domain whether steady-state or transient solution is desired. For steady state calculations solution is marched in time until steady state is reached. For some physical problems, for example the problems of acoustic or electromagnetic scattering, the solution is harmonic and therefore it is more efficient to account for the harmonic nature of the solution at the outset in the existing mathematical framework as follows.

For the scattering problems, since the governing Equations (5) and (7) are linear, an assumption that the incident field is harmonic in time with a frequency ω will result in a time-dependent total-field that will also be harmonic with frequency ω. Thus the governing equations can be recast in frequency domain by the use of the single-frequency assumption:

$$Q = \Re\{\tilde{Q}(x,y)\}e^{-i\omega t}, \tag{16}$$

where tilde denotes a complex quantity and $\Re$ denotes the real part.

By applying the single-frequency assumption, and by recasting the equations in scattered form with pseudo-time derivative, the governing equations become,

$$\frac{\partial \tilde{Q}_S}{\partial t^*} + \frac{\partial \tilde{F}_S}{\partial x} + \frac{\partial \tilde{G}_S}{\partial y} - i\omega \tilde{Q}_S = \tilde{S}, \tag{17}$$

where $\tilde{Q}, \tilde{F}$ and $\tilde{G}$ contain the complex coefficients of Q, F and G. The source term $\tilde{S}$ is defined

$$\tilde{S} = S + i\omega \tilde{Q}_i - \frac{\partial \tilde{F}_i}{\partial x} - \frac{\partial \tilde{G}_i}{\partial y}. \tag{18}$$

The total-field is cast as a sum of the known incident value and the scattered value, $\tilde{Q} = \tilde{Q}_i + \tilde{Q}_S$.

As mentioned before, although a variety of numerical methods exist for solving Equations (1), in what follows we describe a typical method which has been successfully employed by the author and his students in solving Equations (1) and (17). The details of the method are described in (Agarwal & Huh, 1996c). Here we provide some salient features:

2.1. SPATIAL DISCRETIZATION

The set of weakly conservative governing Equations (1) or (17) is solved numerically using a finite-volume node-based scheme. The physical domain in (x, y) is mapped to the computational domain (ξ, η) to allow for arbitrarily shaped bodies. The semi-discrete form of the compact fourth-order method can be written as,

$$\frac{d}{dt}JQ_{ij} + \frac{\mu_\eta^2 \mu_\xi \delta_\xi}{1 + \delta_\xi^2/6} F'_{ij} + \frac{\mu_\xi^2 \mu_\eta \delta_\eta}{1 + \delta_\eta^2/6} G'_{ij} - i\omega JQ_{ij} = JS_{ij} + D_{ij}, \tag{19}$$

where D is the added dissipation term defined as,

$$D_{ij} = v_6 \delta_\xi (\frac{J}{\Delta t} \delta_\xi^5) Q - v_6 \delta_\eta (\frac{J}{\Delta t} \delta_\eta^5) Q \tag{20}$$

and $\Delta \vec{t}$ is the time step when the CFL number, σ_i, is 1. $\Delta\xi$ and $\Delta\eta$ are defined to be 1, and hence have been omitted. The vectors F' and G' are the curvilinear flux vectors and are defined as, $F' = Fy_\eta - Gx_\eta, G' = -Fy_\xi + Gx_\xi$ and J is the determinant of the metric tensor, $J = x_\xi y_\eta -$

$x_\eta y_\xi \cdot \delta$ and μ are the standard difference and averaging operators defined as $\delta_\xi Q_{i+\frac{1}{2},j} = Q_{i+1,j} - Q_{ij}$ and $\mu_\xi Q_{i+\frac{1}{2},j} = Q_{i+1,j} + Q_{i,j}$. The coefficient v_6 is user specified, and normally varies between 0.0008 to 0.001. The discretized form of Equation (1) is the same as (19) except that $-i\omega J Q_{ij}$ is dropped and $\tilde{S}$ is replaced by S.

The spatial discretizations in Equation (19) is based on the classical Pade scheme and is fourth-order accurate on smooth grids. For many physical problems, for example the steady state aerodynamics problems, the second-order discretization on "reasonable" grids is sufficient. However for wave propagation problems, a higher-order discretization becomes necessary to obtain accurate solutions on "reasonable grids" efficiently. The resolution of second-order central differencing with 20 points per wavelength is matched by compact fourth-order differencing with roughly 8 points per wavelength.

2.2. TIME INTEGRATION

A four stage Runge-Kutta scheme is used to integrate the governing Equations (1) in real time and (17) in the pseudo-time plane. The integration method is explicit for Equation (1) and is point implicit for Equation (17) to alleviate the stiffness arising from the term $-i\omega Q$ for large values of ω (Agarwal & Huh, 1996c).

For the two-dimensional system of Equations (17), the time integration is computed as follows:

$$Q_{ij}^0 = Q_{ij}^n$$

$$S_k(Q_{ij}^k - Q_{ij}^0) = -\alpha^k \frac{\Delta t_{ij}}{J}(R_{ij}^{k-1} - D_{ij}^0) \quad k = 1, 4$$

$$Q_{ij}^{n+1} = Q_{ij}^4 \qquad (21)$$

$$R_{ij} = \frac{\mu_\eta^2 \mu_\xi \delta_\xi}{1 + \delta_\xi^2/6}F'_{ij} + \frac{\mu_\xi^2 \mu_\eta \delta_\eta}{1 + \delta_\eta^2/6}G'_{ij} - i\omega J Q_{ij} - J S_{ij}$$

$$\text{Here, } \alpha^1 = \frac{1}{4}, \alpha^2 = \frac{1}{3}, \alpha^3 = \frac{1}{2}, \alpha^4 = 1,$$

and $s_i = 1 + \alpha^i \omega \Delta t$. D_{ij} is the dissipation term described previously. A single evaluation of the dissipative terms is required and the scheme is stable provided that the CFL number, σ_i, does not exceed $2\sqrt{2}$ regardless of the size of ω. By putting $\omega = 0$ in s_i, we obtain the explicit scheme for Equation (1).

2.3. BOUNDARY CONDITIONS

Analytical boundary conditions for the governing partial differential equations are defined by the physics of the problem to be solved. The accurate numerical implementation of the boundary conditions is extremely important because of its influence on the overall solution accuracy. For exterior boundary value problems, the approach to the numerical implementation of the boundary condition in the farfield may determine the size of the computational domain needed for desired solution accuracy. A correct mathematical model requires the implementation of the boundary condition at infinity which is impractial numerically. Therefore a set of well-posed boundary conditions must be placed on a finite-domain and therefore the accurate and efficient implementation of these boundary conditions becomes crucial for a good computational approach.

In the following section, we present a few selected results for a variety of physical problems using the above algorithm. Problem specific numerical issues for both the governing equations and the boundary conditions are explained.

3. Selected Solution Examples

In this section, a solution example for each of the governing equations (with exception of Einstein's Equations of General Relativity) is described. All the examples are based on the work of the author, his colleagues, and students.

3.1. LAUNCH VEHICLE ANALYSIS INCLUDING MAIN ENGINE PLUME

For determining the installed engine performance of a launch vehicle in flight from subsonic to hypersonic Mach numbers, three-dimensional version of the Navier-Stokes Equations (2) and the species Equations (3) and (4) is solved in conjunction with a two-equation $k - \epsilon$ turbulence model and a finite-rate hydrocarbon/air chemistry model (Gielda et al., 1991). Figures 1 and 2 show the grid about the Delta/Thor forebody and internal nozzle. Figure 3 shows the flight trajectory of the vehicle. Figures 4 shows the Mach contours on the vehicle (with engine on) at a flight Mach number of 1.40 and an altitude of 31,700 ft. Figures 5 and 6 show the comparison of numerical prediction and the flight test data for pressure and temperature in the base region of the vehicle at three instances during the flight after take-off. The agreement between the predictions and the experimental data is excellent. The details of the calculation can be found in (Gielda et al., 1991).

3.2. ACOUSTIC RADIATION FROM AN OSCILLATING CYLINDER

In (Agarwal & Huh, 1996c), acoustic Equations (5) are used to compute the radiated sound field from an oscillating circular cylinder in quiescent flow. A novel treatment of radiation boundary conditions (Agarwal & Huh, 1995a) is employed to minimize the size of the computational domain. It is based on the modal analysis of the similarity form of the acoustic equations. Analytical relations are derived to insure that there are no incoming modes in the computational domain. For a circular boundary, this boundary condition is equivalent to exact integral form of the Sommerfeld radiation condition. As shown in Figure 7, no numerical reflection was observed in calculations even by bringing the farfield boundary at a distance of 0.02 diameters from the surface of the cylinder. Accurate solutions were obtained on a 66 x 10 grid in 24 cpu seconds on a Cray X-MP with six-orders of magnitude reduction in residuals as shown in Figure 8. (Agarwal & Huh, 1996a) and (Agarwal & Huh, 1996b) provide the numerical solutions for more complex problems of acoustic radiation and scattering from airfoils in the presence of mean flow.

3.3. TRANSVERSE ELECTRIC (TE) ELECTROMAGNETIC SCATTERING BY AN OGIVE

Maxwell's Equations (6) are solved to compute the electromagnetic scattering from a perfectly conducting ogive due to transverse electric polarization as described in (Shu & Agarwal, 1992). Figure 9 shows lines of constant modulus of the scattered magnetic field intensity. Figure 10 shows the comparison for the computed bistatic radar cross section (RCS) using the CFD-based approach and the method of moments. Agreement is excellent. Again the farfield boundary is only at a distance of 1.2 times the chord from the center of the ogive. (Shu & Agarwal, 1992) and (Shu & Agarwal, 1994) provide details of the calculations for scattering from complex 2- and 3-D shapes with dielectric and lossy coatings.

3.4. SCATTERING OF A BEAM OF ELECTRONS IN THE SPHERICALLY SYMMETRIC POTENTIAL FIELD OF AN ATOM

The Schroedinger Equation (9) is solved to compute the scattering from a beam of electrons in the spherically symmetric potential field of a hydrogen atom given by,

$$\hat{U}(r_0, \theta, 0) = \frac{1}{4\rho r_0^2} \left[\frac{1}{2\rho} \left(\frac{\partial \rho}{\partial \theta} \right)^2 - \frac{1}{\sin\theta} \frac{\partial}{\partial \theta} \left(\sin\theta \frac{\partial \rho}{\partial \theta} \right) \right]. \qquad (22)$$

where r_0 = range of the atomic field. We consider a hydrogen atom in ground state given by,

$$u(r) = e^{-2r/\pi}.$$

The initial and boundary conditions employed for the solution of Equation (9) are $\rho(r,\theta,0) = 1, V_r(r,\theta,0) = \sqrt{2E_k}\sin\theta, V_\theta = \sqrt{2E_k}\cos\theta, \hat{U}(r,\theta,0) = 0$, and $\rho(0,\theta,t) = 0, V_r(r,\theta,t) = \sqrt{2E_k}, V_\theta(r_0,\theta,t) = 0$ respectively. In the farfield $\rho(r_0,\theta,t) = \sigma(\theta)/r_0^2, E_k$ = kinetic energy of the scattered beam of electrons and $\sigma(\theta)$ = differential cross section. As shown in (Landau & Lifshitz, 1959), this problem has an exact analytical solution given by,

$$\frac{d\sigma}{d\theta} = 2\pi \frac{\sin\frac{\theta}{2}\left[1 + E_k \sin\frac{\theta}{2}\right]^2}{\left[1 + 2E_k \sin\frac{\theta}{2}\right]^4}. \tag{23}$$

The CFD-based approach used in solving Equation (9) produces an excellent agreement with Equation (23) as shown in (Agarwal et al., 1992).

3.5. SIMULATION OF ONE-DIMENSIONAL SI $N^+ - N - N^-$ DIODE

Equations (10) and (11) are solved to simulate the electron shock waves in a submicron semiconductor device. The collision terms in Equation (10) are modeled as follows (Gardner et al., 1993). The effects of electron-phonon and electron-impurity collisions are included.

$$\left(\frac{\partial \bar{p}}{\partial t}\right)_c = -\frac{\bar{p}}{\tau_p}$$

$$\left(\frac{\partial W}{\partial t}\right)_c = -\frac{W - W_0}{\tau_W}$$

$$\tau_p = \tau_{p0}\frac{T_0}{T} \tag{24}$$

$$\tau_W = \frac{\tau_p}{2}\left(1 + \frac{\frac{3}{2}T}{\frac{1}{2}mv_s^2}\right)$$

where

τ_{p0}	=	low-energy momentum relaxation time
T_0	=	semi-conductor lattice temperature
v_s	=	$v_s(T_0)$ = saturation velocity
K	=	$K_0\tau_{p0}nT_o/m$, and K_0 = a positive constant

The boundary conditions employed in the calculations are (see Figure 13):

1. charge neutral contacts $(n = N)$ with $dT/dx = 0$ at x_{min} and x_{max}, and

2. assuming a bias V across the device, where n_i = the intrinsic electron concentration

$$e\phi(x_{min}) = T \ln (n/n_i), \text{ and}$$
$$e\phi(x_{max}) = T \ln (n/n_i) + eV.$$

The calculations are performed using the following physical parameters (Gardner et al., 1993):

Si $n^+ - n - n^-$ diode
0.25 micron source, 0.25 micron channel, 0.25 micron domain
n^+ region: doping density $N = 10^{18}$ cm^{-3}
n region: doping density $N = 10^{15}$ cm^{-3}
$T_0 = 77K = 0.00665 ev, m = 0.24 m_e, m_e$ = electron mass
$v_s = 1.2 \times 10^7$ cm/sec, $n_i = 2.84 \times 10^{20}$ cm^{-3}, $\tau_{p0} = 1.67$ picoseconds
low field electron mobility, $\mu_{n0} = \frac{e\tau_{p0}}{m} = 12,200$ cm$^2/V$
$\varepsilon = 11.7, \gamma = \frac{5}{3}$ for a polytropic gas, $V = 1$ volt
$c = \sqrt{\frac{T}{m}}$ = sound speed

Figure 11 shows the doping profile used in the simulation. Figures 12 and 13 show the computations for electron temperature and electron velocity using the numerical method described in this paper and the ENO method reported in (Gardner et al., 1993). There are some minor differences in two sets of calculations. Again the unified CFD based approach was effectively employed in the solution of Equations (10) and (11) as described in (Agarwal, 1995).

3.6. NUMERICAL SOLUTION OF A RELATIVISTIC SHOCK

Gas dynamics models for the analysis of reactor dynamics of high energy collisions between heavy ions such as atomic nuclei require the inclusion of relativistic effects in the conservation equations of mass, momentum, and energy given by Equations (12) and (13). As described in (Agarwal & Reddy, 1996), these equations have been solved using the numerical algorithm described in this paper for a relativistic shock tube problem and a shock model problem. Figures 14 and 15 show the comparison of results reported in (Agarwal & Reddy, 1996) with those reported in (Schneider et al., 1993). The two calculations are in good agreement.

3.7. TORSION OF A CIRCULAR CYLINDRICAL SHAFT

To transmit a torque from one place to another, a shaft is employed. The problem is to solve the three-dimensional version of the Equations of elasticity (15) to obtain the stress distribution in the shaft. The degree of difficulty to solve this problem depends on the geometry of the shaft. For a noncircular shaft, numerical solution is required. Here we consider the simple example of a circular shaft shown in Figure 16. The boundary conditions on the circumference C are

$$x\tau_{xx} + y\tau_{xy} = 0, x\tau_{yx} + y\tau_{yy} = 0, \text{and } x\tau_{zx} + y\tau_{zy} = 0. \tag{25}$$

The conditions on the stresses at the end are that they are equipollent to a torque T on the cylinder given by,

$$T = \frac{\pi a^4 G\phi}{2}, \tag{26}$$

where α = radius of the cylinder, G = shear modulus, and ϕ = twist per unit length of the cylinder. This problem has an exact solution for the displacements α, β, and γ given by (Fung, 1977),

$$\alpha = -\phi z y, \beta = \phi z x, \gamma = 0. \tag{27}$$

The stress components are given by,

$$\tau_{xx} = \tau_{yy} = \tau_{zz} = \tau_{xy} = 0, \text{and } \tau_{xz} = -G\phi y, \tau_{yz} = G\phi x. \tag{28}$$

The numerical code described in (Riddel & Agarwal, 1994) reproduces the analytical solutions given by Equations (26)-(28).

4. Conclusions

Numerical solutions of governing equations of mathematical physics from eight different disciplines have been considered. It has been shown that most of these equations can be written in a hyperbolic conservation law form. A unified CFD-based approach has been successfully applied to solve these equations. It has been demonstrated that a single computational code with minor changes reflecting the particular numerical requirements (gridding, order-of-the scheme, implementation of the boundary conditions) of a given problem can be successfully employed to solve a wide variety of problems in computational physics.

References

Agarwal, R.K., "A CFD-Based Approach for Hydrodynamic Model of Semiconductor Device Simulation Equations," unpublished report, Wichita State University, 1995.

Agarwal, R.K. and Huh, K.S., "Acoustic Radiation Due to Gust-Airfoil Interaction in Compressible Flow," AIAA Paper 96-1755, 1996a.

Agarwal, R.K. and Huh, K.S., "Acoustic Radiation from Oscillating Rigid Bodies in Mean Compressible Flow," *Computational Acoustics*, ASME Fluids Engineering Conference, San Diego, CA, 1996b.

Agarwal, R.K. and Huh, K.S., "A Dispersion-Relation-Preserving Fourth-Order Compact Time-Domain/Frequency-Domain Finite-Volume Method for Computational Acoustics," AIAA Paper 96-0277, 1996c.

Agarwal, R.K. and Huh, K.S., "A Novel Formulation of Farfield Boundary Condition for Computational Acoustics," *Computational Acoustics*, ASME FED-Vol. 219, 1995a, pp. 35-40.

Agarwal, R.K. and Huh, K.S., "Scattering of Sound by Rigid Bodies in Arbitrary Flow," *Computational Fluid Dynamics Review 1995*, John Wiley, M. Hafez, and K. Oshima, eds., 1995b, pp. 797-820.

Agarwal, R.K., Huh, K.S., and Shu, M., "A CFD-Based Approach for the Solution of Acoustics, Maxwell, and Schroedinger Equations for Scattering Problem," *Thirteenth International Conference on Numerical Methods in Fluid Dynamics, Lecture Notes in Physics*, Vol. 414, Springer Verlag, M. Napolitano and F. Sabetta, eds., 1992, pp. 370-375.

Agarwal, R.K. and Reddy, S.K., "Exact Determination of Eigenvalues and Numerical Solution of Relativistic Equations of Gas Dynamics," submitted for publication in *Journal of Computational Physics*, 1996.

Anderson, D.A., Tannehill, J.C., and Pletcher, R.H., *Computational Fluid Dynamics and Heat Transfer*, Taylor and Francis, Bristol, 1984.

Arnowitt, R., Deser, S., and Misner, C.W., *Gravitation: An Introduction to Current Research*, L. Witten, ed., John Wiley, New York, 1962.

Fatemi, E., Jerome, J.W., and Osher, S., "Solution of the Hydrodynamics Device Model Using High-Order Non-Oscillatory Shock Capturing Algorithm," *IEEE Transactions on Computer-Aided Design of Integrated Circuits and Systems*, Vol. 10, 1991, pp. 232-244.

Fung, Y.C., *A First Course in Continuum Mechanics*, Prentice Hall, New Jersey, 1977.

Gardner, C.L., Jerome, J.W., and Shu, C.W., "The ENO Method for the Hydrodynamic Model for Semiconductor Device Simulation," *High Performance Computing 1993: Grand Challenges in Computer Simulation*, San Diego, CA, 1993, pp. 96-101.

Gielda, T.P., Pavish, D.L., Deese, J.E., and Agarwal, R.K., "Three-Dimensional Delta/Thor Launch Vehicle Analysis Including Main Engine Plume," AIAA Paper 91-3338, 1991.

Gielda, T.P., Walter, T.M., and Agarwal, R.K., "Single Stage Rocket Performance: Prediction and Test," *International Journal of Computational Fluid Dynamics*, Vol. 2, 1994, pp. 83-110.

Landau, L. and Lifshitz, E.M., *Quantum Mechanics*, Pergamon Press, 1959.

Riddel, S. and Agarwal, R.K., "A CFD-Based Approach for the Solution of Equations of Elasticity," unpublished report, Washington University in St. Louis, 1994.

Schneider, V., Katscher, U., Rischke, D.M., Waldhauser, B., and Marhun, J.A., "New Algorithms for Ultra-Relativistic Numerical Hydrodynamics," *Journal of Computational Physics*, Vol. 105, 1993, pp. 92-107.

Seidel, E. and Suen, Wai-Mo, "Numerical Relativity," *International Journal of Modern Physics C*, Vol. 5, 1994, pp. 181-187.

Shu, M. and Agarwal, R.K., "A Compact Higher-Order Finite-Volume Time-Domain/Frequency Domain Method for Electromagnetic Scattering," AIAA Paper 92-0453, 1992.

Shu, M. and Agarwal, R.K., "A Spatially Compact Solver for Electromagnetic Scattering," *Numerical Methods for the Solution of Maxwell Equations*, R. Lohner, J. Periaux, and H. Steve, eds., John Wiley, 1994.

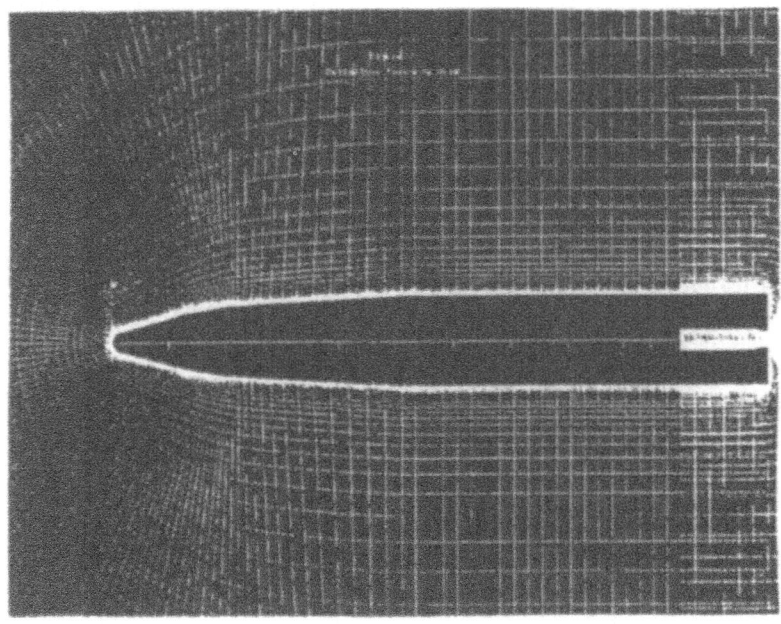

Figure 1.   Computational grid about the Delta/Thor forebody and internal nozzle.

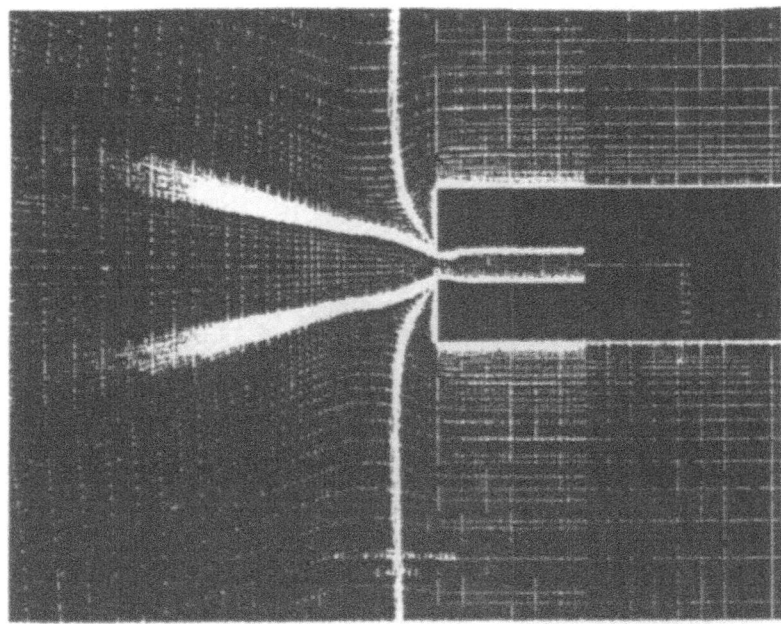

Figure 2. Expanded view of the Delta/Thor computational grid in the nozzle and base region.

Delta/Thor Flight 419 Reconstructed Trajectory

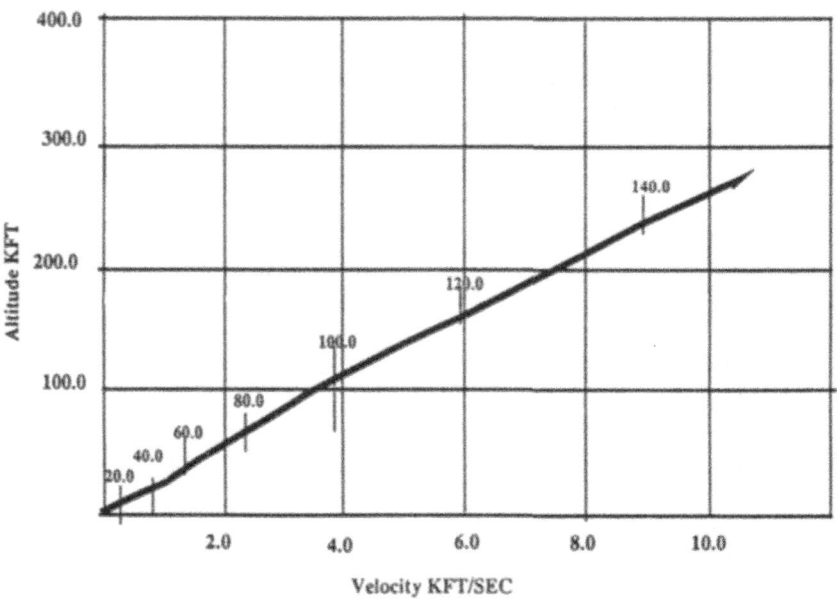

Figure 3. Flight trajectory of Delta/Thor vehicle.

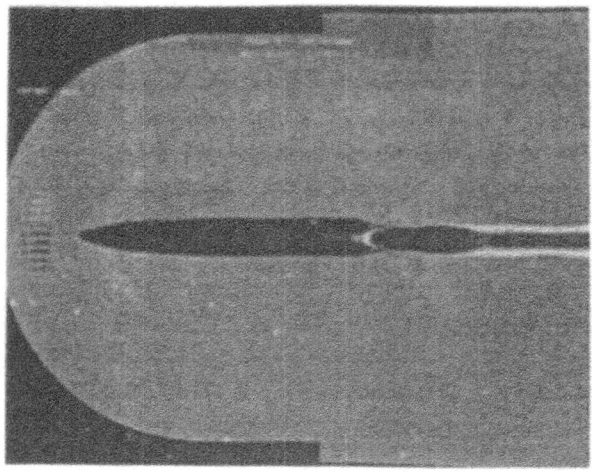

Figure 4. Computed Mach contours about the Delta/Thor ELV, $M_\infty = 1.40$, alt=31, 700.0 ft.

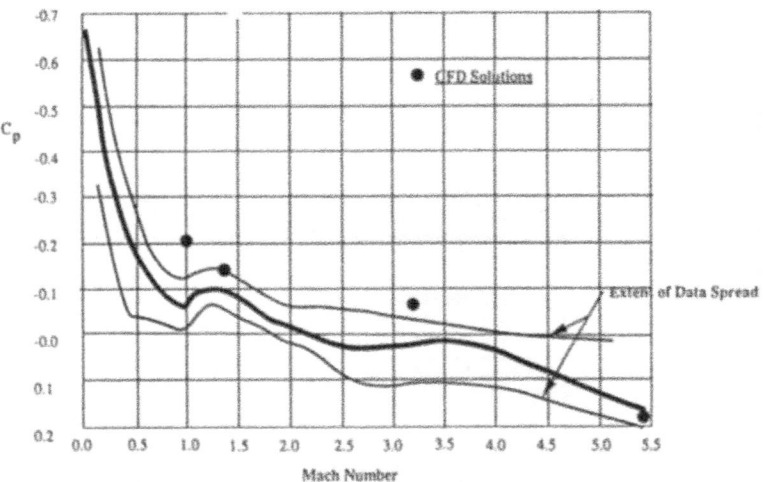

Figure 5. Comparison of computed and measured base pressure coefficient for Delta/Thor full-up flight vehicle.

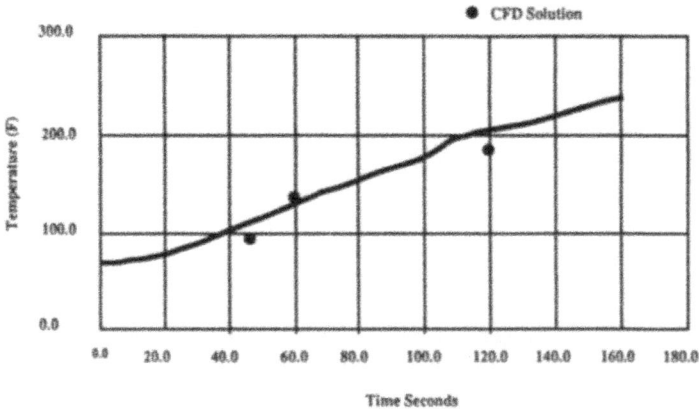

Figure 6. Delta/Thor base surface temperature history.

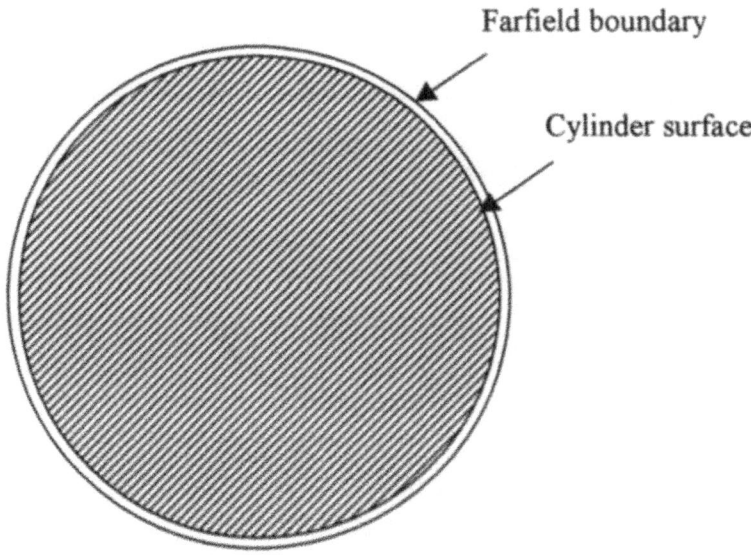

Figure 7. Acoustic radiation from an oscillating cylinder; $D = \lambda$, farfield boundary $D0 = 1.02\lambda, kD = 2\pi, 65 \times 10$ grid.

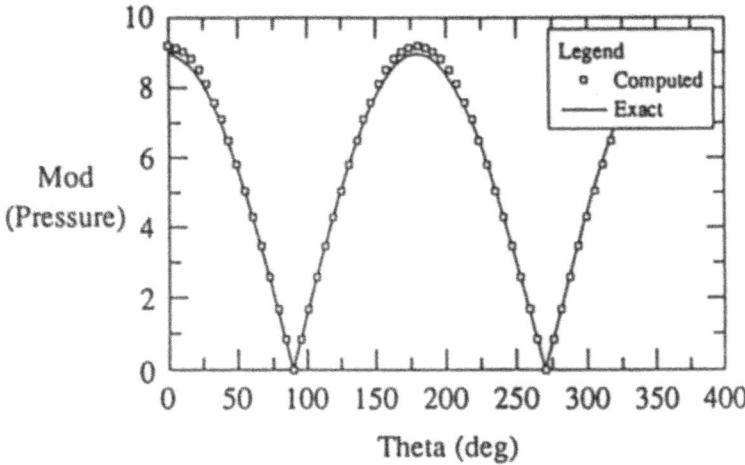

Figure 8. Amplitude of pressure on the surface of the oscillating cylinder of Fig. 7, $\lambda = 1$.

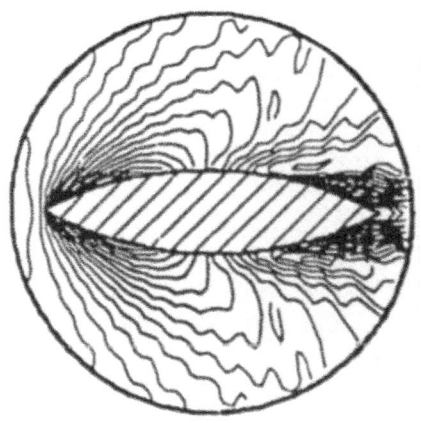

Ka = 16 π

Grid size: 257 × 100

Diameter of
farfield boundary = 1.2 × chord

Figure 9. TE scattering from an ogive.

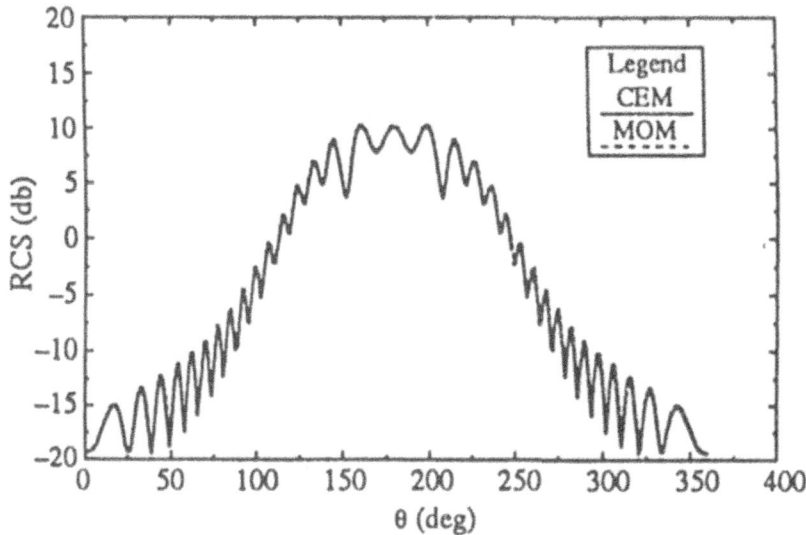

Figure 10. Bistatic RCS of the ogive in Fig. 9 due to TE scattering.

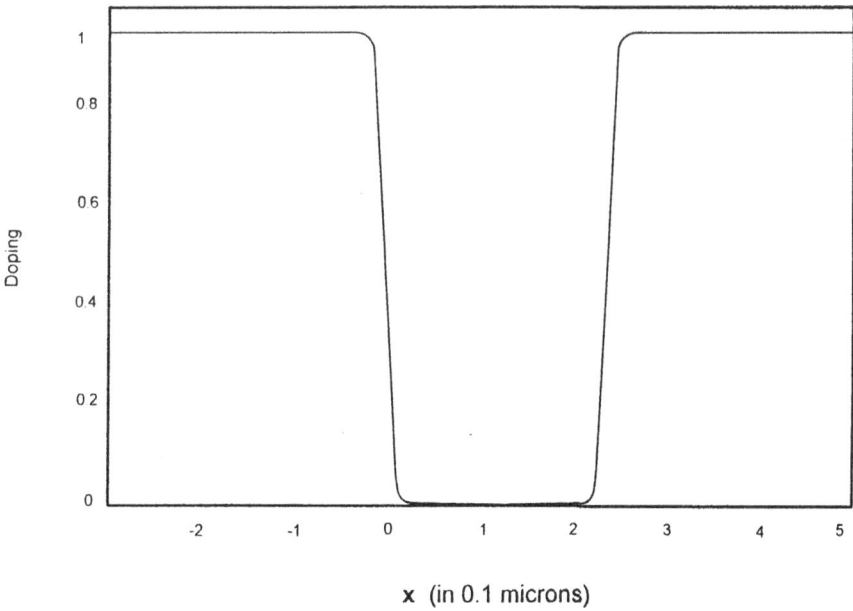

Figure 11. Doping profile in 10^{-18} cm^{-3} for simulation of 1-D Si n$^+$-n-n$^-$ diode (0.25 micron channel).

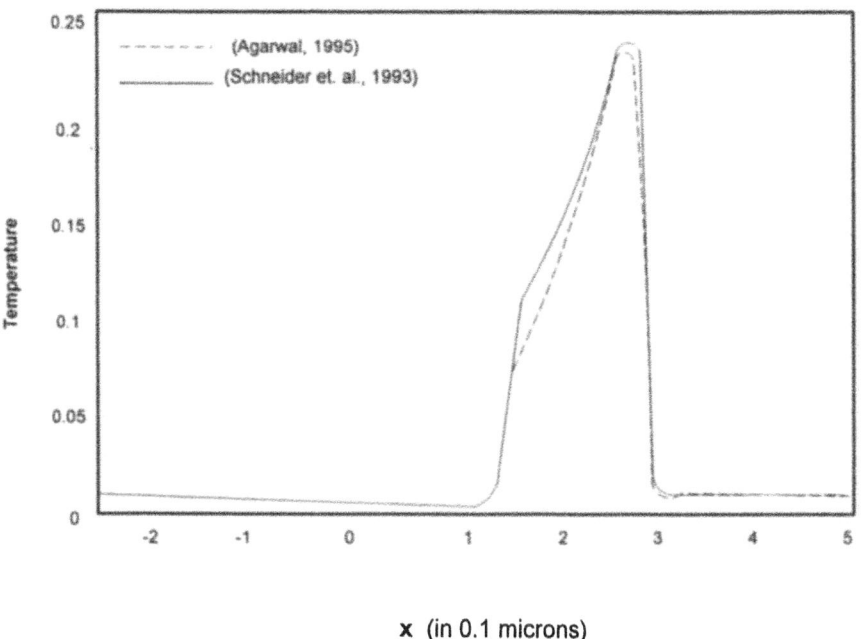

x (in 0.1 microns)

Figure 12.   Electron temperature in *ev* for the device simulation of Fig. 11.

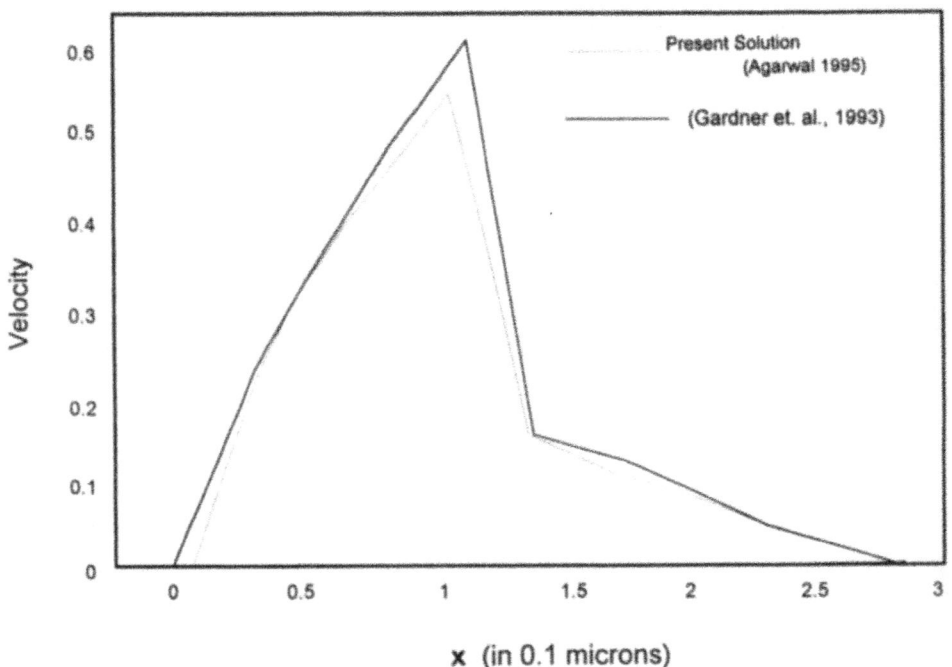

x (in 0.1 microns)

Figure 13.   Electron velocities in 10^8 cm/s for the device simulation of Fig. 11.

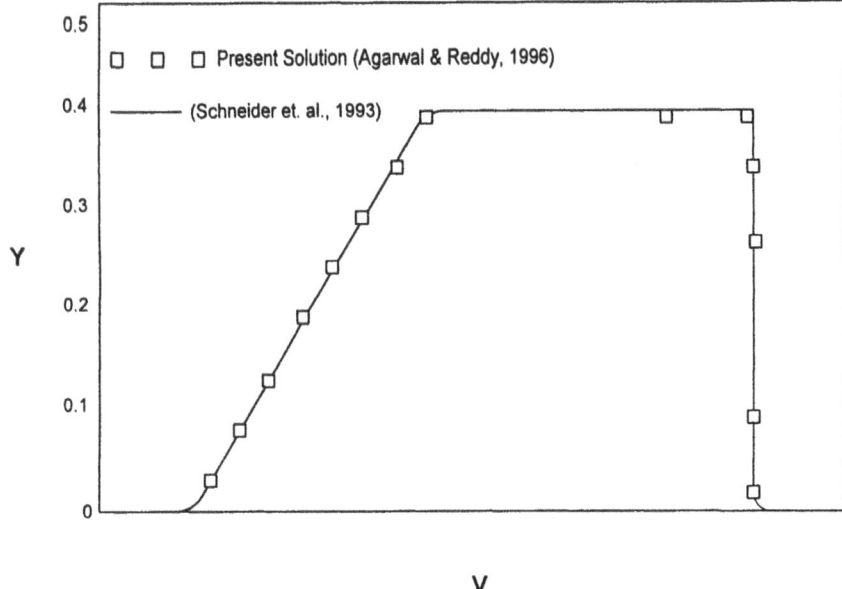

Figure 14. Relativistic shock model (Schneider *et al.*, 1993), V=Velocity.

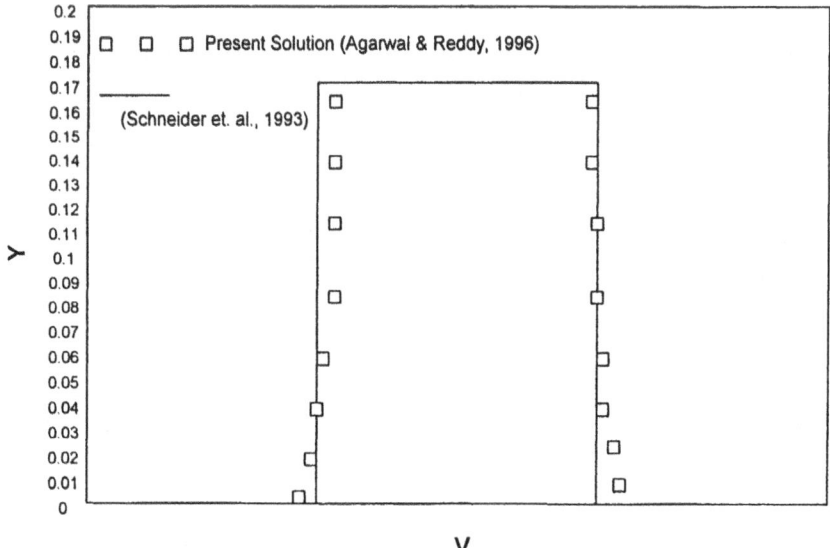

Figure 15. Relativistic shock tube problem (Schneider *et al.*, 1993), V=Velocity.

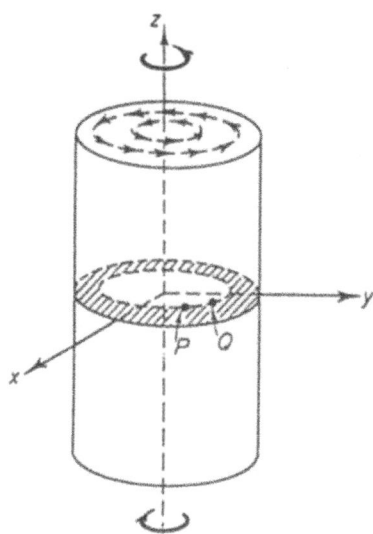

Figure 16.   Torsion of a circular shaft.

SECOND ORDER GODUNOV SCHEMES FOR 2D AND 3D MHD EQUATIONS AND DIVERGENCE-FREE CONDITION

WENLONG DAI AND PAUL R. WOODWARD
Laboratory for Computational Science and Engineering
University of Minnesota
200 Union Street S.E., Minneapolis, MN 55455

Abstract.
A finite difference scheme for 1D, 2D and 3D magnetohydrodynamical (MHD) equations is proposed in this paper. The approximate MHD Riemann solver developed in the scheme is based on characteristic formulations. Both the conservation laws for mass, momentum, energy and magnetic field, and the divergence-free condition for the magnetic field are exactly satisfied in the proposed scheme. The scheme is second order accurate in both space and time, and is simple compared to existing Godunov schemes for MHD flows. The correctness and robustness of the scheme are shown through numerical examples. The approach proposed in this paper to maintain the divergence-free condition may be applied to other dimensionally split and unsplit Godunov schemes for MHD flows.

1. Introduction

During the last ten years, effort has been put on the development of Godunov schemes for Magnetohydrodynamical (MHD) equations. (Brio & Wu, 1988) developed an upwind scheme for the one-dimensional (1D) MHD equations. (Zachary & Colella, 1992) and (Zachary et al., 1994) applied the general principle for hyperbolic systems of conservation laws to one- and multi-dimensional MHD equations. Dai and Woodward developed a nonlinear Riemann solver (1994) and extended the piecewise parabolic method to the multi-dimensional MHD equations (Dai & Woodward, 1994b). (Powell, 1994) and (Powell et al., 1995) developed a Roe-type Riemann solver and an upwind scheme for MHD equations using eight wave families. (Ryu & Jones, 1995), (Ryu et al., 1995) and (Balsara, 1996) applied TVD method to the MHD equations.

V. Venkatakrishnan et al (eds.), Barriers and Challenges in Computational Fluid Dynamics, 283-297.
© 1998 *Kluwer Academic Publishers.*

These Godunov schemes for MHD equations may obtain good results for some of MHD problems, but questions remain in these schemes, which have to be answered. In all these Godunov schemes for multi-dimensional MHD flows, either the conservation laws or the divergence-free condition for the magnetic field is satisfied only to the accuracy of schemes. In order to maintain the divergence-free condition, one typical approach used in the existing Godunov schemes is to add a "cleaning-up" procedure after each time step (Zachary *et al.*, 1994; Ryu *et al.*, 1995; Balsara, 1996). When the cleaning-up is applied to supersonic MHD flows, new issues arise. First, a fraction of the total CPU time has to be spent for the Poison equation after each time step. Second, to solve the Poison equation needs global information, and the existing parallel computers are not favorable to the global dependence of information. Third, direct derivatives have to be evaluated for the evaluation of the divergence of the magnetic field, which may introduce $O(1)$ truncation errors near MHD discontinuities. More seriously, the cleaning-up may destroy the conservative property of a Godunov scheme. As we know, the intermediate results for the magnetic field obtained before the cleaning-up is applied, together with the solutions for other variables, exactly satisfy the conservation laws. After the cleaning-up is applied, the magnetic field is modified, the conservation laws for the total energy and the three components of magnetic field are satisfied only to the accuracy of schemes, and truncation errors may be $O(1)$ near MHD discontinuities. This kind of $O(1)$ truncation errors may be easily detected when a strong MHD shock propagating obliquely in a 2D domain is simulated using a non-conservative scheme. We would like to mention that Stone and Norman incorporated a constrained transport method (Evans & Hawley, 1988) in their MHD algorithms (1992) in which the divergence-free condition is exactly satisfied, although the conservation laws are satisfied only to the accuracy of the algorithm.

Therefore, developing numerical schemes for supersonic MHD flows, in which both the conservation laws for all conservative quantities and the divergence-free condition are exactly satisfied, is an open question. In this paper, we propose a second order finite difference scheme for 1D, 2D and 3D MHD equations, in which an approximate MHD Riemann solver is developed. The scheme is very simple. In the proposed scheme, both the conservation laws and the divergence-free condition are exactly satisfied.

2. Numerical Representations

The set of MHD equations may be written in the form of conservation laws:

$$\frac{\partial \mathbf{U}}{\partial t} + \frac{\partial \mathbf{F}_x}{\partial x} + \frac{\partial \mathbf{F}_y}{\partial y} + \frac{\partial \mathbf{F}_z}{\partial z} = 0, \tag{1}$$

$$\mathbf{U} \equiv \begin{pmatrix} \rho \\ \rho u_x \\ \rho u_y \\ \rho u_z \\ \rho e \\ B_x \\ B_y \\ B_z \end{pmatrix}, \quad \mathbf{F}_x(\mathbf{U}) \equiv \begin{pmatrix} \rho u_x \\ \rho u_x u_x + P_{xx} \\ \rho u_x u_y + P_{xy} \\ \rho u_x u_z + P_{xz} \\ \rho u_x \epsilon + u_x P_{xx} + u_y P_{xy} + u_z P_{xz} \\ 0 \\ \Omega_z \\ -\Omega_y \end{pmatrix}.$$

Here $e \equiv \epsilon + \mathbf{u}^2/2 + \mathbf{B}^2/(8\pi\rho)$, $P_{xx} \equiv p + (B_y^2 + B_z^2 - B_x^2)/(8\pi)$, $P_{yy} \equiv p + (B_z^2 + B_x^2 - B_y^2)/(8\pi)$, $P_{zz} \equiv p + (B_x^2 + B_y^2 - B_z^2)/(8\pi)$, $P_{xy} \equiv -B_x B_y/(4\pi)$, $P_{xz} \equiv -B_x B_z/(4\pi)$, and $P_{yz} \equiv -B_y B_z/(4\pi)$. In the basic equation, ρ, $\mathbf{u}$, p, $\mathbf{B}$ and ϵ are the mass density, flow velocity, thermal pressure, magnetic field and specific internal energy, and $\mathbf{\Omega} \equiv \mathbf{u} \times \mathbf{B}$. We assume the γ-law for the equation of state.

Consider a numerical grid $\{x_i, y_j, z_k\}$ in a 3D domain. Integrating Eq.(1) in a grid cell $x_i \le x \le x_{i+1}$, $y_j \le y \le y_{j+1}$, $z_k \le z \le z_{k+1}$ and over a time step $0 \le t \le \Delta t$ yields

$$\begin{aligned} \mathbf{U}_{i,j,k}(\Delta t) = \mathbf{U}_{i,j,k}(0) \; &+ \; \frac{\Delta t}{\Delta x_i}[\overline{\mathbf{F}}_{xj,k}(x_i) - \overline{\mathbf{F}}_{xj,k}(x_{i+1})] \\ &+ \; \frac{\Delta t}{\Delta y_j}[\overline{\mathbf{F}}_{yi,k}(y_j) - \overline{\mathbf{F}}_{yi,k}(y_{j+1})] \\ &+ \; \frac{\Delta t}{\Delta z_k}[\overline{\mathbf{F}}_{zi,j}(z_k) - \overline{\mathbf{F}}_{zi,j}(z_{k+1})], \quad (2) \end{aligned}$$

$$\mathbf{U}_{i,j,k}(t) \equiv \frac{1}{\Delta x_i \Delta y_j \Delta z_k} \int_{z_k}^{z_{k+1}} \int_{y_j}^{y_{j+1}} \int_{x_i}^{x_{i+1}} \mathbf{U}(t, x, y, z) dx \, dy \, dz, \quad (3)$$

$$\overline{\mathbf{F}}_{xj,k}(x_i) \equiv \frac{1}{\Delta t \Delta y_j \Delta z_k} \int_0^{\Delta t} \int_{z_k}^{z_{k+1}} \int_{y_j}^{y_{j+1}} \mathbf{F}_x[\mathbf{U}(t, x_i, y, z)] dy \, dz \, dt,$$

and $\overline{\mathbf{F}}_{yi,k}(y_j)$ and $\overline{\mathbf{F}}_{zi,j}(z_k)$ have similar definitions.

Although the set of Eqs.(2) is a natural one to represent the conservation laws, it is not the convenient one for the divergence-free condition of the magnetic field. Notice that the conservation laws for the magnetic field in an orthogonal coordinate system has a special feature. For example, the conservation law for B_x doesn't involve derivatives in the x-direction. It is this feature that makes it possible that eight conservation laws and the divergence-free condition may be simultaneously exactly satisfied in a system of eight equations. We write $\mathbf{U}$ in a form

$$\mathbf{U} \equiv \begin{pmatrix} \mathbf{W} \\ \mathbf{B} \end{pmatrix}.$$

In our scheme to be presented in the paper, the flow variables $\mathbf{W}$ are represented by volume-averages, Eq.(3), while the magnetic field is represented at interfaces of grid cells:

$$b_{xi,j,k} \equiv \frac{1}{\triangle y_j \triangle z_k} \int_{z_k}^{z_{k+1}} \int_{y_j}^{y_{j+1}} B_x(x_i, y, z) dy dz, \qquad (4)$$

$$b_{yi,j,k} \equiv \frac{1}{\triangle x_j \triangle z_k} \int_{z_k}^{z_{k+1}} \int_{x_i}^{x_{i+1}} B_y(x, y_j, z) dx dz, \qquad (5)$$

$$b_{zi,j,k} \equiv \frac{1}{\triangle x_j \triangle y_j} \int_{y_j}^{y_{j+1}} \int_{x_i}^{x_{i+1}} B_z(x, y, z_k) dx dy. \qquad (6)$$

Notice that different components of the magnetic field are represented at different sets of interfaces. Considering these averages at the interfaces, we define a vector $\mathbf{V}_{i,j,k}$

$$\mathbf{V}_{i,j,k} \equiv \left(\begin{array}{c} \mathbf{W}_{i,j,k} \\ \mathbf{b}_{i,j,k} \end{array} \right), \qquad (7)$$

whose elements have mixed definitions Eqs.(3-6). In our scheme, $\mathbf{V}_{i,j,k}$ is considered as independent variables which have to be updated, but cell-averaged $\mathbf{B}_{i,j,k}$ is considered only as an intermediate variable for an approximate calculation of the time-averaged fluxes needed in the scheme. The value of $\mathbf{B}_{i,j,k}$ may be approximately calculated from $\mathbf{b}_{i,j,k}$ through interpolations.

We would like to mention two points here. First, $\mathbf{b}_{i,j,k}$ is well defined even if there exists a MHD discontinuity (a fast shock, or slow shock, or rotational discontinuity, or contact discontinuity) at any interface of grid cells. Second, we are actually using a staggered grid, but it is different from traditional staggered grids in which flow variables are defined at different sets of grid points.

In our difference representation of the magnetic field, the net flux across six interfaces of a grid cell is

$$\oint_S \mathbf{B} \cdot ds = \triangle y_j \triangle z_k (b_{xi+1,j,k} - b_{xi,j,k}) \;\; + \;\; \triangle x_i \triangle z_k (b_{yi,j+1,k} - b_{yi,j,k})$$
$$+ \;\; \triangle x_i \triangle y_j (b_{zi,j,k+1} - b_{zi,j,k}).$$

Here S stands for the surface of the grid cell $x_i \leq x \leq x_{i+1}$, $y_j \leq y \leq y_{j+1}$ and $z_k \leq z \leq z_{k+1}$. The divergence-free condition is exactly satisfied if and only if the expression above is exactly vanishing for each grid cell.

We should mention that for a 2D problem $(\partial/\partial z = 0)$, $b_{zi,j,k} = B_{zi,j,k}$, and the divergence-free condition becomes

$$\triangle y_j (b_{xi+1,j,k} - b_{xi,j,k}) + \triangle x_i (b_{yi,j+1,k} - b_{yi,j,k}) = 0.$$

3. Numerical Schemes

As we said before, the averages of the magnetic field at three sets of interfaces, $\mathbf{b}_{i,j,k}$, are considered as independent variables in our scheme. We integrate the induction equation for B_x (or B_y, or B_z) at the interface $x = x_i$ (or $y = y_j$, or $z = z_k$) of a grid cell $x_i \leq x \leq x_{i+1}$, $y_j \leq y \leq y_{j+1}$ and $z_k \leq z \leq z_{k+1}$. The integrations give us

$$
\begin{aligned}
b_{xi,j,k}(\Delta t) = b_{xi,j,k}(0) \quad &+ \quad \frac{\Delta t}{\Delta y_j}[\overline{\Omega}_{zk}(x_i, y_{j+1}) - \overline{\Omega}_{zk}(x_i, y_j)] \\
&- \frac{\Delta t}{\Delta z_k}[\overline{\Omega}_{yj}(x_i, z_{k+1}) - \overline{\Omega}_{yj}(x_i, z_k)],
\end{aligned} \tag{8}
$$

$$
\begin{aligned}
b_{yi,j,k}(\Delta t) = b_{yi,j,k}(0) \quad &+ \quad \frac{\Delta t}{\Delta z_k}[\overline{\Omega}_{xi}(y_j, z_{k+1}) - \overline{\Omega}_{xi}(y_j, z_k)] \\
&- \frac{\Delta t}{\Delta x_i}[\overline{\Omega}_{zk}(x_{i+1}, y_j) - \overline{\Omega}_{zk}(x_i, y_j)],
\end{aligned} \tag{9}
$$

$$
\begin{aligned}
b_{zi,j,k}(\Delta t) = b_{zi,j,k}(0) \quad &+ \quad \frac{\Delta t}{\Delta x_i}[\overline{\Omega}_{yj}(x_{i+1}, z_k) - \overline{\Omega}_{yj}(x_i, z_k)], \\
&- \frac{\Delta t}{\Delta y_j}[\overline{\Omega}_{xi}(y_{j+1}, z_k) - \overline{\Omega}_{xi}(y_j, z_k)],
\end{aligned} \tag{10}
$$

$$
\overline{\Omega}_{xi}(y_j, z_k) \equiv \frac{1}{\Delta t \Delta x_i} \int_0^{\Delta t} \int_{x_i}^{x_{i+1}} \Omega_x(x, y_j, z_k) dx\, dt,
$$

$$
\overline{\Omega}_{yj}(x_i, z_k) \equiv \frac{1}{\Delta t \Delta y_j} \int_0^{\Delta t} \int_{y_j}^{y_{j+1}} \Omega_y(x_i, y, z_k) dy\, dt,
$$

$$
\overline{\Omega}_{zk}(x_i, y_j) \equiv \frac{1}{\Delta t \Delta z_k} \int_0^{\Delta t} \int_{z_k}^{z_{k+1}} \Omega_z(x_i, y_j, z) dz\, dt.
$$

We should mention that Eqs.(8-10) are exact. From this set of difference equations, it is easy to verify that the net magnetic flux across six interfaces of a grid cell is exactly conserved, i.e.,

$$
\oint_S \mathbf{B}(\Delta t) \cdot \mathbf{ds} = \oint_S \mathbf{B}(0) \cdot \mathbf{ds}. \tag{11}
$$

For the 2D situation $(\partial/\partial z = 0)$, $b_{zi,j,k}$ is exactly same defined as $B_{zi,j,k}$. $b_{xi,j}$ and $b_{yi,j}$ are updated according to the equations

$$
b_{xi,j}(\Delta t) = b_{xi,j}(0) + \frac{\Delta t}{\Delta y_j}[\overline{\Omega}_z(x_i, y_{j+1}) - \overline{\Omega}_z(x_i, y_j)], \tag{12}
$$

$$b_{yi,j}(\Delta t) = b_{yi,j}(0) + \frac{\Delta t}{\Delta x_i}[\overline{\Omega}_z(x_{i+1}, y_j) - \overline{\Omega}_z(x_i, y_j)], \qquad (13)$$

$$\overline{\Omega}_z(x_i, y_j) \equiv \frac{1}{\Delta t}\int_0^{\Delta t} \Omega_z(x_i, y_j)dt.$$

Again Eqs.(12,13) are exact. We would like to emphasize that in order to exactly maintain the divergence-free condition, it is necessary to update the magnetic field in the unsplit manner.

Our current algorithm for 2D and 3D MHD equations is based on a dimensionally split technique to update $\mathbf{U}_{i,j,k}$. The approximate MHD Riemann solver is based on characteristic formulations, and it is an extension of the Riemann solver in (Dai & Woodward, 1995) to its single-step Eulerian version. The approximate MHD Riemann solver is very simple.

Our 1D functioning code starts from the cell-averages of a set of variables $(\rho, \rho u_x, \rho u_y, \rho u_z, \rho e, B_x, B_y, B_z)$. For purely 1D MHD flows, $\mathbf{B}_{i,j,k}$ is exactly the same as $\mathbf{b}_{i,j,k}$, but $\mathbf{B}_{i,j,k}$ is approximately calculated from $\mathbf{b}_{i,j,k}$ in our multi-dimensional scheme. Other variables, such as thermal pressure and internal energy, are approximately derived from this set of variables. For each variable, interpolations are used to determine the structure of the variable inside each grid cell. Although more sophisticated interpolations may be used, we use the linear interpolation in this paper for the cell structure.

After the interpolation, we have to find the effective left and right states for the Riemann problem arising from each interface between grid cells. Unlike Godunov schemes for Lagrangian hydrodynamics, the left (or right) state in a single-step Eulerian scheme may come from the cell structure of the right (or left) cell of the interface for supersonic flows. Consider the Riemann problem arising at the interface $x = x_i$. In order to find domains from which the left and right states come, we first find the estimate for the time-averaged u_x at the interface: $u_x = (u_{xi} + u_{xi-1})/2$. Different waves have different domains of dependence during a time step. Suppose c_f, c_a and c_s being the averages of the fast, Alfven and slow wave speeds in the $(i-1)th$ and ith cells. If $(u_x + c_f)$ is non-negative, we approximately consider the average over the domain $x_i - \Delta t(u_x + c_f) \leq x \leq x_i$ as the left state for the fast wave. Otherwise we consider the average over the domain $x_i \leq x \leq x_i - \Delta t(u_x + c_f)$ as the the left state for the fast wave. Slow and Alfven waves are similarly treated.

After we obtain the effective left and right states, we solve the Riemann problem to approximately calculate the time-averaged values at the interface. The time-averaged density at the interface is approximately calculated through the characteristic formulation

$$\bar{\rho} = \rho^{(0)}[\frac{\bar{p}}{p^{(0)}}]^{1/\gamma}.$$

Here the superscript $^{(0)}$ stands for a domain-average over the domain between x_i and $(x_i - u_x \triangle t)$ and $\bar{p}$ is the time-averaged value of p at the interface.

As stated before, in our current algorithm, the updating of $\mathbf{U}_{i,j,k}$ is based on a dimensionally split technique (Strang, 1968). Each time step of a multi-dimensional problem may be broken down into 1D passes in which derivatives in other dimensions are set to zero. Therefore, the dimensionally split technique may be written in the form

$$\mathbf{U}_{i,j,k}(2\triangle t) = L^x_{\triangle t} L^y_{\triangle t} L^z_{\triangle t} L^z_{\triangle t} L^y_{\triangle t} L^x_{\triangle t} \mathbf{U}_{i,j,k}(0). \qquad (14)$$

Here $L^x_{\triangle t}$ is a 1D operator in the x-direction with the time step $\triangle t$.

In our current algorithm, this dimensionally split technique is used to update the conserved quantities, $\mathbf{W}_{i,j,k}$ and $\mathbf{B}_{i,j,k}$. But the solution for $\mathbf{B}_{i,j,k}$ obtained from Eq.(2) doesn't exactly satisfy the divergence-free condition. Therefore, in our scheme, the solution $\mathbf{B}_{i,j,k}$ is used only as an intermediate result which is used to calculate the time-averaged fluxes needed when $\mathbf{W}_{i,j,k}$ and $\mathbf{b}_{i,j,k}$ are updated.

We would like to mention two points here. First, according to the basic equation for the magnetic field (i.e., the last three equations contained in Eq.(1), which are called the induction equation), B_x (or B_y or B_z) remains unchanged in a sweep in the x-direction (or y-direction, or z-direction). Second, in a purely 1D problem, for example, in the x-direction, B_x is a constant, but B_x varies with x in a sweep in the x-direction in the multi-dimensional scheme. $b_{xi,j,k}$ is used as B_x in the Riemann problem arising from the interface, $x = x_i$.

As a summary, in our current algorithm, $\mathbf{V}_{i,j,k}$ is updated in the form

$$\mathbf{V}_{i,j,k}(2\triangle t) = D_{\triangle t} L^x_{\triangle t} L^y_{\triangle t} L^z_{\triangle t} D_{\triangle t} L^z_{\triangle t} L^y_{\triangle t} L^x_{\triangle t} \mathbf{V}_{i,j,k}(0). \qquad (15)$$

Here the operator $D_{\triangle t}$ stands for Eqs.(8-10).

The remaining task in our scheme is to approximately evaluate the time-averaged $\mathbf{\Omega}$ needed in Eqs.(8-10) for the 3D situation and needed in Eqs.(12,13) for the 2D situation. In the second order of accuracy,

$$\overline{\Omega}_{xi} \approx \Omega_x(\tilde{\mathbf{u}}^*_i(y_j, z_k), \tilde{\mathbf{B}}^*_i(y_j, z_k)), \qquad (16)$$

$$\overline{\Omega}_{yi} \approx \Omega_y(\tilde{\mathbf{u}}^*_j(x_i, z_k), \tilde{\mathbf{B}}^*_j(x_i, z_k)), \qquad (17)$$

$$\overline{\Omega}_{zk} \approx \Omega_y(\tilde{\mathbf{u}}^*_k(x_i, y_j), \tilde{\mathbf{B}}^*_k(x_i, y_j)). \qquad (18)$$

Here $\tilde{\mathbf{u}}_i^*(y_j, z_k) \equiv [\tilde{\mathbf{u}}_i(y_j, z_k, \triangle t) + \tilde{\mathbf{u}}_i(y_j, z_k, 0)]/2$, $\tilde{\mathbf{u}}_j^*(x_i, z_k) \equiv [\tilde{\mathbf{u}}_j(x_i, z_k, \triangle t) + \tilde{\mathbf{u}}_j(x_i, z_k, 0)]/2$, $\tilde{\mathbf{u}}_k^*(x_i, y_j) \equiv [\tilde{\mathbf{u}}_k(x_i, y_j, \triangle t) + \tilde{\mathbf{u}}_k(x_i, y_j, 0)]/2$, and $\tilde{\mathbf{u}}_i(y_j, z_k)$ (or $\tilde{\mathbf{u}}_j(x_i, z_k)$, or $\tilde{\mathbf{u}}_k(x_i, y_j)$) is the line-averaged value of $\mathbf{u}$ on the line $x_{i-1} < x < x_i$, $y = y_j$ and $z = z_k$ (or the line $y_{j-1} < y < y_j$, $x = x_i$, $z = z_k$, or the line $z_{k-1} < z < z_k$, $x = x_i$, $y = y_j$). $\tilde{\mathbf{B}}_i^*$, $\tilde{\mathbf{B}}_j^*$ and $\tilde{\mathbf{B}}_k^*$ in Eqs.(16-18) have similar meanings. Through the dimensionally split approach, we have already obtained the updated cell-averages, $\mathbf{u}_{i,j,k}(\triangle t)$ and $\mathbf{B}_{i,j,k}(\triangle t)$. Interpolations may be used to approximately calculate these line-averaged values. Thus, the magnetic field $\mathbf{b}_{i,j,k}$ may be updated through Eqs.(8-10).

We would like to mention again that $\mathbf{B}_{i,j,k}$ is used only as an intermediate result. After $\mathbf{b}_{i,j,k}$ is updated, $\mathbf{B}_{i,j,k}$ may be approximately calculated from $\mathbf{b}_{i,j,k}$, for example, $B_{xi,j,k} \approx (b_{xi+1,j,k} + b_{xi,j,k})/2$.

For the 2D situation, The value of $\overline{\Omega}_z$ in Eqs.(12,13) may be approximately evaluated:

$$\overline{\Omega}_z(x_i, y_j) \approx \Omega_z(\overline{\mathbf{u}}(x_i, y_j), \overline{\mathbf{B}}(x_i, y_j)). \tag{19}$$

$\overline{\mathbf{u}}(x_i, y_j)$ and $\overline{\mathbf{B}}(x_i, y_j)$ may be approximately calculated through

$$\overline{\mathbf{u}}(x_i, y_j) \approx \frac{1}{2}[\mathbf{u}(x_i, y_j, \triangle t) + \mathbf{u}(x_i, y_j, 0)], \tag{20}$$

$$\overline{\mathbf{B}}(x_i, y_j) \approx \frac{1}{2}[\mathbf{B}(x_i, y_j, \triangle t) + \mathbf{B}(x_i, y_j, 0)]. \tag{21}$$

The point values $\mathbf{u}(x_i, y_j)$ and $\mathbf{B}(x_i, y_j)$ here may be approximately obtained, for example, $\mathbf{u}(x_i, y_j) \approx (\mathbf{u}_{i,j} + \mathbf{u}_{i-1,j} + \mathbf{u}_{i,j-1} + \mathbf{u}_{i-1,j-1})/4$.

4. Numerical Examples

The numerical scheme described in this paper has been tested for some 1D and 2D examples for its correctness, accuracy and robustness. The correctness of the scheme are tested through the propagation of smooth fast, slow and Alfven waves and shock-tube problems in both 1D and 2D domains. For these test problems, we know the "exact solutions". We compare our numerical solutions with the exact ones, and find that they have a good agreement. For the accuracy of the scheme, we carried out a few convergence studies for smooth waves, for which we know the exact solutions.

In this paper, we present only four of 2D test problems we carried out. Uniform grids and 5/3 for γ are used in the four examples. In the first two examples, a single wave propagating along a ξ-direction is set up. The ξ-direction makes an angle α ($\equiv 30°$) with the x-axis. The fast wave speed in the direction of wave propagation is very close to unity. The simulations

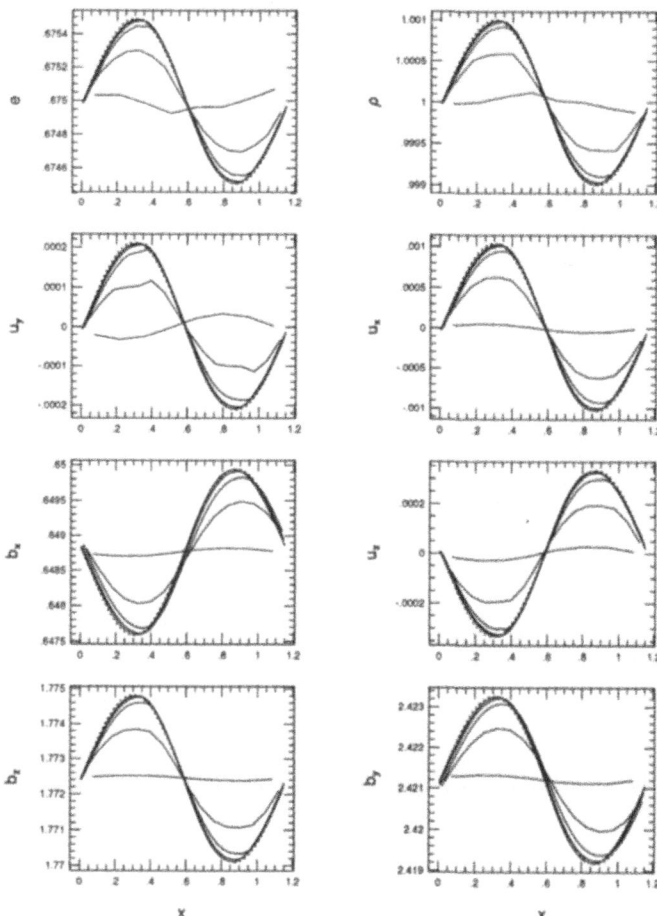

Figure 1. A fast wave propagating in a 2D domain $(L_x - 0) \times (L_y - 0)$. The dashed lines are the initial profiles along the line $y = L_y/2$, and solid lines are the results at $t = 10$ obtained when five grids are used, which contain 8×8, 16×16, 32×32, 64×64 and 128×128 grid cells.

are performed in the domain $(L_x - 0) \times (L_y - 0)$ with periodic boundary condition, where $L_x = 1/cos\alpha$ and $L_y = 1/sin\alpha$. The plots in Figs.1 and 2 for the first two examples are the profiles along $y = L_y/2$ and the dashed there are initial profiles. The first example is to test the accuracy of the scheme. The solid lines in Fig.1 are the simulation results at $t = 10$ when 8×8, 16×16, 32×32, 64×64 and 128×128 grid cells are used. The

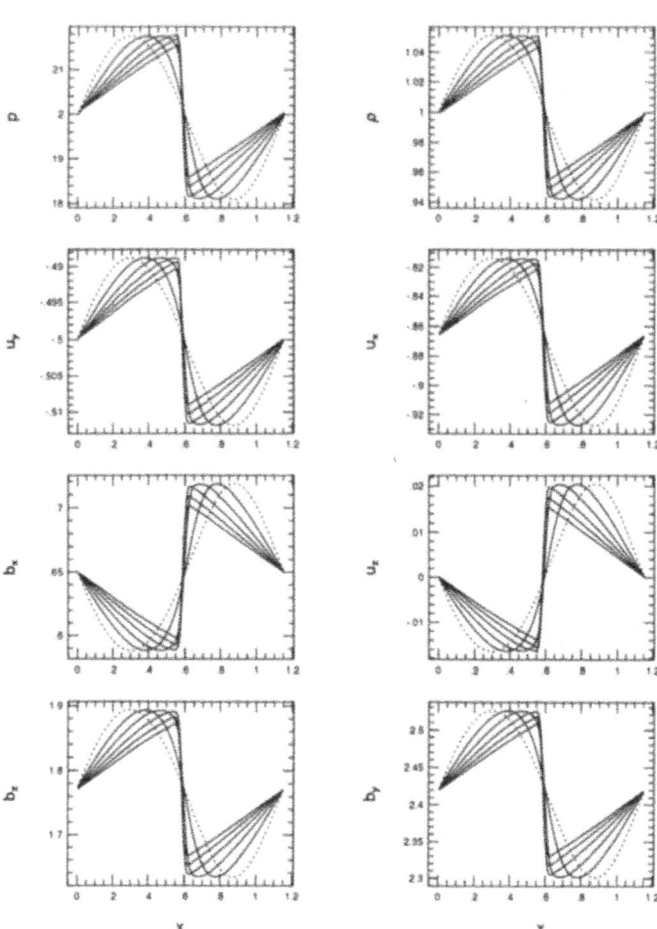

Figure 2.   A nonlinear fast wave propagating in a 2D domain $(L_x - 0) \times (L_y - 0)$. The dashed lines are the initial profiles along the line $y = L_y/2$. The solid lines are the profiles along the line at $t = 1, 2, 3, 4, 5$.

second example is for the propagation of a nonlinear fast wave along the ξ-direction. Figure 2 gives the profiles along $y = L_y/2$ at $t = 1.0, 2.0, 3.0,$ 4.0 and 5.0. 128×128 grid cells are used in this example.

The third one is a 2D Riemann problem with continuation boundary conditions in both x- and y-directions. The initial four states for $(\rho, p, u_x,$ $u_y, u_z)$ is $(1, 1, 0.75, -0.5, 0)$ for $x > 0$ and $y > 0$, $(2, 1, 0.75, 0.5, 0)$ for $x < 0$ and $y > 0$, $(1, 1, -0.75, 0.5, 0)$ for $x < 0$ and $y < 0$, and $(3, 1,$ -0.75,-0.5, 0)$ for $x > 0$ and $y < 0$. The initial magnetic field is uniform, **B** $= (2, 0, 1)$. A mesh containing 256×256 grid cells is used. Figure 3 shows the mass density and Fig.4 gives the directions of the magnetic field on the

the (x,y)-plane at $t = 0.7001$ for the third example. The length of a arrow in Fig.4 is proportional to the strength of the part of the field, (B_x, B_y). The early stage of this problem without the magnetic field was numerically studied in (Schulz-Rinne *et al.*, 1993).

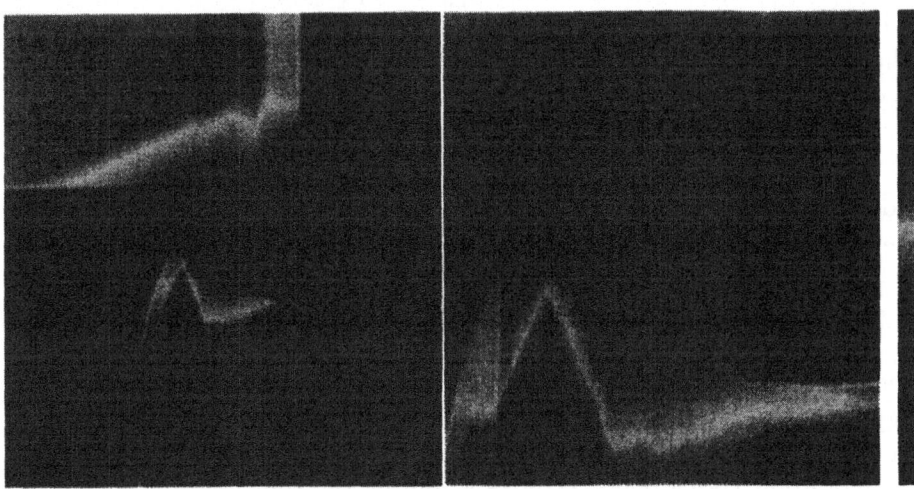

Figure 3. The images of the mass density in the 2D Riemann problem.

The fourth example is the interaction between a magnetosonic shock and a denser cloud. The problem is of interests to the community of astrophysics and space physics. The simulation for this problem is performed in a $(2 - 0) \times (1 - 0)$ domain. Initially, a magnetosonic shock propagating in the x-direction with a Mach number (≈ 10) is set up in the simulation domain. The shock front is initially located at $x = 0.325$. The pre-shock state for $(\rho, p, u_x, u_y, u_z, B_x, B_y, B_z)$ is $(1, 1, 0, 0, 0, 0, 2, 2)$, and the post-shock state is $(3.86859e+00, 1.67345e+02, 1.12536e+01, 0, 0, 0, 7.73718e+00, 7.73718e+00)$. In the front of the shock, there is initially a circular cloud, whose center is located at $(x, y) = (0.55, 0.5)$. The radius of the cloud is 0.175. The cloud is twenty times denser than the plasma at the pre-shock state. Figure 5 shows the mass density at four instants. and Fig.6 gives the directions of the magnetic field at the (x,y)-plane at $t = 0.08510$.

5. Conclusions and Discussions

In this paper, we have proposed a second order finite difference scheme for MHD equations. The scheme is of the Godunov-type, and is based on an approximate MHD Riemann solver. The MHD Riemann solver is based on the characteristic formulations of MHD equations. In the proposed scheme, the mass, momentum, energy and the three components of magnetic field are exactly conserved, and the divergence-free condition is exactly satisfied.

Compared to existing Godunov schemes for MHD equations, the scheme proposed in this paper is simple.

The approach proposed in this paper for the divergence-free condition may be applied to other dimensionally split and unsplit Godunov schemes for MHD equations. The operator $L^x_{\triangle t}$ in Eq.(14) may be a 1D operator in other schemes for MHD equations. If Eqs.(8-10) are used in an unsplit scheme $L_{\triangle t}$, the resulting scheme may be written in the form

$$\mathbf{V}_{i,j,k}(\triangle t) = D_{\triangle t} L_{\triangle t} \mathbf{V}_{i,j,k}(0).$$

An approximate Riemann solver for fully multi-dimensional Riemann problems arising from grid points will be useful for the calculation of time-averaged values needed in Eqs.(8-10,12,13).

6. Acknowledgments

The work presented here has been supported by the Department of Energy through grants DE-FG02-87ER25035 and DE-FG02-94ER25207, by the National Science Foundation through grant ASC-9309829, by NASA through grant USRA/5555-23/NASA, and by the University of Minnesota through its Minnesota Supercomputer Institute.

References

Balsara, D., "Total variation diminishing scheme for magnetohydrodynamics," *J. Comput. Phys.*, submitted, 1996.

Brio, M. and Wu, C.C., "An upwind differencing scheme for the equations of ideal magnetohydrodynamics," *J. Comput. Phys.*, Vol. 75, 1988, pp. 400-422.

Dai, W. and Woodward, P.R., "An approximate Riemann solver for ideal magnetohydrodynamics," *J. Comput. Phys.*, Vol. 111, 1994, pp. 354-372.

Dai, W. and Woodward, P.R., "A simple Riemann solver and high-order Godunov schemes for hyperbolic systems of conservation laws," *J. of Comput. Phys.*, Vol. 121, 1995, pp. 51-65.

Dai, W. and Woodward, P.R., "Extension of the piecewise parabolic method (PPM) to multidimensional ideal magnetohydrodynamics," *J. Comput. Phys.*, Vol. 115, 1994, pp. 485-514.

Evans, C. and Hawley, J.F., "Simulation of Magnetohydrodynamic flows: A constrained transport method," *Astrophys. J.*, Vol. 332, 1988, pp. 659-677.

Powell, K.G., "An approximate Riemann solver for magnetohydrodynamics (that works in more than one dimension)," *ICASE Report No. 94-24*, 1994.

Powell, K.G., Roe, P.L., Myong, R.S., Gombosi, T., and Zeeuw, D.De, "An upwind scheme for magnetohydrodynamics," *AIAA Paper*, 95-1704-CP, 1995.

Ryu, D. and Jones, T.W., "Numerical magnetohydrodynamics in astrophysics: Algorithm and tests for one-dimensional flow," *Astrophys. J.*, Vol. 442, 1995, pp. 228-256.

Ryu, D., Jones, T.W., and Frank, A., "Numerical magnetohydrodynamics in astrophysics: Algorithm and tests for multidimensional flow," *Astrophys. J.*, Vol. 452, 1995, pp. 785-796.

Schulz-Rinne, C.W., Collins, J.P., and Glaz, H.M., "Numerical solution of the Riemann problem for two-dimensional gas dynamics," *SIAM J. Sci. Comput.*, Vol. 14, 1993, pp. 1394-1414.

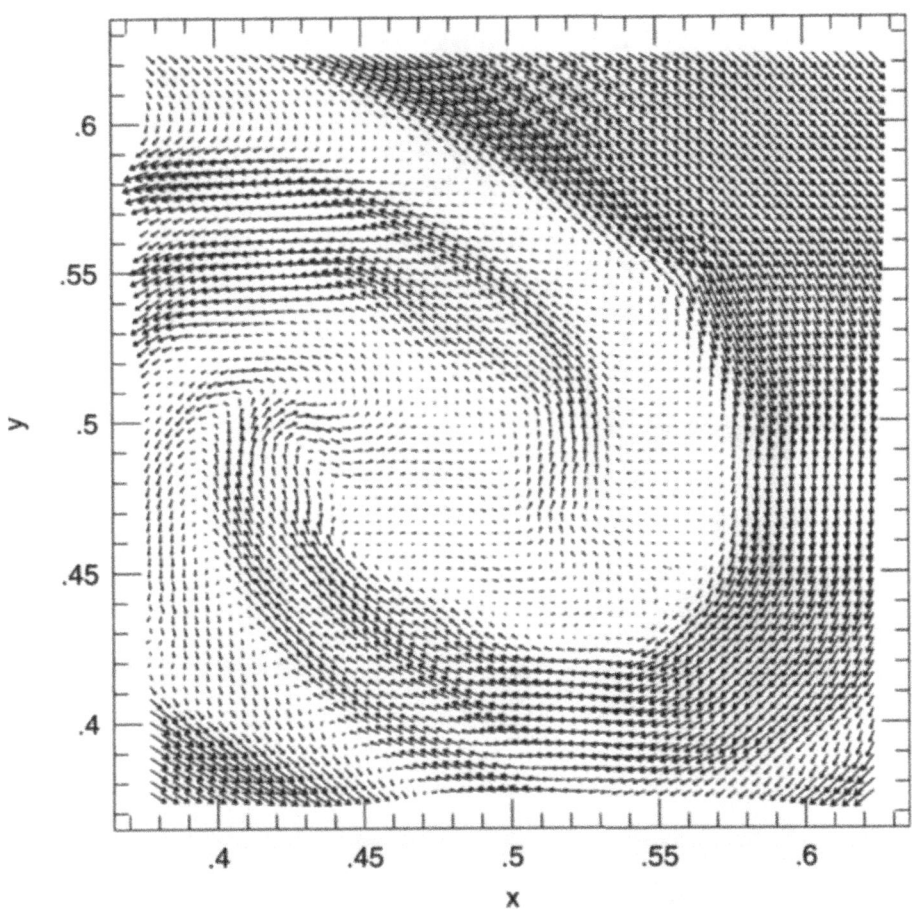

Figure 4. The direction of the part of the magnetic field, (B_x, B_y), at $t = 0.7001$ in the 2D Riemann problem.

Stone, J.M. and Norman, M.L., "ZEUS-2D: A radiation magnetohydrodynamics code for astrophysical flows in two space dimensions. II. The magnetohydrodynamics algorithms and tests," *Astrophys. J. Supl.*, Vol. 80, 1992, pp. 791-818.

Strang, G., "On construction and comparison of differential schemes," *SIAM J. Numer. Anal.*, Vol. 5, 1968, pp. 506-517.

Zachary, A.L. and Colella, P., "A high-order Godunov method for the equations of ideal magnetohydrodynamics," *J. Comput. Phys.*, Vol. 99, 1992, pp. 341-347.

Zachary, A.L., Malagoli, A., and Colella, P., "A higher-order Godunov method for multidimensional ideal magnetohydrodynamics," *SIAM J. Sci. Comput.*, Vol. 15, 1994, pp. 263-380.

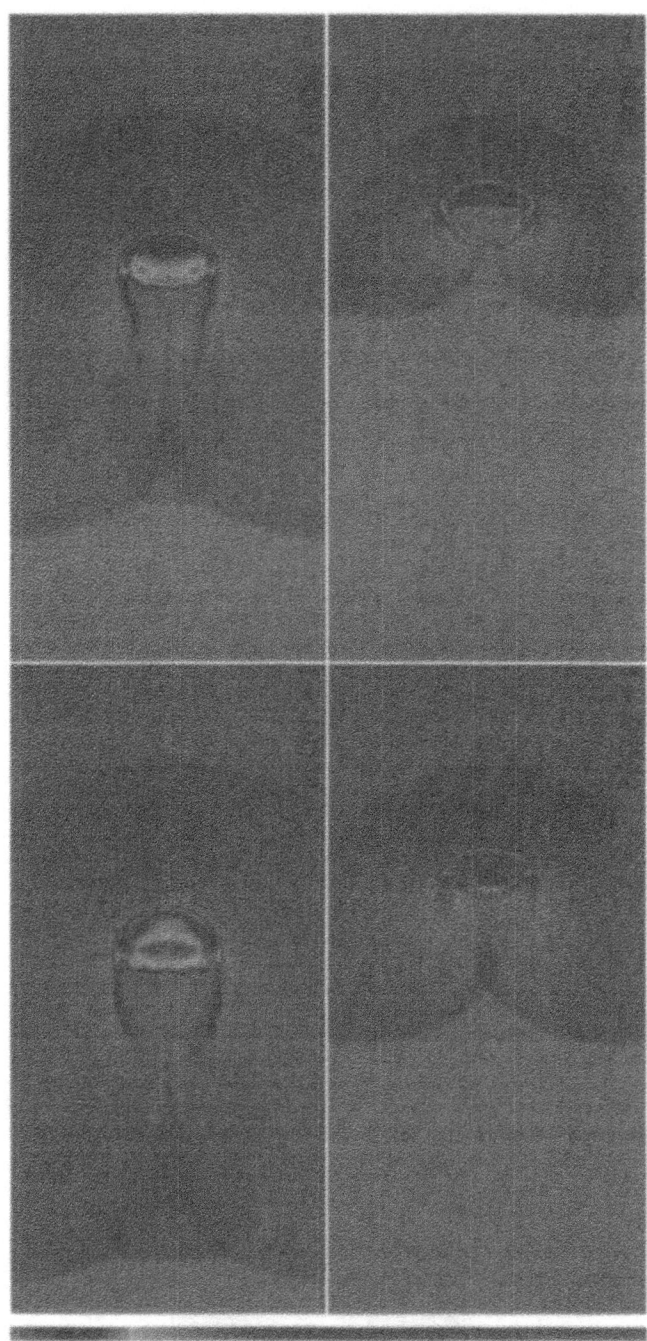

Figure 5. The mass density in the interaction between a magnetosonic shock and a denser cloud.

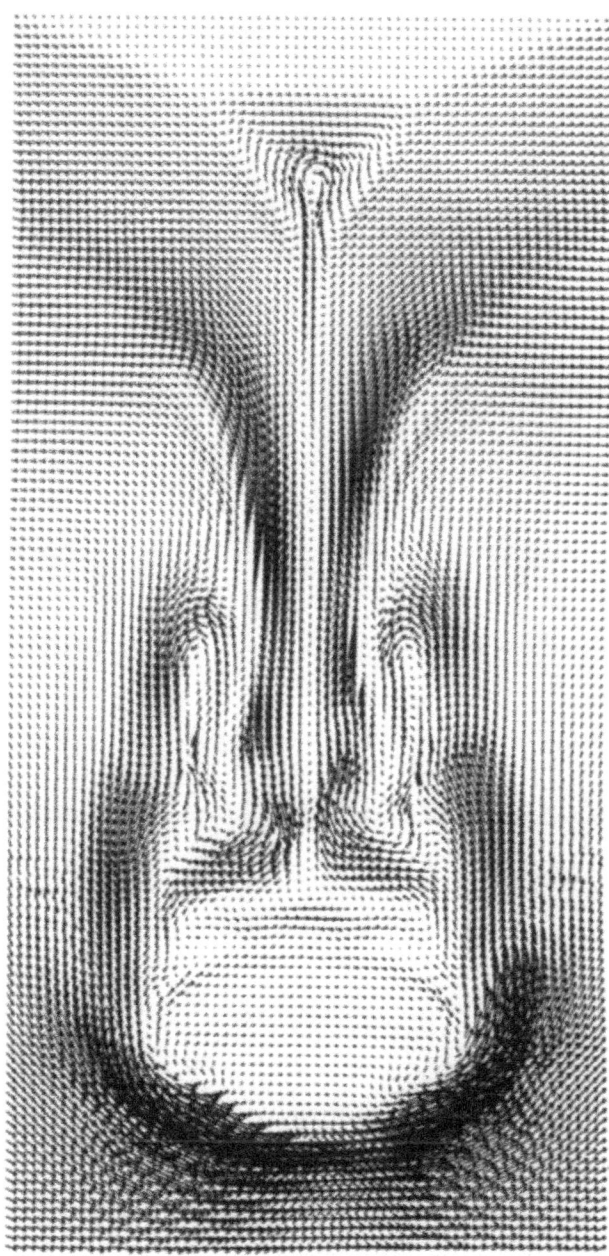

Figure 6. The part of the magnetic field, (B_x, B_y), in a part of the simulation domain, $0.582 < x < 1.582$ and $0.25 < y < 0.75$ at $t = 0.08510$.

ON MULTIDIMENSIONAL POSITIVELY CONSERVATIVE HIGH-RESOLUTION SCHEMES

TIMUR LINDE AND PHILIP L. ROE
W.M. Keck Foundation Laboratory
for Computational Fluid Dynamics
Department of Aerospace Engineering
The University of Michigan
Ann Arbor, MI 48109

1. Introduction

There is often a need to compute flows containing regions in which the total energy is overwhelmingly dominated by the kinetic energy mode. This poses a difficulty if the equations are solved in conservation form. Pressure, being a difference of two large quantities, may be contaminated by various types of numerical error, and its value may become negative. It is also possible for a particular scheme to produce negative density even if an exact solution does not contain vacuum zones. As soon as density or pressure become negative a computation fails. When the problem occurs, it can be usually postponed by lowering the time step. However, one needs a scheme that preserves positivity for all time, under conditions not much more severe that the usual CFL restriction. This problem has recently received growing attention in the literature.

Einfeldt *et al.* (1992) were the first to introduce the class of positively conservative schemes, the schemes that always generate physically admissible solutions from physical data. For one-dimensional flows they proved that the Godunov scheme (Godunov, 1959) is positively conservative, but any Godunov-type scheme based on a linearized Riemann problem must sometimes fail to produce physically meaningful results. Roe's scheme (Roe, 1981) is a linearization-type scheme, hence it is not positively conservative. It is interesting to note that another popular scheme, Osher's scheme (Osher & Solomon, 1982), is not positively conservative either since it can be shown that it fails to compute interactions of strong shocks. Einfeldt *et al.* (1992) proved that the HLLE scheme (Einfeldt, 1988) and the Lax-Friedrichs

V. Venkatakrishnan et al (eds.), Barriers and Challenges in Computational Fluid Dynamics, 299-313.
© 1998 *Kluwer Academic Publishers.*

scheme, to which the HLLE scheme reduces for a specific choice of the numerical signal velocities, are positively conservative. They also showed that a less diffusive modification of the HLLE scheme, the HLLEM scheme, does not have this property. This is unfortunate since in the quest for robustness one seems to face the dilemma of using either the rather expensive Godunov scheme or an economical but far more diffusive scheme. It is therefore natural that the design of a practical accurate positively conservative scheme is becoming a research issue.

Recently an improvement to the HLLEM scheme has been suggested. Charrier et al. (1993) showed that with an optimal choice of the antidiffusion coefficients the HLLEM scheme can be made both positively conservative and fairly accurate. The resulting HLLEMR scheme, however, uses a rather involved optimization step. A very detailed description of the HLLEMR scheme can be found in (Flandrin, 1995). Xu et al. (1995) have demonstrated the ability of their Boltzmann scheme to give accurate results for problems with strong rarefaction waves. Liou (1996) has recently presented some remarkable positivity properties of the AUSM-family of schemes. Donat and Marquina (1996) have designed a new combined Roe-Lax-Friedrichs scheme. They observed that the first-order version of their scheme exhibits the positive conservation property, but its third-order extension fails to compute the test cases described in Einfeldt et al. (1992).

This is not very surprising, because most of the previous work has been restricted to one-dimensional first-order accurate schemes, and it is not obvious under what conditions a multidimensional high-resolution version of a particular scheme can be guaranteed to remain positively conservative. The determination of such conditions is the main goal of the present paper.

2. Preliminaries

We consider the Euler equations describing the flows of inviscid fluids. In conservation form the equations are

$$\frac{\partial \mathbf{U}}{\partial t} + \nabla \cdot \mathcal{F}(\mathbf{U}) = 0, \qquad \mathcal{F} = (\mathbf{F}_x, \mathbf{F}_y, \mathbf{F}_z), \qquad (1)$$

where

$$\mathbf{U} = \begin{pmatrix} \rho \\ m_x \\ m_y \\ m_z \\ E \end{pmatrix}, \qquad \mathbf{F}_x(\mathbf{U}) = \begin{pmatrix} m_x \\ m_x^2/\rho + p \\ m_x m_y/\rho \\ m_x m_z/\rho \\ m_x(E+p)/\rho \end{pmatrix}, \qquad (2)$$

and $\mathbf{F}_y(\mathbf{U})$ and $\mathbf{F}_z(\mathbf{U})$ are defined in a similar way. Naturally, ρ is the energy of a gas per unit volume. We will use the ideal gas model, therefore the equation of state is $p = (\gamma - 1)(E - (m_x^2 + m_y^2 + m_z^2)/2\rho)$, where γ is the adiabatic constant.

Following (Einfeldt $et\ al.$, 1992) we define the set of physically admissible states by

$$G = \left\{ \mathbf{U} \mid \rho > 0 \quad \text{and} \quad E - (m_x^2 + m_y^2 + m_z^2)/2\rho > 0 \right\}. \tag{3}$$

By definition this set contains all the states with positive density and pressure. It is straightforward to show that G is a convex set. Moreover, we can see that both inequalities in (3) involve homogeneous functions of $\mathbf{U}$ of degree one. Therefore, if some state $\mathbf{U}_0 \in G$, then for any $\alpha > 0$, $\alpha\mathbf{U}_0 \in G$. Hence G is an open convex cone. Then it is trivial to prove that for any $\mathbf{U}_i \in G$, where $i = 1, \ldots, n$, and for any $\alpha_i \geq 0$ such that $\sum_{i=1}^{n} \alpha_i > 0$, $\sum_{i=1}^{n} \alpha_i \mathbf{U}_i \in G$. Thus any non-negative non-trivial linear combination of admissible states is always an admissible state, and this statement is not restricted to convex combinations (for which $\sum_{i=1}^{n} \alpha_i = 1$). Physically this means that a mixture of various quantities of a gas will obviously have positive density and pressure. Also, we simply observe that an admissible state remains admissible under a change of coordinates. Repeated use of the above two observations will suffice for most of our analysis.

We assume that a positively conservative one-dimensional first-order accurate Godunov-type scheme is available, and our task will be to extend it to more dimensions and higher order. In the following we do not need to know the details of this scheme; our proof applies to any present or future method. We only assume that the flux function is consistent and rotationally invariant. Also, we require that this scheme must produce physical solutions under a CFL condition $\Delta t < k\Delta x/\lambda$ with k of order unity, where λ is a characteristic velocity which determines the allowable time step. Since the expansion fans are of main concern in the present problem, a good choice of this velocity would be $\lambda(\mathbf{U}_L, \mathbf{U}_R) = \max\{|u_L| + a_L, |u_R| + a_R\}$, where a is the sound speed.

At this point we would like to address the issue of accuracy. Note that if the pressure of a gas is very small compared to the kinetic energy, it may be of the same order as the numerical errors associated with the computation of the total energy. Hence the actual value of the numerical pressure may not be very accurate. Whether this is acceptable may depend on the context. For the present we merely require that the solution belongs to the set of physically meaningful states at every point for all times. Thus we prove that our codes will be robust but not necessarily accurate. We hope that the desired accuracy can be achieved via refinement.

3. First-order Multidimensional Schemes

Let us consider a computational grid consisting of arbitrary cells in any number of dimensions. For given cell i let us denote the set of its neighbors by ω_i. Then with piecewise constant data in each cell a typical first-order accurate finite volume numerical scheme can be written as

$$U_i^{k+1} = U_i^k - \frac{\Delta t}{V_i} \sum_{j \in \omega_i} \tilde{\mathbf{F}}_{\mathbf{n}_{ij}}(U_i^k, U_j^k) S_{ij}, \qquad (4)$$

where $\tilde{\mathbf{F}}_{\mathbf{n}_{ij}}(U_i, U_j)$ is the numerical flux normal to face ij, $\mathbf{n}_{ij}$ is the normal vector pointing from cell i to cell j, S_{ij} is the area of face ij, V_i is the volume of cell i, Δt is the time step, and k denotes the time level. For all i let us choose some α_{ij} such that $0 < \alpha_{ij} < 1$, $\sum_{j \in \omega_i} \alpha_{ij} = 1$. Subsequently we will make a specific choice of α_{ij} that will maximize the allowable time step. Since $\sum_{j \in \omega_i} \mathbf{F}_{\mathbf{n}_{ij}}(U_i^k) S_{ij} = 0$, we can rewrite Equation 4 as follows,

$$U_i^{k+1} = \sum_{j \in \omega_i} \alpha_{ij} U_i^k - \frac{\Delta t}{V_i} \sum_{j \in \omega_i} \left[\tilde{\mathbf{F}}_{\mathbf{n}_{ij}}(U_i^k, U_j^k) - \mathbf{F}_{\mathbf{n}_{ij}}(U_i^k) \right] S_{ij}$$

$$= \sum_{j \in \omega_i} \alpha_{ij} \left(U_i^k - \frac{\Delta t S_{ij}}{\alpha_{ij} V_i} \left[\tilde{\mathbf{F}}_{\mathbf{n}_{ij}}(U_i^k, U_j^k) - \mathbf{F}_{\mathbf{n}_{ij}}(U_i^k) \right] \right)$$

$$= \sum_{j \in \omega_i} \alpha_{ij} T_{ij}^{-1} T_{ij} \left(U_i^k - \frac{\Delta t S_{ij}}{\alpha_{ij} V_i} \left[\tilde{\mathbf{F}}_{\mathbf{n}_{ij}}(U_i^k, U_j^k) - \mathbf{F}_{\mathbf{n}_{ij}}(U_i^k) \right] \right).$$

In the last step T_{ij} is an orthogonal matrix that defines a local rotation of state U from the coordinate frame $\{x, y, z\}$ to a face oriented frame $\{\mathbf{n}_{ij}, \tau_{1ij}, \tau_{2ij}\}$, i.e. $T_{ij} : (\rho, m_x, m_y, m_z, E)^T \mapsto (\rho, m_{\mathbf{n}_{ij}}, m_{\tau_{1ij}}, m_{\tau_{2ij}}, E)^T$, where τ_{1ij} and τ_{2ij} are two arbitrarily chosen tangential vectors.

Let us denote $\mathbf{W}_{i_{ij}} = T_{ij} U_i$ and $\mathbf{W}_{j_{ij}} = T_{ij} U_j$. Since T_{ij} is orthogonal then $U_i \in G$ implies $\mathbf{W}_{i_{ij}} \in G$ and vice versa. It is also easy to verify that for a Godunov-type scheme that is invariant under rotations

$$T_{ij} \tilde{\mathbf{F}}_{\mathbf{n}_{ij}}(U_i, U_j) = \tilde{\mathbf{F}}_x(\mathbf{W}_{i_{ij}}, \mathbf{W}_{j_{ij}}). \qquad (5)$$

Then we can write

$$U_i^{k+1} = \sum_{j \in \omega_i} \alpha_{ij} T_{ij}^{-1} \mathbf{W}_{i_{ij}}^{k+1}, \qquad (6)$$

so that the new state is a superposition of states that could have arisen from some one-dimensional calculation,

$$\mathbf{W}_{i_{ij}}^{k+1} = \mathbf{W}_{i_{ij}}^k - \frac{\Delta t S_{ij}}{\alpha_{ij} V_i} \left[\tilde{\mathbf{F}}_x(\mathbf{W}_{i_{ij}}^k, \mathbf{W}_{j_{ij}}^k) - \mathbf{F}_x(\mathbf{W}_{i_{ij}}^k) \right]. \qquad (7)$$

If all $\mathbf{W}_{i_{ij}}^{k+1} \in G$, then $U_i^{k+1} \in G$, and the scheme will be positively conservative.

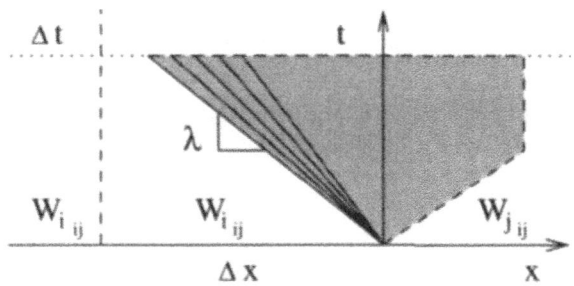

Figure 1. Equivalent one-dimensional problem.

Equation 7 clearly describes an equivalent one-dimensional problem for $\mathbf{W}_{i_{ij}}$ on a grid with the cell size $\triangle x = \alpha_{ij} V_i / S_{ij}$. In this equivalent problem (see Figure 1) the left neighbor has the same state as the cell under consideration; a trivial Riemann problem that will be solved exactly by any consistent Riemann solver. Therefore $\mathbf{W}_{i_{ij}}^{k+1}$ will be physically admissible if

$$\triangle t \leq \frac{\alpha_{ij} V_i}{\lambda(\mathbf{W}_{i_{ij}}^k, \mathbf{W}_{j_{ij}}^k) S_{ij}}. \tag{8}$$

If we use the definitions of $W_{i_{ij}}$ and λ, the above inequality becomes,

$$\triangle t \max(|u_{\mathbf{n}_{ij}i}^k| + a_i^k, |u_{\mathbf{n}_{ij}j}^k| + a_j^k) S_{ij} \leq \alpha_{ij} V_i. \tag{9}$$

Now recall that coefficients α_{ij} have been left undefined. Let us choose them in a way that maximizes the allowable time step. To achieve this we require that for a given cell all the inequalities in (9) are violated simultaneously. Then for each i we can add the inequalities to obtain

$$\triangle t \leq \frac{V_i}{\sum_{j \in \omega_i} \max(|u_{\mathbf{n}_{ij}i}^k| + a_i^k, |u_{\mathbf{n}_{ij}j}^k| + a_j^k) S_{ij}}, \tag{10}$$

where we used $\sum_{j \in \omega_i} \alpha_{ij} = 1$. If condition 10 is satisfied for all the cells, the scheme will be positively conservative. Thus *given a first-order one-dimensional positively conservative scheme we can always build a first-order multidimensional positively conservative scheme, provided, as for the Euler equations. that the set of physically admissible states is convex.*

Condition 10 is not too restrictive, and it closely resembles the CFL restriction on the time step that comes from a linear stability analysis (see e.g. Barth (1990)). Hence we can conclude that in a sense positive conservation is equivalent to nonlinear stability. Since the allowable time step is always finite, there is no limit to the time for which the calculation can be continued. However, this does not necessarily imply that the answer will be correct. There are certain situations, as documented by Quirk (1994), where even the exact Riemann solution leads to a limited form of instability, usually when shocks are almost aligned with the grid. We do not address this issue here.

4. High-resolution Multidimensional Schemes

4.1. RECONSTRUCTION AND TIME EVOLUTION

Let us now consider a numerical grid consisting of arbitrary convex cells[1]. We will extend the previous analysis to cover the case where the state in each cell need not be uniform, although we assume that the spatial variation does not take the states beyond the admissible limits. In order to improve both spatial and time resolution of a numerical scheme one needs to employ some sort of reconstruction strategy as well as a time evolution technique more elaborate than Forward Euler. In the following we will consider a simple formally second-order accurate finite volume scheme.

First of all, we need to choose a set of variables to be reconstructed. Both the primitive and the conservative variables can used for that. Numerical experience shows that it is often preferable to work with the primitive variables since then many physical constraints, for example positivity of pressure, are relatively straightforward to enforce. However, this creates the problem of recovery of one set of variables from another one. Commonly a direct mapping of the cell averaged conservative variables onto the primitive variables that represent the centroid values is used. It can be shown (van Leer, 1979) that with the piecewise linear reconstruction this technique is second-order accurate. However, the generated second-order error is virtually uncontrollable. If the density or the pressure in a cell are very low this may take the state out of the set of physically admissible states. To eliminate this problem we will reconstruct the conservative variables.

There are various ways to compute gradients in the cells (for example, see Barth (1990)). For the purpose of simplicity and generality we will use the least squares approach. It is clear that in addition to spurious oscillations reconstruction may also lead to negative densities or pressures. To prevent these from happening our scheme will have a limiter, whose computation will be discussed later.

We will use the classical Heun (Gear, 1971) scheme to compute time evolution. Shu and Osher (1988) showed that this scheme is TVD stable.

---

[1]The requirement that the cells must be convex is not very restrictive and in practice it is usually satisfied. Cartesian methods may be an exception since there a body must be cut out of the grid, and the "cut-cells" may not be convex.

Also, as will be seen, this scheme nicely fits the splitting method described in the previous section.

Let us denote the linear extrapolation of the state in cell i to the centroid of face ij by $\mathbf{U}_{ij}$. Then we can write

$$\mathbf{U}_{ij} = \mathbf{U}_i + \phi_i(\nabla\mathbf{U})_i \cdot (\mathbf{r}_{ij} - \mathbf{r}_i), \tag{11}$$

where $\mathbf{r_i}$ is the centroid of cell i, $\mathbf{r_{ij}}$ is the centroid of face ij, $(\nabla\mathbf{U})_i$ is the gradient matrix and ϕ_i is the single scalar limiter. With these notations the overall scheme can be written as follows,

$$\begin{aligned}
\mathbf{U}_i^{k+\frac{1}{2}} &= \mathbf{U}_i^k - \Delta t\,\mathbf{Res}[\mathbf{U}^k, i] \\
\mathbf{U}_i^{k+1} &= \tfrac{1}{2}\mathbf{U}_i^k + \tfrac{1}{2}(\mathbf{U}_i^{k+\frac{1}{2}} - \Delta t\,\mathbf{Res}[\mathbf{U}^{k+\frac{1}{2}}, i]),
\end{aligned} \tag{12}$$

where

$$\mathbf{Res}[\mathbf{U}^k, i] = \frac{1}{V_i} \sum_{j\in\omega_i} \tilde{\mathbf{F}}_{\mathbf{n}_{ij}}(\mathbf{U}_{ij}^k, \mathbf{U}_{ji}^k)S_{ij}. \tag{13}$$

Clearly, we need to ensure that both $\mathbf{U}_i^{k+\frac{1}{2}}$ and $\mathbf{U}_i^{k+1}$ belong to G. Since the right hand sides of both equations in (12) are very similar, it is enough to consider only the first of them, and the second one can be analyzed in exactly the same way. Let us therefore study the following generic equation,

$$\mathbf{U}^i = \mathbf{U}_i - \frac{\Delta t}{V_i} \sum_{j\in\omega_i} \tilde{\mathbf{F}}_{\mathbf{n}_{ij}}(\mathbf{U}_{ij}, \mathbf{U}_{ji})S_{ij}, \tag{14}$$

where the superscript denotes the updated state.

As before, we would like to express the centroid value of the state in a cell in terms of the corresponding face-centroid states, i.e. for each i we want to find a set of coefficients ξ_{ij}, $0 < \xi_{ij} < 1$, such that

$$\sum_{j\in\omega_i} \xi_{ij} = 1 \quad \text{and} \quad \sum_{j\in\omega_i} \xi_{ij}\mathbf{U}_{ij} = \mathbf{U}_i. \tag{15}$$

Note, since generally $\phi_i(\nabla\mathbf{U})_i \neq \mathbf{0}$, the choice of these coefficients is not arbitrary, and Equation 11 shows that we must require

$$\sum_{j\in\omega_i} \xi_{ij}(\mathbf{r}_{ij} - \mathbf{r}_i) = 0 \quad \text{or} \quad \mathbf{r}_i = \sum_{j\in\omega_i} \xi_{ij}\mathbf{r}_{ij}. \tag{16}$$

Thus coefficients ξ_{ij} are related to the geometry of the cells. Unless a cell is a simplex the above equations admit multiple solutions. However, since Equations 16 relate the centroid of the cell to the centroids of its faces, a very natural solution can be found. By dividing a two-dimensional cell into triangles with a common vertex at the centroid, or a three-dimensional cell into pyramids, we can show that all the requirements are satisfied by

$$\xi_{ij} = \delta_{dim} \frac{(\mathbf{r}_{ij} - \mathbf{r}_i) \cdot \mathbf{n}_{ij} S_{ij}}{V_i}, \quad \text{where} \quad \delta_{dim} = 1/\text{dimension}. \quad (17)$$

This formula shows why we require the convexity of the cells. If a cell is not convex, its centroid may not belong to the cell. Then some of its coefficients ξ_{ij} will be negative. But second-order accuracy will be questionable in that cell anyway, and by setting the limiter to zero we can return to the first-order case for which the shape of the cell is irrelevant.

Let us now choose some $\alpha_{ij} > 0$ and β_{ij} such that $\alpha_{ij} + \beta_{ij} = 1$ (note, we do not require $\beta_{ij} > 0$). The actual values of these coefficients will be determined later. We need this additional splitting because now in general $\sum_{j\in\omega_i} \mathbf{F}_{\mathbf{n}_{ij}}(\mathbf{U}_{ij})S_{ij} \neq 0$. Then Equation 14 becomes

$$\mathbf{U}^i = \sum_{j\in\omega_i} \xi_{ij}(\alpha_{ij} + \beta_{ij})\mathbf{U}_{ij} - \frac{\Delta t}{V_i} \sum_{j\in\omega_i} \tilde{\mathbf{F}}_{\mathbf{n}_{ij}}(\mathbf{U}_{ij}, \mathbf{U}_{ji})S_{ij}$$

$$+ \frac{\Delta t}{V_i} \sum_{j\in\omega_i} \mathbf{F}_{\mathbf{n}_{ij}}(\mathbf{U}_{ij})S_{ij} - \frac{\Delta t}{V_i} \sum_{j\in\omega_i} \mathbf{F}_{\mathbf{n}_{ij}}(\mathbf{U}_{ij})S_{ij}$$

$$= \sum_{j\in\omega_i} \xi_{ij}\alpha_{ij} \left(\mathbf{U}_{ij} - \frac{\Delta t S_{ij}}{\xi_{ij}\alpha_{ij}V_i}\left[\tilde{\mathbf{F}}_{\mathbf{n}_{ij}}(\mathbf{U}_{ij}, \mathbf{U}_{ji}) - \mathbf{F}_{\mathbf{n}_{ij}}(\mathbf{U}_{ij})\right]\right)$$

$$+ \sum_{j\in\omega_i} \xi_{ij} \left(\beta_{ij}\mathbf{U}_{ij} - \frac{\Delta t S_{ij}}{\xi_{ij}V_i}\mathbf{F}_{\mathbf{n}_{ij}}(\mathbf{U}_{ij})\right)$$

$$= \sum_{j\in\omega_i} \xi_{ij}\alpha_{ij}T_{ij}^{-1} \left(\mathbf{W}_{ij_{ij}} - \frac{\Delta t S_{ij}}{\xi_{ij}\alpha_{ij}V_i}\left[\tilde{\mathbf{F}}_x(\mathbf{W}_{ij_{ij}}, \mathbf{W}_{ji_{ij}}) - \mathbf{F}_x(\mathbf{W}_{ij_{ij}})\right]\right)$$

$$+ \sum_{j\in\omega_i} \xi_{ij}T_{ij}^{-1} \left(\beta_{ij}\mathbf{W}_{ij_{ij}} - \frac{\Delta t S_{ij}}{\xi_{ij}V_i}\mathbf{F}_x(\mathbf{W}_{ij_{ij}})\right),$$

where as previously we introduced $\mathbf{W}_{ij_{ij}} = T_{ij}\mathbf{U}_{ij}$ and $\mathbf{W}_{ji_{ij}} = T_{ij}\mathbf{U}_{ji}$. If all the terms in the parentheses in the last equation belong to the set of physically admissible states, the scheme will be positively conservative.

The terms in the first sum above describe equivalent one-dimensional problems for $\mathbf{W}_{ij_{ij}}$. Thus as before we conclude that they will belong to G if all of the initial states $\mathbf{W}_{ij_{ij}}$ belong to G and

$$\Delta t \max(|u_{\mathbf{n}_{ij}ij}| + a_{ij}, |u_{\mathbf{n}_{ij}ji}| + a_{ji})S_{ij} \leq \xi_{ij}\alpha_{ij}V_i. \quad (18)$$

The terms in the second sum can be computed explicitly, and it is straightforward (see Appendix A) to show that they will be in G as long as

$$\Delta t \left(u_{\mathbf{n}_{ij}ij} + \sqrt{\frac{\gamma - 1}{2\gamma}}a_{ij}\right)S_{ij} \leq \xi_{ij}\beta_{ij}V_i. \quad (19)$$

Now we need to maximize the time step. Since α_{ij} are the only adjustible parameters, it is clear that for each cell all the inequalities in (18) and (19) can not fail simultaneously. But they can fail together per face. Then for each $j \in \omega_i$ we can add (18) and (19), and the time step becomes

$$\Delta t \leq \min_j \frac{\xi_{ij} V_i}{\left(\max(|u_{\mathbf{n}_{ij}ij}| + a_{ij}, |u_{\mathbf{n}_{ij}ji}| + a_{ji}) + u_{\mathbf{n}_{ij}ij} + \sqrt{\frac{\gamma-1}{2\gamma}a_{ij}}\right) S_{ij}}. \tag{20}$$

Finally, if we make use of Equation 17, we can rewrite the above equation in the following form,

$$\Delta t \leq \min_j \frac{\delta_{dim}(\mathbf{r}_{ij} - \mathbf{r}_i) \cdot \mathbf{n_{ij}}}{\max(|u_{\mathbf{n}_{ij}ij}| + a_{ij}, |u_{\mathbf{n}_{ij}ji}| + a_{ji}) + u_{\mathbf{n}_{ij}ij} + \sqrt{\frac{\gamma-1}{2\gamma}a_{ij}}}. \tag{21}$$

Equation 21 must be applied to both stages of the time-stepping scheme, and the smallest time step will determine the overall time step. If the second stage happens to be more restrictive then the first one, the first stage may be recomputed using this smaller time step, and the whole iteration can be repeated. This proves that *given a first-order one-dimensional positively conservative scheme and a computational grid consisting of convex cells we can always build a multidimensional formally second-order accurate scheme provided that the set of physically admissible states is convex.*

The time step given by (21) is not overly restraining. In fact, since $u + \sqrt{\frac{\gamma-1}{2\gamma}}a < |u| + a$, we can see that on a fairly regular grid the time step for a second-order scheme will be roughly half the time step for a first-order scheme under the same flow conditions. Note also that this time step is increased if $(\mathbf{r}_{ij} - \mathbf{r}_i)$ is nearly parallel to $\mathbf{n}_{ij}$; this defines a sense in which the grid should be as regular as possible.

4.2. LIMITING

The main purpose of limiting is to eliminate spurious oscillations; it must also ensure the positivity of density and pressure everywhere in the cells[2]. By the assumptions the cells are convex, hence any point inside a cell is a convex linear combination of its vertices. Then the state at that point is the same linear combination of the vertex states. Thus the positivity of density and pressure only at the cell vertices will suffice.

We will limit the primitive variables using an approach similar to the one suggested in (Barth, 1990). The idea is to find for each cell i the largest possible value of the limiter (up to some target value) that will keep the

---

[2]This is actually too restrictive. It is usually sufficient to preserve positivity only at the face centroids. We will, however, present strict positivity conditions here.

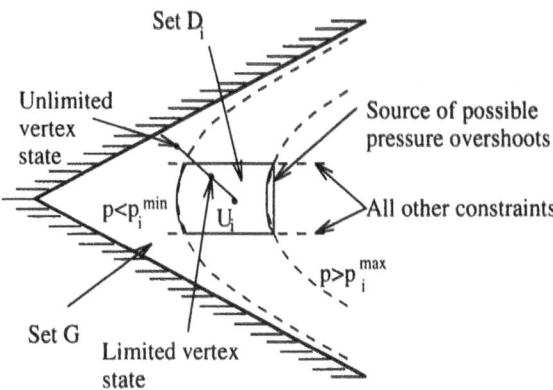

Figure 2. Schematic representation of limiting in the state space.

primitive variables at the vertices within a certain desired range. This range is determined by the maximum and minimum neighboring primitive centroid values, including the centroid of the cell itself. Let us denote them by ρ_i^{max}, ρ_i^{min}, u_i^{max}, u_i^{min}, etc. Then in the space of the conservative variables the desired range defines the following subset of G,

$$
D_i = \left\{ \mathbf{U} \in G \;\middle|\;
\begin{array}{ccccc}
\rho_i^{min} & < & \rho & < & \rho_i^{max} \\
u_i^{min} & < & u & < & u_i^{max} \\
v_i^{min} & < & v & < & v_i^{max} \\
w_i^{min} & < & w & < & w_i^{max} \\
p_i^{min} & < & p & < & p_i^{max}
\end{array}
\right\}. \tag{22}
$$

We will choose the limiter such that the vertex states, U_i^{ver}, belong to the convex hull of this set and if possible to the set itself.

Set D_i is not convex, but its non-convexity comes only from the last inequality in (22). In fact, while equations like $\rho = \rho_*$ or $u = u_*$, where an asterisk means some constant value, define hyperplanes in the state space, the equation $p = p_*$ defines a more complicated hypersurface in that space. Note, however, that the set of states whose pressure exceeds some positive value p_*, $G_{p_*} = \{\mathbf{U} \in G | p > p_*\}$, is convex. Hence only the inequality $p < p_i^{max}$ in (22) leads to the non-convexity of set D_i. This means that the convex hull of set D_i consists of all those states whose density and the velocity components are in the desired range, but whose pressure may exceed p_i^{max}. This result is illustrated in Figure 2. The fact that pressure in cell i may overshoot p_i^{max} is not perfect from the point of view of monotonicity[3], but it is quite favorable from the point of view of positivity.

---

[3] Our experience, however, shows that in practice possible pressure overshoots do not appear in the numerical solutions.

It is not hard to limit density and the velocity components since density is a linear function and the velocity components are monotone functions of the limiter. For each cell we set $\phi_i = 1$, then for each vertex of the cell we compute the *limited* vertex states $\mathbf{U}_i^{ver}(\phi_i)$ always using the *current* value of the limiter, and use these values to repeatedly update the limiter,

$$\phi_i \Leftarrow \begin{cases} \phi_i \dfrac{\rho_i^{max} - \rho_i}{\rho_i^{ver}(\phi_i) - \rho_i} & \text{if } \rho_i^{ver}(\phi_i) > \rho_{max} \\[2ex] \phi_i \dfrac{\rho_i^{min} - \rho_i}{\rho_i^{ver}(\phi_i) - \rho_i} & \text{if } \rho_i^{ver}(\phi_i) < \rho_{min} \end{cases} \tag{23}$$

then (remember, the updated value of ϕ_i should already be used here)

$$\phi_i \Leftarrow \begin{cases} \phi_i \dfrac{\rho_i(u_i^{max} - u_i)}{\rho_i(u_i^{max} - u_i) + \rho_i^{ver}(\phi_i)(u_i^{ver}(\phi_i) - u_i^{max})} & \text{if } u_i^{ver}(\phi_i) > u_{max} \\[2ex] \phi_i \dfrac{\rho_i(u_i^{min} - u_i)}{\rho_i(u_i^{min} - u_i) + \rho_i^{ver}(\phi_i)(u_i^{ver}(\phi_i) - u_i^{min})} & \text{if } u_i^{ver}(\phi_i) < u_{min}. \end{cases} \tag{24}$$

The other velocity components contribute to the limiter computation in exactly the same way.

Not surprisingly pressure is the hardest variable to limit. We were not able to find elegant explicit criteria to limit pressure, therefore in this work we used a simple iterative procedure to do it. If $p_i^{ver} \notin [p_i^{min}, p_i^{max}]$ the procedure finds a value of the limiter that brings pressure into the desired range. Note, however, that p_i^{ver} is not a monotone function of the limiter. This means that further decreasing of the limiter may take pressure at the vertices out of $[p_i^{min}, p_i^{max}]$. In that case one could repeat the procedure to check whether some extra limiting is required. However, we do not believe that that would be necessary because this situation is not very likely to happen in a real computation. Moreover, even if it does happen, pressure, as we demonstrated above, will never become smaller then p_i^{min}, therefore it will never become negative.

5. Results

To demonstrate the robustness of the proposed method we apply it to test problems for which Roe's scheme would fail after the very first time step. In all our computations we set $\gamma = 1.4$, and we use the HLLE flux function, though we could use any other positively conservative flux function.

We first study a symmetric $M = 5$ double rarefaction wave problem. The initial distribution for this problem is $\rho_L = \rho_R = 7$, $u_L = -1$, $u_R = 1$, $p_L = p_R = 0.2$, and the y and z velocity components are set to zero. Note that the resulting problem is more challenging than the one suggested in Einfeldt *et al.* (1992) since for $\gamma = 1.4$ this initial distribution leads to vacuum at $x = 0$. Figure 3 shows the resulting first- and second-order accurate distributions of density and pressure at $t = 0.6$. The computational grid

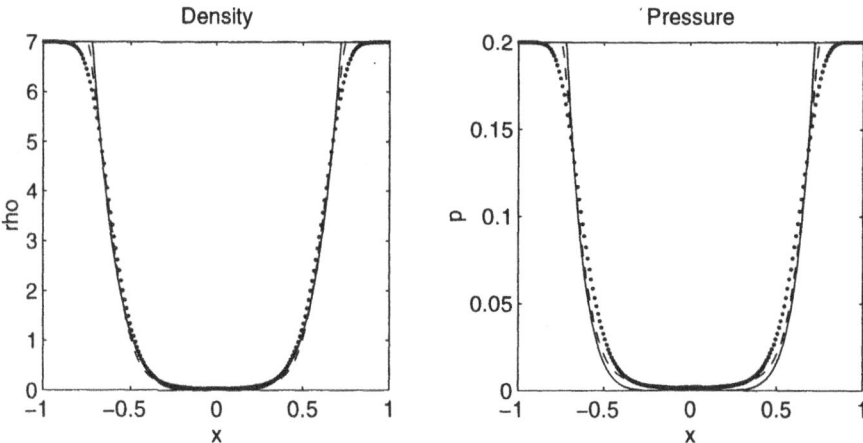

Figure 3.   Double rarefaction problem. Solutions are — exact, $\cdots$ first-order and - - - second-order.

has 200 points. Although both schemes can run at the maximum allowable time step, we find that accuracy is improved by setting the time step to half its maximum value.

We can see that both schemes produce a good approximation to the exact density distribution, but they noticeably overestimate the value of pressure in the middle of the expansion fan. This illustrates our comment that an admissible solution is not necessarily an accurate one.

We also apply our scheme to a corner flow problem shown on Figure 4. The initial conditions are $\rho = 7$, $u = 1$, $v = 0$, $p = 0.8$ everywhere, which corresponds to a sudden $M = 2.5$ motion of a gas relatively to a 90° degree corner. Although these are not realistic initial conditions this is a good test since very often numerical computations start with a supersonic flow everywhere in a computational domain that contains sharp corners. Moreover, it can be easily verified that for $\gamma = 1.4$ the maximum Mach number in a Prandtl-Meyer fan that can turn around a right angled corner without cavitating is about 2.56. Hence after a long enough time, when the flow becomes self-similar in the vicinity of the corner, we expect to observe a very strong rarefaction fan emanating from the corner[4].

The results from the first- and the second-order schemes obtained on a 400x400 grid are presented on Figure 4. The maximum allowable time steps were used in both simulations, and the figure shows the density contours at $t = 0.6$. It is clear that the initial conditions result in a sudden expansion at the vertical wall. The initial Mach number, however, is not high enough

---

[4]Note that in the common 90° corner shock diffraction problem the maximum Mach number that can be used for the downstream post-shock state is only about 1.89. Hence one can not obtain an extreme rarefaction at the corner in that problem.

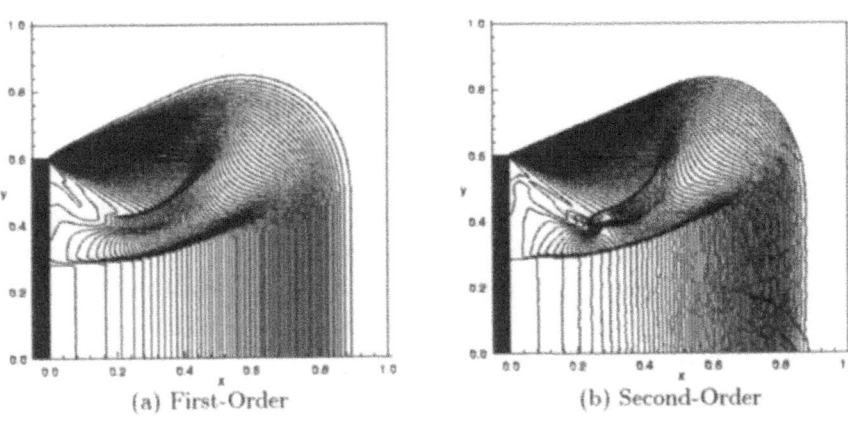

(a) First-Order   (b) Second-Order

Figure 4. Numerical solutions to the corner flow problem. Density contours are shown.

to produce vacuum at that wall. We can also observe that an expansion fan forms at the corner. It is seen that the second-order calculation provides a very good resolution of the shocks. Also, the feature between the shocks is a contact discontinuity: it is badly resolved in the first-order calculation because of the dissipative nature of the HLLE flux, but the second-order reconstruction proposed here is able to define it very well in the second-order code.

At the corner, the values of both density and pressure drop by three orders of magnitude, and the maximum Mach number obtained there is about 10. This value is lower than the Mach number is supposed to be in the corner region, which is probably the result of inaccuracies associated with the computation of pressure. An elaborate corner treatment, for example, as suggested by Woodward and Colella (1984) may improve the situation, but this is an issue of numerical resolution which we will not discuss here.

6. Conclusions

We presented in this paper an extension of first-order one-dimensional positively conservative schemes to more dimensions and higher order. We showed that the positive conservation property can be guaranteed for a fairly general class of schemes provided that certain conditions are met. Among them are the convexity of the set of physically admissible states and the availability of a first-order one-dimensional positively conservative scheme. We also demonstrated that the resulting restrictions on the time step are compatible to the linear stability requirements. In this sense positive conservation is analogous to nonlinear stability.

At the same time we understand that our method has potential for refinement. We derived a set of sufficient conditions that ensure positive conservation for a scheme, and in practice many of them can be relaxed or better estimates can be found. Also, the issue of reconstruction and limiting needs a more profound analysis. Nevertheless, it is pleasing that a robust scheme can be built at a relatively inexpensive price. Of course, some accuracy is lost due to extra safety considerations, but in crash-prone computations one needs to compromise between accuracy and robustness.

Finally, we would like to point out that the technique described in this paper is very general and can be applied to other systems of equations. For example, in magnetohydrodynamics (MHD) there is frequently a need to compute flows of highly rarefied plasmas. It is straightforward to show that the set of physically admissible states in MHD is convex, and an exact Riemann solver would be positively conservative. However, we do not yet have available an economical one-dimensional solver with this property. It would be very useful to design one.

7. Acknowledgments

This work was supported by a Doctoral Fellowship from the François-Xavier Bagnoud Foundation.

A. Appendix

Suppose that $\mathbf{U} \in G$. If we let $\mu = \frac{\xi \beta V}{\Delta t S}$, then we need to find when

$$\mu \mathbf{U} - \mathbf{F}_x(\mathbf{U}) \in G.$$

Since $\rho > 0$, the first of Conditions 3 requires that

$$\mu - u > 0,$$

and the second requires that

$$2(\mu \rho - \rho u)(\mu E - u(E+p)) > (\mu \rho u - \rho u^2 - p)^2 + (\mu \rho v - \rho uv)^2 + (\mu \rho w - \rho uw)^2.$$

It is easy to show that the above inequality reduces to

$$(\mu - u)^2 - \frac{\gamma - 1}{2\gamma} a^2 > 0.$$

Since $\mu - u > 0$, this gives us

$$u + \sqrt{\frac{\gamma - 1}{2\gamma}} a < \mu,$$

or

$$\Delta t \left(u + \sqrt{\frac{\gamma - 1}{2\gamma}} a \right) S < \xi \beta V.$$

References

Barth, T.J., "On unstructured grids and solvers," In *Computational Fluid Dynamics*, Lecture Series 1990-03, VKI, March 1990.

Charrier, P., Dubroca, B., and Flandrin, L., "Un solveur de Riemann approché pour l'"etude d'"ecoulements ypersoniques bidimensionnels, *C.R. Acad. Sci. Paris*, Vol. 317, 1993, pp. 1083-1086.

Donat, R. and Marquina, A., "Capturing shock reflections: An improved flux formula," *J. Comput. Phys.*, Vol. 125, No. 1, April 1996, pp. 42-58.

Einfeldt, B., Munz, C.D., Roe, P.L., and Sj'"ogreen, B., "On Godunov-type methods near low densities," *J. Comput. Phys.*, Vol. 92, 1992, pp. 273-295.

Einfeldt, B., "On Godunov-type methods for gas dynamics," *SIAM J. Numer. Anal.*, Vol. 25, No. 2, April 1988, pp. 294-318.

Flandrin, L., "Méthodes 'cell-centered' pour l'approximation des équations d'Euler et de Navier-Stokes sur des maillages non structurés," Ph.D. thesis, l'Universite Bordeaux I, December 1995.

Gear, C.W., "Numerical Initial Value Problems in Ordinary Differential Equations," Prentice-Hall, Englewood Cliffs, NJ, 1971.

Godunov, S.K., "A difference scheme for numerical computation of discontinuous solutions of hydrodynamic equations," *Mat. Sb.*, Vol. 47, No. 3, 1959, pp. 271-306 (in Russian).

Liou, M.-S., "Further progress in numerical flux scheme," In *15th International Conference on Numerical Methods in Fluid Dynamics*, Monterey, CA, June 1996.

Osher, S. and Solomon, F., "Upwind difference schemes for hyperbolic systems of conservation laws," *Math. Comput.*, Vol. 38, No. 158, April 1982, pp. 339-374.

Quirk, J.J., "A contribution to the Great Riemann solver debate, *Int. J. Numer. Methods Fluids*, Vol. 18, No. 6, 1994, pp. 555-574. Also ICASE Report No. 92-64.

Roe, P.L., "Approximate Riemann solvers, parameter vectors, and difference schemes," *J. Comput. Phys.*, Vol. 43, 1981, pp. 357-372.

Shu, C.-W. and Osher, S., "Efficient implementation of essentially non-oscillatory shock-capturing schemes," *J. Comput. Phys.*, Vol. 77, No. 2, August 1988, pp. 439-471.

van Leer, B., "Towards the ultimate conservative difference scheme. V. A second-order sequel to Godunov's method," *J. Comput. Phys.*, Vol. 32, 1979, pp. 101-136.

Woodward, P. and Colella, P., "The numerical simulation of two-dimensional fluid flow with strong shocks," *J. Comput. Phys.*, Vol. 54, 1984, pp. 115-173.

Xu, K., Martinelli, L., and Jameson, A., "Gas-kinetic finite volume methods, flux-vector splitting, and artificial diffusion," *J. Comput. Phys.*, Vol. 120, No. 1, September 1995, pp. 48-65.

A NEW SCHEME FOR THE SOLUTIONS OF MULTIDIMENSIONAL MHD EQUATIONS

NECDET ASLAN

Marmara University
Fen-Ed. Physics Department
81040 Göztepe İstanbul
Turkey

AND

TERRY KAMMASH

University of Michigan
Nuclear Engineering Department
Ann Arbor, MI 48109

Abstract.

In this paper, the solution of generalized system of hyperbolic equations by means of upwind, limited, second order accurate fluxes including a new sonic fix is presented. The new sonic fix introduced here utilizes a dissipation term embedded directly in the fluxes and it is totally based on physical grounds producing the correct decay rate of sonic gradients. In addition to the sonic fix, the effects of the source term on the flux limiters are also introduced. The resulting scheme is applied to a variety of test problems resulting from the solutions of Euler's and magneto-hydrodynamic, MHD equations. To eliminate the divergence problem, a new implementation of a recently introduced scheme for the MHD equations which includes a divergence wave and a source related to $\vec{\nabla} \cdot \vec{B}$ is introduced. The numerical test results obtained with this new scheme are in excellent agreement with the previous ones and they show that the scheme presented here is robust, accurate, and entropy satisfying by producing very sharp contact discontinuities and shocks without post shock oscillations and divergence errors.

V. Venkatakrishnan et al (eds.), Barriers and Challenges in Computational Fluid Dynamics, 315-333.
© 1998 *Kluwer Academic Publishers.*

1. Introduction

The behaviour of the hyperbolic equations near sonic points is very critical. While the algorithms including a reconstruction stage of primitive quantities are less affected from the sonic points, the algorithms using flux limiters may lead to unphysical expansion shocks if the sonic points are not handled correctly. In this case the algorithm may not converge to the correct entropy satisfying solution. Treatments of sonic points had been developed so far by many investigators (see (Harten, 1983), (Sweby, 1984), (van Leer *et al.*, 1983), (LeVeque, 1988)). The most striking sonic fix was introduced lately by Roe (1992) for the solutions of Euler's equations where both of the states around the sonic interfaces are modified to obtain correct decay rate of sonic gradients. This idea was applied to the MHD equations by Aslan (1993) and very satisfactory results were obtained with the requirement of no structure coefficents addressed in (Zachary & Colella, 1992). The new sonic fix introduced here follows a similar idea given in (Roe, 1992) but it differs in the sense that it is embedded directly in fluxes and that it requires no additional modification of the left and right states around the sonic interface as done in (Roe, 1992). As in (LeVeque, 1988) and (Roe, 1992), it relies on the physical grounds and it leads to the correct amount of sonic flux transfer between the related states producing a correct decay rate of sonic gradients.

Sec.2 describes the basic equations for resistive MHD equations. These equations are written in the generalized curvilinear coordinates in a form suitible for the splitting schemes (see (Aslan, 1993) and (Glaister, 1987)). In Sec.3, the finite volume method for two dimensions is presented. This section also explains the new sonic fix that is used in the fluxes. How the 2D algorithm is implemented is also described in this section. Sec.4 gives one and two dimensional results obtained with the secheme described in Sec.3. The results show excellent performance of the new sonic fix and the new divergence source along with the divergence wave. Finally, the conclusion is given in Sec.5.

2. Basic Equations

The resistive MHD is the simplest fluid model that can describe the macroscopic behaviour of the plasma interacting with external and internal fields

$$\frac{\partial \rho}{\partial t} + \vec{\nabla} \cdot (\rho \vec{V}) = 0 \tag{1}$$

$$\rho(\frac{\partial \vec{V}}{\partial t} + \vec{V} \cdot \vec{\nabla}\vec{V}) = -\vec{\nabla}P + \frac{1}{c}\vec{J} \times \vec{B} \tag{2}$$

$$\frac{\partial \rho \epsilon}{\partial t} + \vec{\nabla} \cdot (\rho \epsilon \vec{V}) + P \vec{\nabla} \cdot \vec{V} - \eta \vec{J} \cdot \vec{J} = 0 \tag{3}$$

$$\vec{\nabla} \times \vec{B} = \frac{4\pi}{c} \vec{J} \tag{4}$$

$$\vec{\nabla} \times \vec{E} = -\frac{1}{c} \frac{\partial \vec{B}}{\partial t} \tag{5}$$

$$\eta \vec{J} = \vec{E} + \frac{1}{c} \vec{V} \times \vec{B} \tag{6}$$

$$\vec{\nabla} \cdot \vec{B} = 0 \tag{7}$$

where ρ is the density, $\vec{V}$ velocity, $P = (\gamma - 1)\rho \epsilon$ pressure, $\vec{J}$ current density, ϵ internal energy, $\vec{B}$ magnetic field, $\vec{E}$ electric field, and η is the resistivity and $\gamma = c_p / c_v$ is the ratio of specific heats. Notice the absence of $q\vec{E}$ term in the Lorentz force (second term on the right of (2)). This term is dropped by the assumption of local charge neutrality ($q = 0$). The last term in Eq.(3) is due to the ohmic heating (i.e., the heating of resistive fluid due to the current flowing in it).

To solve these equations numerically, one can use the following normalization:

$$\rho' = \frac{\rho}{\rho_0}, \quad \vec{V}' = \frac{\vec{V}}{V_0}, \quad \vec{B}' = \frac{\vec{B}}{B_0}, \quad P' = \frac{P}{P_0}, \quad \epsilon' = \frac{\epsilon}{\epsilon_0}, \quad \vec{x}' = \frac{\vec{x}}{x_0}, \quad t' = \frac{t}{t_0} \tag{8}$$

where primed quantities denote normalized variables and subscript "0" refers to the characteristic values which satisfy the following relations

$$V_0 = \frac{x_0}{t_0}, \quad P_0 = B_0^2 = (\gamma - 1)\rho_0 \epsilon_0, \quad J_0 = \frac{E_0}{\eta_0} = \frac{c}{4\pi} \frac{B_0}{x_0}, \quad S = \frac{x_0 V_0}{c^2 \eta_0} = \frac{\tau_R}{\tau_A} \tag{9}$$

where S is the magnetic Reynolds number which is a measure of the ratio of the resistive diffusion time scale τ_R to the Alfven transit time scale τ_A. Normalizing the MHD equations (Eqs: (1- 7)) and rearranging via the vectoral identities one gets their conservative forms

$$\frac{\partial \rho}{\partial t} + \vec{\nabla} \cdot [\rho \vec{V}] = 0 \tag{10}$$

$$\frac{\partial \rho \vec{V}}{\partial t} + \vec{\nabla} \cdot [\rho \vec{V} \vec{V} - \frac{\vec{B}\vec{B}}{4\pi} + \bar{\bar{I}}(P + \frac{B^2}{8\pi})] = -\frac{\vec{B}}{4\pi} \vec{\nabla} \cdot \vec{B} \tag{11}$$

$$\frac{\partial E}{\partial t} + \vec{\nabla} \cdot [(E + P^*)\vec{V} - \frac{\vec{B}}{4\pi}\vec{B} \cdot \vec{V} + \frac{\eta}{S} \frac{(\vec{\nabla} \times \vec{B}) \times \vec{B}}{(4\pi)^2}] = -\frac{\vec{B} \cdot \vec{V}}{4\pi} \vec{\nabla} \cdot \vec{B} \tag{12}$$

$$\frac{\partial \vec{B}}{\partial t} + \vec{\nabla} \cdot [\vec{V}\vec{B} - \vec{B}\vec{V} + \frac{\eta}{S}(\vec{\nabla}\vec{B}^T - \vec{\nabla}\vec{B})] = -\vec{V}\vec{\nabla} \cdot \vec{B} \qquad (13)$$

where primes are dropped. Here $P^* = P + \frac{B^2}{8\pi}$ and $E = \frac{1}{2}\rho V^2 + \frac{P}{\gamma-1} + \frac{B^2}{8\pi}$ are called as the total pressure and the total energy respectively. Notice the appearence of the source terms related to $\vec{\nabla} \cdot \vec{B}$. These source terms are naturally required in order the MHD equations given by Eqs.(1)-(7) to be identical to the Eqs.(10-13). The existence of the source terms may seem to be in conflict with Eq.(7); however; (remaining very small in magnitude) they are required for only numerical purposes. The existence of the source vector brings about a divergence wave which appears only in multi-dimensional problems. This wave succesfully convects out the local increase of $\vec{\nabla} \cdot \vec{B}$ and the divergence source acts as a stabilizing factor. This issue was first addressed by Aslan and Roe (1993) and first implemented by Powell (1994) and then by Powell and Gombasi (1994) and recently by Aslan (1996).

When MHD equations are used to investigate the laboratory plasmas, usually curvilinear coordinates are required (i.e., toroidal, spherical etc.). To write the MHD equations in generalized curvilinear coordinates, (x_1, x_2, x_3), the following definitions can be used:

$$ds^2 = h_1^2 dx_1^2 + h_2^2 dx_2^2 + h_3^2 dx_3^2 \ , \quad dv = h_1 h_2 h_3 dx_1 dx_2 dx_3 \ , \quad g = (h_1 h_2 h_3)^2 \qquad (14)$$

where ds is the area element, dv is the volume element, g is the metric, and h_i is the scale factor along the coordinate axis x_i. In what follows, the scale factors and coordinate directions are given for several curvilinear geometries often encountered in MHD simulations:

Cartesian Geometry:

$$h_1 = h_2 = h_3 = 1 \ ; \quad x_1 = x, \quad x_2 = y, \quad x_3 = z \qquad (15)$$

Cylindrical Geometry:

$$h_1 = 1, h_2 = r, h_3 = 1 \ ; \quad x_1 = r, \quad x_2 = \theta, \quad x_3 = z \qquad (16)$$

Spherical Geometry:

$$h_1 = 1, h_2 = r, h_3 = rsin(\theta) \ ; \quad x_1 = r, \quad x_2 = \theta, \quad x_3 = \phi \qquad (17)$$

Toroidal Geometry:

$$h_1 = 1, h_2 = r, h_3 = R + rcos(\theta) \ ; \quad x_1 = r, \quad x_2 = \theta, \quad x_3 = \phi \qquad (18)$$

where R is the major axis radius.

The operators used in curvilinear geometries are given by

$$\vec{\nabla}P = \sum_i \frac{\hat{e}_i}{h_i}\frac{\partial P}{\partial x_i} \tag{19}$$

$$\vec{\nabla}\cdot\vec{F} = \sum_i \frac{1}{\sqrt{g}}\frac{\partial}{\partial x_i}\left(\sqrt{g}\frac{F_i}{h_i}\right) \tag{20}$$

$$\vec{\nabla}\times\vec{B} = \sum_i \frac{h_i\hat{e}_i}{\sqrt{g}}\sum_{l,m}\epsilon_{ilm}\frac{\partial h_m B_m}{\partial x_l} \tag{21}$$

$$(\vec{\nabla}\cdot\overline{\overline{T}})_j = \sum_i \left[\frac{1}{\sqrt{g}}\frac{\partial}{\partial x_i}(\sqrt{g}\frac{T_{ij}}{h_i}) + \sum_{k\neq j}\frac{T_{ik}}{h_i h_k}\left\{\begin{array}{c} j \\ ki \end{array}\right\}\right] \tag{22}$$

where $\hat{e}_i$ is the unit vector,

$$\epsilon_{ijk}\hat{e}_k = \hat{e}_i \times \hat{e}_j = \left\{\begin{array}{cc} +1 & ijk : \text{odd permutation} \\ -1 & ijk : \text{even permutation} \\ 0 & \text{otherwise} \end{array}\right\} \tag{23}$$

is the Levi-Civita symbol, and

$$\left\{\begin{array}{c} i \\ ii \end{array}\right\} = \frac{1}{h_i}\frac{\partial h_i}{\partial x_i}, \quad \left\{\begin{array}{c} i \\ ij \end{array}\right\} = \left\{\begin{array}{c} i \\ ji \end{array}\right\} = \frac{1}{h_i}\frac{\partial h_i}{\partial x_j}, \quad \left\{\begin{array}{c} j \\ ii \end{array}\right\} = -\frac{h_i}{h_j^2}\frac{\partial h_i}{\partial x_j} \tag{24}$$

are the Christoffel symbols which are related to the change in the direction of unit vectors if one moves in space.

Using these definitions the MHD equations can be written as (Aslan, 1993)

$$h_1 h_2 h_3\frac{\partial\vec{U}}{\partial t}+\frac{\partial}{\partial x_1}[h_2 h_3\vec{F}_1]+\frac{\partial}{\partial x_2}[h_1 h_3\vec{F}_2]+\frac{\partial}{\partial x_3}[h_1 h_2\vec{F}_3]+\vec{S}_1+\vec{S}_2+\vec{S}_3 = \vec{S}_{div} \tag{25}$$

where $\vec{U} = [\rho, \rho V_1, \rho V_2, \rho V_3, B_1, B_2, B_3, E]^T$ is the state vector, $\vec{F}_i$ and $\vec{S}_i$ are the flux and source vectors along the coordinate x_i, and $\vec{S}_{div}$ is the source vector given by

$$\vec{F}_1 = \begin{bmatrix} \rho V_1 \\ \rho V_1^2 - B_1^2/4\pi + P^* \\ \rho V_1 V_2 - B_1 B_2/4\pi \\ \rho V_1 V_3 - B_1 B_3/4\pi \\ 0 \\ V_1 B_2 - B_1 V_2 + \frac{\eta}{S}(\frac{1}{h_2}\frac{\partial B_1}{\partial x_2} - \frac{1}{h_1}\frac{\partial B_2}{\partial x_1}) \\ V_1 B_3 - B_1 V_3 + \frac{\eta}{S}(\frac{1}{h_3}\frac{\partial B_1}{\partial x_3} - \frac{1}{h_1}\frac{\partial B_3}{\partial x_1}) \\ (E + P^*)V_1 - \frac{B_1}{4\pi}\vec{V}\cdot\vec{B} + \\ \frac{\eta}{(4\pi)^2 S}[\frac{B_2}{\partial h_2}\frac{\partial B_1}{\partial x_2} + \frac{B_3}{\partial h_3}\frac{\partial B_1}{\partial x_3} - \frac{B_2}{\partial h_1}\frac{\partial B_2}{\partial x_1} - \frac{B_3}{\partial h_1}\frac{\partial B_3}{\partial x_1}] \end{bmatrix}, \quad \vec{S}_{div} = -\begin{bmatrix} 0 \\ B_1/4\pi \\ B_2/4\pi \\ B_3/4\pi \\ V_1 \\ V_2 \\ V_3 \\ \frac{\vec{V}\cdot\vec{B}}{4\pi} \end{bmatrix}\vec{\nabla}\cdot\vec{B} \tag{26}$$

$$
\vec{S}_1 =
\begin{bmatrix}
0 \\
-P^* \frac{\partial}{\partial x_1}(h_2 h_3) - h_3 \frac{\partial h_2}{\partial x_1}(\rho V_2^2 - B_2^2/4\pi) - h_2 \frac{\partial h_3}{\partial x_1}(\rho V_3^2 - B_3^2/4\pi) \\
h_3 \frac{\partial h_2}{\partial x_1}(\rho V_2 V_1 - B_2 B_1/4\pi) \\
h_2 \frac{\partial h_3}{\partial x_1}(\rho V_3 V_1 - B_3 B_1/4\pi) \\
0 \\
h_3 \frac{\partial h_2}{\partial x_1}[V_2 B_1 - B_2 V_1 + \frac{\eta}{S}(\frac{1}{h_1}\frac{\partial B_2}{\partial x_1} - \frac{1}{h_2}\frac{\partial B_1}{\partial x_2})] \\
h_2 \frac{\partial h_3}{\partial x_1}[V_3 B_1 - B_3 V_1 + \frac{\eta}{S}(\frac{1}{h_1}\frac{\partial B_3}{\partial x_1} - \frac{1}{h_3}\frac{\partial B_1}{\partial x_3})] \\
0
\end{bmatrix}
\tag{27}
$$

Note that in order to get $\vec{F}_2$ and $\vec{S}_2$ from $\vec{F}_1$ and $\vec{S}_1$ given above, interchange the indices 1 with 2 and interchange the rows 2 with 3 and 6 with 7. Also notice that excluding the resistive terms, $\vec{S}_i$ is written such that it has derivatives only with respect to x_i.

This form of MHD equations therefore leads to a structure which can be used by any splitting scheme. If the equilibrium is sought as well as the transients one needs to use finite volume type schemes. The following section includes the numerical representation of such a scheme which includes the finite volume method with quadrilateral or triangular cells and with the new sonic fix and source parameters embedded directly into the fluxes.

3. Numerics with New Sonic Fix and Source

When the differential form of the MHD equations is integrated over the finite volume in 2D cartesian geometry, one gets (Aslan, 1995), (Aslan, 1996b)

$$
\int \int_v \int \vec{U}_t \, dA \, dt + \int \int_v \int (\vec{F}_x - \vec{G}_y) \, dA \, dt = \int \int_v \int \vec{S} \, dA \, dt \tag{28}
$$

where $\vec{G}_y$ is equivalent to $\vec{F}_2$ in Eq.(25) and $\vec{S}_i = 0$ since there exists no curvature in the coordinates. Defining the average state and source vectors as $< \vec{U}^n > = 1/A \int_A \int \vec{U}^n \, dA$ and $< \vec{S}_{div}^{n+1/2} > = 1/A \int_A \int \vec{S}_{div}^{n+1/2} \, dA$ Eq.(28) leads to

$$
< \vec{U}_{ij}^{n+1} > = < \vec{U}_{ij}^n > - \frac{\Delta t}{A_{ij}} \sum_{k=1}^{3 \, or \, 4} \vec{\mathcal{F}}_n^k \cdot \Delta s_k + \Delta t < \vec{S}_{ij}^{n+1/2} > \tag{29}
$$

where n is the time level, Δt is the time increment, A_{ij} is the area of the cell (i, j) and $\vec{\mathcal{F}}_n^k$ is the normal flux to its kth face ($k : 1-4$ for quadrilateral cells, $k : 1 - 3$ for triangular cells) whose length is Δs_k. In order to implement the 2D line integral for the fluxes, the cartesian components of vectoral quantities (i.e., velocity and magnetic fields) are rotated to get the normal and tangential components (see Fig.1):

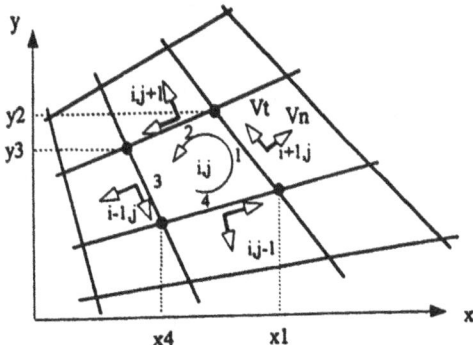

Figure 1.   The mesh structure using the quadrilateral cells

$$\vec{V}_n^k = \vec{V}_x^k cos(\phi_k) + \vec{V}_y^k sin(\phi_k) , \quad \vec{V}_T^k = -\vec{V}_x^k sin(\phi_k) + \vec{V}_y^k cos(\phi_k) \quad (30)$$

$$sin(\phi_k) = \frac{x_k - x_{k+1}}{\Delta s_k} , \quad cos(\phi_k) = \frac{y_{k+1} - y_k}{\Delta s_k} ,$$

$$\Delta s_k = \sqrt{(x_k - x_{k+1})^2 + (y_{k+1} - y_k)^2}. \quad (31)$$

These quantities are then used to obtain the limited, second order, and normal fluxes at each interface from the formula (Aslan & Kammash, 1996)

$$\vec{\mathcal{F}}_n = (\overline{\overline{I}} - \mathcal{K}^*)\frac{\vec{F}_L + \vec{F}_R}{2} - \frac{1}{2}\sum_{k=1}^{8}[(1 - \kappa^*)|\hat{\lambda}_k|\hat{\alpha}_k \vec{r}_k$$

$$- \frac{1}{2}\sum_k \left[\hat{\lambda}_k \left((\nu_k - \sigma_k(1 - \kappa^*))\hat{\alpha}_k - \Delta t \hat{\beta}_k \right) \vec{r}_k \right]$$

$$\Phi(\hat{\theta}_u) \quad (32)$$

where $\Phi(\hat{\theta}_u)$ is the flux limiter and for instance θ_u at interface $i + \frac{1}{2}, j$ is defined as

$$\hat{\theta}_u^k = \frac{(1 - \sigma_k)\hat{b}_{i+\frac{3}{2},j}^k + (1 + \sigma_k)\hat{b}_{i-\frac{1}{2}}^k, j}{2\hat{b}_{i+\frac{1}{2}}^k, j},$$

$$\hat{b}_{i+\frac{1}{2},j}^k = [\frac{\hat{\lambda}_k}{2} \left((\nu_k - \sigma_k(1 - \kappa^*))\hat{\alpha}_k - \Delta t \hat{\beta}_k \right) \vec{r}_k]_{i+\frac{1}{2},j} \quad (33)$$

In Eq.(30), $\vec{F}_L$ and $\vec{F}_R$ are the fluxes computed from the left and right states, $\vec{U}_L$ and $\vec{U}_R$ (where L refers to the lower cell index). The hats refer to Roe averaged interface quantities (see (Aslan, 1995) and (Aslan, 1996b)). Here, $\hat{\lambda}$, $\hat{r}$, and $\hat{\alpha}$ are the eigenvalues, eigenvectors, and wave strengths of 8 possible waves in the MHD (fast±, Alfven±, Slow±, Entropy, Divergence) obtained from the Jacobian, A, of $\vec{F}(\vec{U})$. σ_k and ν_k are defined as the sign of $\hat{\lambda}_k$ and the local Courant number. Also $\hat{\beta}$ is the strength of the source obtained by projecting the source onto the right eigenvectors. The wave and source strengths satisfy

$$\vec{U}_R - \vec{U}_L = \sum_k \hat{\alpha}_k \vec{r}_k \ , \quad \vec{F}_R - \vec{F}_L = \sum_k \hat{\lambda}_k \hat{\alpha}_k \vec{r}_k \ , \quad \hat{\vec{S}} = \sum_k \hat{\beta}_k \vec{r}_k \quad (34)$$

where the first two equations are defined as the local R-H conditions. The terms $\mathcal{K}^*$ and κ^* are defined as sonic fix matrix and parameter respectively. Having been recently introduced by Aslan and Kammash (1996), the origin of this fix can be explained shortly as follows:

Assume that the system of hyperbolic equations in one dimension

$$U_t + F_x = U_t + AU_x = 0 \quad (35)$$

is expanded into second order Taylor's series by

$$U^{n+1} = U^n + \Delta t U_t + \frac{\Delta t^2}{2} U_{tt} \quad (36)$$

Using (35) one can obtain

$$\vec{S}_t = -\vec{F}_x \ , \quad \vec{U}_{tt} = (-\vec{F}_x)_t = A_x \vec{F}_x + A\vec{F}_{xx} = (A\vec{F}_x)_x \quad (37)$$

where $A = \partial \vec{F}/\partial \vec{U}$ is the Jacobian of $\vec{F}$. When examined, the above equations show that the only second order contribution comes from the term $A_x \vec{F}_x$ at sonic points, $A \to 0$. Inserting an additional $A_x \vec{F}_x$ into the second order part of (36) will then create extra dissipation at sonic points. Thus, taking $\vec{U}_{tt} = (A\vec{F}_x)_x + A_x^* \vec{F}_x$, Eq.(36) will lead to

$$\vec{U}^{n+1} = \vec{U}^n + \Delta t(\mathcal{K}^* - I)\vec{F}_x + \frac{\Delta t^2}{2}(A\vec{F}_x)_x \quad (38)$$

where the term $\mathcal{K}^* \vec{F}_x$ should be used only at sonic points (see (Aslan & Kammash, 1996) for details). This leads to a scheme in which the sonic fix is embedded directly into the fluxes so that no extra modification of the states around the sonic interface is required at sonic points as done in (Roe, 1992). Assume that $i + \frac{1}{2}, j$ is the sonic interface (i.e., $A \to 0$, $A_R > 0$, $A_L < 0$). A careful examination shows that the term $\frac{\Delta t}{\Delta x}[\mathcal{K}^* \vec{F}(\vec{U})]_{i+1/2}$ will be transferred from $\vec{U}_{i+1}$ to $\vec{U}_i$ across this interface. The fact that this transfer succesfully eliminates the errors at the sonic interfaces has been shown by Aslan and Kammash (1996) on a variety of test problems. To give an example, consider the following transfer term across the sonic interface (in MHD) from the right momentum to the left one

$$(\frac{\Delta t}{\Delta x})^2 \frac{3 - \gamma}{2}(\hat{P} + \frac{\hat{B}_\perp^2}{8\pi} - \frac{\hat{B}_x^2}{8\pi})\Delta V_x + \text{small terms} \tag{39}$$

This term is related to the total perpendicular pressure and the change in V_x (which has a local maximum at unphysical expansion shocks). As will be shown later, while fixing the sonic point by removing the expansion shocks, the effect of this term decays in time.

In this paper, Eq.(29) is solved in one and two dimensional cartesian geometry. The divergence source vanishes in 1D and the numerical implementation of 1D equations is trivial. The 2D algorithm is implemented as follows:

Because the source term at $t^{n+1/2}$ is unknown, (29) is split into two steps. First the update $\vec{U}^*$ is found from

$$< \vec{U}_{i,j}^* >=< \vec{U}_{i,j}^n > - \frac{\Delta t}{A_{i,j}} \sum_{k=1}^{4} \vec{\mathcal{F}}_n^k \cdot \Delta s_k + \Delta t < \vec{S}_{i,j}^n > \tag{40}$$

and from this update a new source term $(\vec{S}^*)^{n+1}$ is calculated and it is used to correct the updated state

$$< \vec{U}_{i,j}^{n+1} >=< \vec{U}_{i,j}^* > + \frac{\Delta t}{2}[(\vec{S}_{i,j}^*)^{n+1} - < \vec{S}_{i,j}^n >] . \tag{41}$$

The numerical tests showed that most of the time the correction step (41) has no significant effect on the divergence source.

4. Test Problems

The performance of the new sonic fix is checked with Roe's 1D sonic test problem (Roe, 1992) for the Euler's equations. In this problem, the solution domain is separated into left and right regions and the initial conditions: $B = 0$, $V = 0$, $\rho_L = P_L = 100$ and $\rho_R = P_R = 1$ with $\gamma = 1.4$ are assumed. Fig.2. shows the result for density without and with the new sonic fix on a mesh of 100 points; where 80 timesteps have been taken with $\frac{\Delta t}{\Delta x} = 0.2$. Without the sonic fix (Fig.2a), the scheme leads to a noticeable expansion shock while as Fig.2b shows, the new sonic fix works well and eliminates the expansion shock.

The next test problem is a high Mach number problem introduced for MHD by Brio and Wu (1988). The initial conditions are given by $W_L = [1, 0, 0, 0, 0, \sqrt{4\pi}, 0, 1000]$ and $W_R = [0.125, 0, 0, 0, 0, -\sqrt{4\pi}, 0, 0.1]$ with $\gamma =$

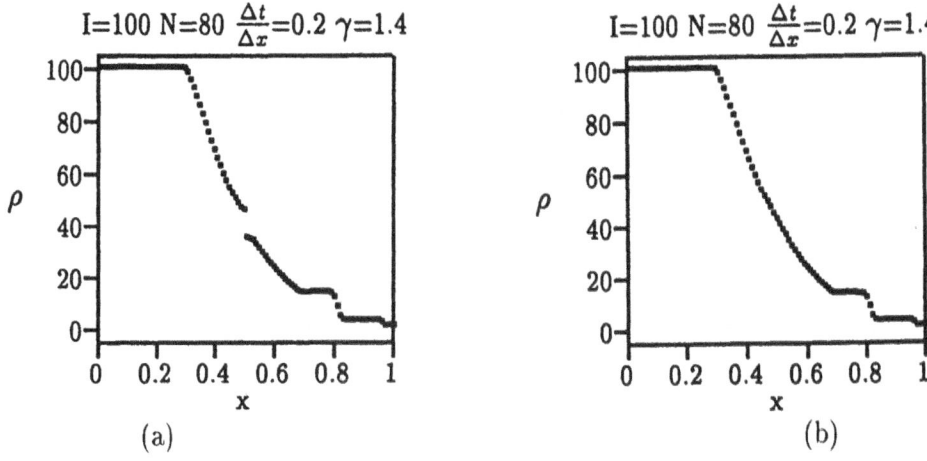

Figure 2.   The density plots for Roe's strong sonic problem.

1.4, where W is the primitive state defined as $W = [\rho, V_x, V_y, V_z, B_x, B_y, B_z, P]$. A grid of 800 points is taken and the results for the density without and with the new sonic fix are displayed in Fig.3 at t=0.0063. Again, the

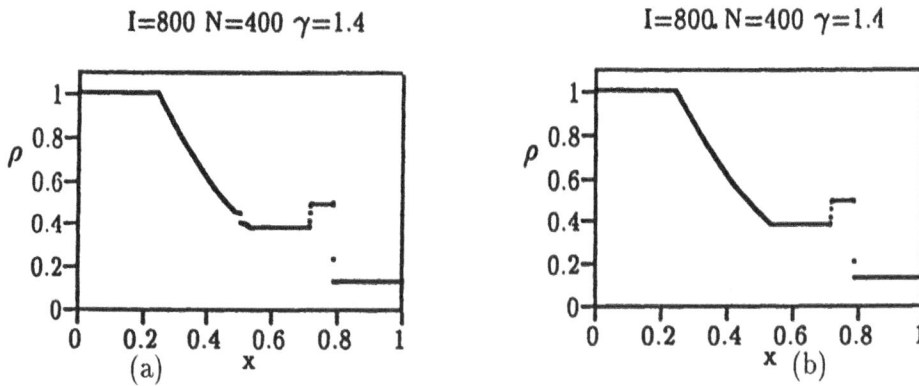

Figure 3.   The density plots for MHD High Mach number problem. (a) without the fix, (b) with the new fix

result with new sonic fix is excellent showing that it works succesfully for the MHD as well. Fig.4 shows the normalized magnitude of the transferred quantity for the x momentum at the sonic interface. This value is related to Eq.(39) and decays in time as stated earlier.

   The first 2D test problem is the $15°$ wedge flow with no magnetic field. In this problem, a Mach=2 flow enters into a rectangle domain from its left face and exits from the right face. The lower boundary includes a $15°$ wedge which produces an expansion corner and gives rise to a shock which reflects off from the upper reflected surface. The density and pressure are set to 1 and $\gamma = 1.4$ throuogut the grid and $V_x = 2$ is assumed as the initial

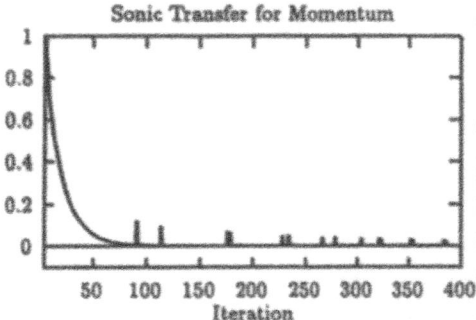

Figure 4. The time history of the magnitude of the sonic transfer between the left and right momenta across the sonic interface

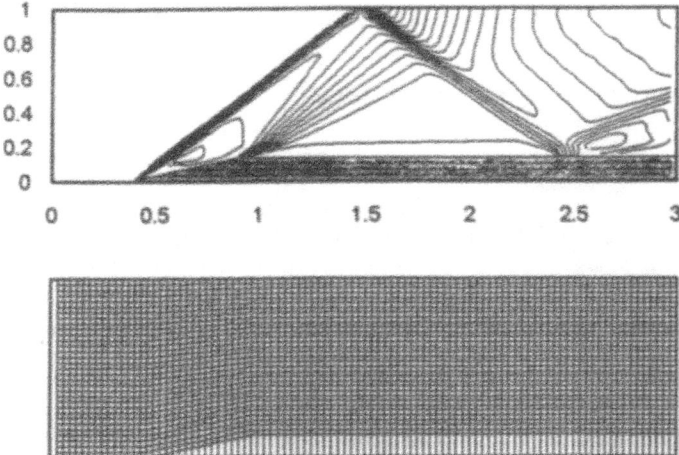

Figure 5. Density contours and grid used for the 15° wedge flow. This result is obtained with the new sonic fix.

conditions. The steady-state solution obtained and a cartesian 96×36 grid used are shown in Fig.5. This result shows that the new sonic fix works well for the problems including more complex geometry. The only problem with this solution is that the shocks diffuse slightly after reflection. As will be shown later, this diffusion can be reduced significantly by applying a 2D telescoping property.

The next 2D test problem was constructed by the author such that a $29°$ reflected shock is the equilibrium solution across a cartesian tube. The states at the left W_L and upper boundaries W_R are specified and the lower and right boundaries are taken as reflective and outgoing respectively. To find the state on the right of the $29°$ shock, the stationary R-H conditions (i.e., $\Delta \vec{F}_n = 0$) are solved in the normal direction, provided that the left

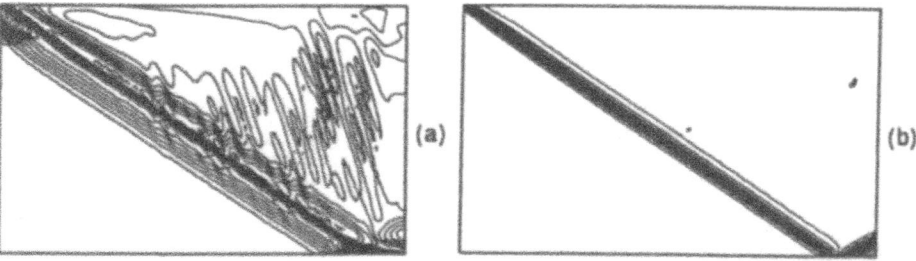

Figure 6. Regular High-Mach reflection problem. The graphs show B_x contours obtained by (a) 7x7 eigen-system without the divergence source and (b) 8x8 eigen-system with the source.

state is known. These states are given by

$$W_L = [1, 2.9, 0, 0, \sqrt{\pi}, 0, 0, 1/\gamma]^T$$

$$W_R = [1.460, 2.716, -0.405, 0, 2.424, -0.361, 0, 1.223]^T \qquad (42)$$

with $\gamma = 1.4$. The solution includes a stationary discontinuity in the magnetic field across the shock which gives rise to a surface current flowing along the infinitesimally thin layer of the shock in z direction. This problem is solved on a cartesian grid with x:[0,2] , y:[0,1] and W_L is assumed throughout the grid as the initial condition. It is expected that the time dependent problem reaches the defined equilibrium state (42) with the divergence condition on the magnetic field is preserved. Since the solution is known analytically, this is a very good test problem to check the accuracy and performance of any numerical method in solving the MHD equations. A 50×25 grid is used along with entropy satisfying second order fluxes limited by Superbee limiter. When the divergence cleaning is not done for the magnetic field, using the 7×7 eigen-system is invalid in multi-dimensional MHD. As seen from the B_x contours given in Fig.6a, the non-physical magnetic islands on each side of the shock have emerged. This result was obtained with 7×7 eigen-system with no source. Once these islands are formed, the non-physical magnetic monopoles are created and the numerical iterations become unstable. Fig.6b gives the result on the same grid obtained with the 8×8 eigen-system and divergence source. As seen, the solution is excellent since there exists no problems arising from the divergence condition and that the shock compression remains uniform (even after reflection). As a remark, it is noted that using a rotated Riemann solver would improve these results which is the main objective of the subsequent papers.

The next problem is an instantaneous high beta ($\beta = P/(B^2/8\pi)$) explosion of circular energetic plasma into the free space with reflective lower boundary. The initial condition is specified within and outside of the circular region whose initial radius is chosen as 0.24 on a 80×80 grid with $x : [-1, 1]$, $y : [-0.8, 1.2]$. The initial states are defined as

$$W_{in} = [20, 0, 0, 0, 0, 0, 1, 50]^T , \quad W_{out} = [1, 0, 0, 0, 0, 0, 1, 1]^T \qquad (43)$$

with $\gamma = 1.4$ so that the interior region includes a 20 times denser plasma with 50 times stronger pressure. A uniform weak magnetic field (magnitude=1) is given initially in the z direction. A safety factor of 0.5 is chosen for the time step and the results are shown in Figs.7(b-h) as the density contours at time steps 100, 200, 300, 400, 500, 600, 900. Fig.7a shows the initial density contour which has just began to expand symmetrically outwards. The initial discontinuity leads to a shock and a contact discontinuity behind it. The reflection of the shock front from the lower boundary is shown in Fig.7b. The reflected shock moves upward interacting with oncoming contact discontinuity and the lower density region as shown in Fig.7c. Fig.7d shows the result as the shock sweeps through the low density region with a flattened front. It catches the upper boundary of the slowly expanding low density region from which a secondary shock emerges towards the lower boundary as shown in Fig.7e. Moving downwards, this secondary shock front diverges turning into a circular shape and creates circulating regions on both sides as seen on Fig.7g. Finally, the secondary shock hits the lower boundary and is reflected back while the circulating regions lead to a mushroom formation as shown in Fig.7h. When these figures are examined it is seen that the shock thickness remains the same in each direction at all times. This was accomplished by utilizing the 2D telescoping property as stated earlier. This result displays the importance of guaranteeing the conservation in the solution domain.

The last application is the preliminary simulation of the internal kink instability of the Tokamak plasmas in cartesian geometry. When the plasma column in a Tokamak bends, the magnetic field lines are concentrated at the inner side while they are diluted at the other side. The imbalance in the magnetic fluxes leads to greater magnetic pressure at the inner side which forces the column to bend further. The plasma is unstable under this kind of bending and the instability is called as the kink-type instability. To eliminate this instability a magnetic field should be applied along the minor axis of the column to create a tension that tries to stretch it to a straight line stabilizing the instability. In order to simulate this kind of instability, usually a consistent equilibrium configuration is assumed and it is perturbed to follow the time behaviour of the plasma. If the plasma is

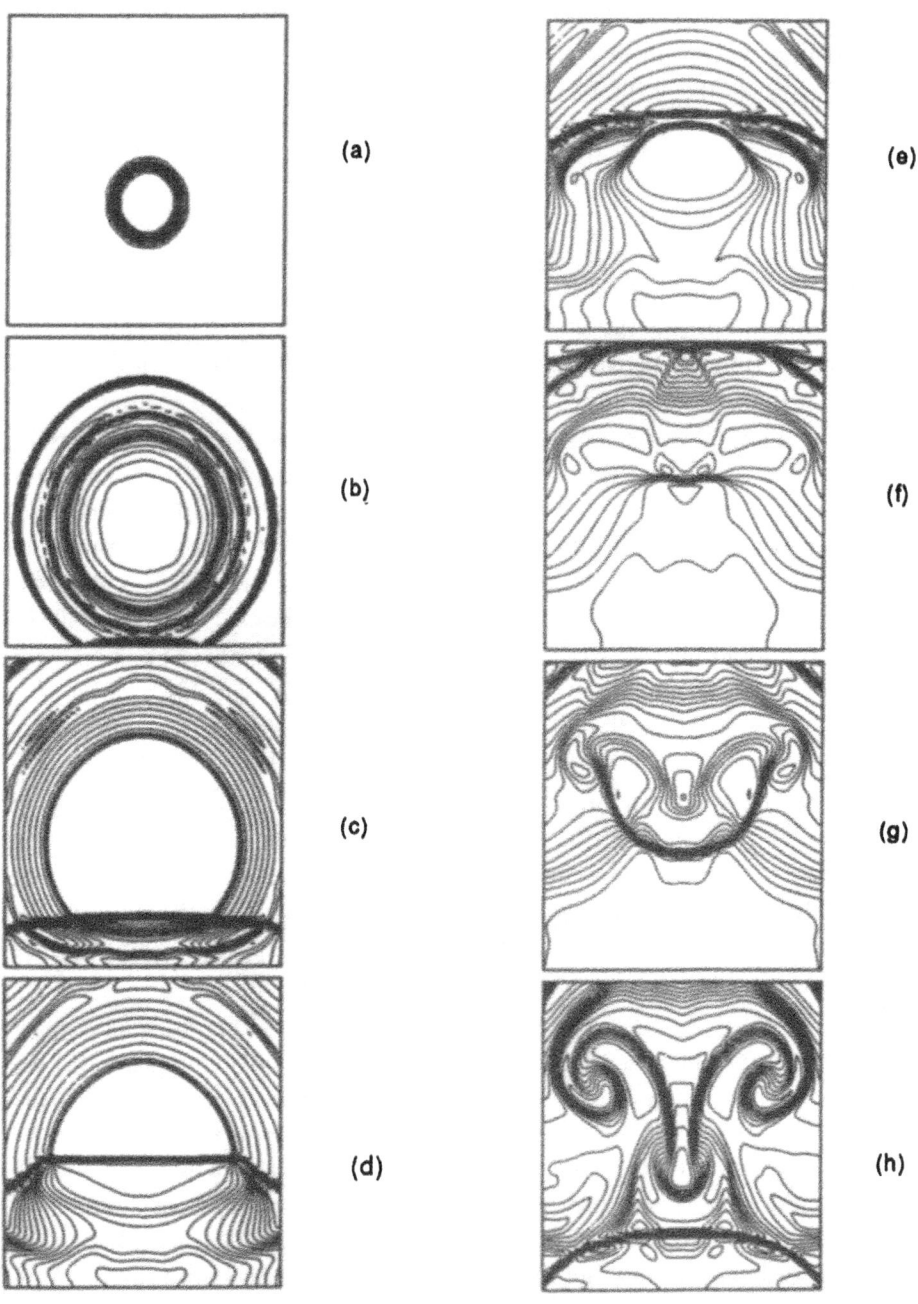

Figure 7. Circular explosion problem. The figures denote the density contours at different times.

unstable against these perturbations, the solutions will exhibit an exponential growth. Otherwise, the perturbation will decay and the configuration will lead to its real equilibrium state.

Since the development of a 3D MHD code (by the author) including curvilinear geometries is still in process, the toroidal geometry is not used here to simulate such a plasma. Instead, the plasma is solved on cartesian geometry with a similar kink-type perturbation. Since the problem is solved only in 2D cartesian geometry, $[x, y]$, it is impossible to examine the effect of bending. If there exists no bending, the plasma will be stable regardless of the value of the magnetic field in perpendicular direction, B_z. In what follows, an equilibrium state which is a solution of $\vec{\nabla}P + [\vec{B} \times (\vec{\nabla} \times \vec{B})]/4\pi = 0$ with $\vec{V} = 0$ (the steady-state form of Eq.2) and $\rho = 1$ is given

$$
\begin{aligned}
B_x &= -\psi\cos(\frac{\pi x}{2})\sin(\frac{\pi y}{2}) \\
B_y &= \psi\sin(\frac{\pi x}{2})\cos(\frac{\pi y}{2}) \\
B_z &= 0 \\
P &= \frac{\psi^2}{4\pi}\cos^2(\frac{\pi y}{2})\sin^2(\frac{\pi x}{2})
\end{aligned}
\tag{44}
$$

This equilibrium state is perturbed with the following velocity perturbation:

$$
\begin{aligned}
V_x &= c\frac{xy}{|xy|}\sin^2(\pi x)\sin^2(\pi y) \\
V_y &= c\chi\cos^2(\frac{3\pi}{2}x)\sin^2(\frac{\pi}{2}y)
\end{aligned}
\tag{45}
$$

where $c \approx 10^{-4}$ is a small number and $\chi = 1$ if $|x| \leq 1/3$ and $\chi = -1$ otherwise. This perturbation generates two counter-rotating vortices in $[x, y]$ plane. See Fig.8 for the initial velocity perturbation.

In simulating the problem in toroidal geometry, the twisting of these vortices is provided by multiplying these velocities with a sinusoidal function depending on the axial direction (Niu, 1988) In the problem considered here the plasma is expected to be stable against this perturbation. Thus, the growth rate given by

$$
\tilde{\gamma} = \frac{1}{2}\frac{d(\ln f)}{dt}
\tag{46}
$$

is expected to decay in time. Here $f = \frac{1}{2}\rho V^2$ is taken as the kinetic energy. To solve this problem numerically, a 54×54 grid is chosen with $x, y : [-1, 1]$ and $\gamma = 5/3$ is taken. The decay in the growth rate is shown in Fig.9a. The time histories of the total, magnetic, and kinetic energies are shown (from top to the bottom) in Fig.9b. The time dependencies of the pressure profile are shown in Fig.10. Fig.10a shows the initial pressure field on the $z = 0$

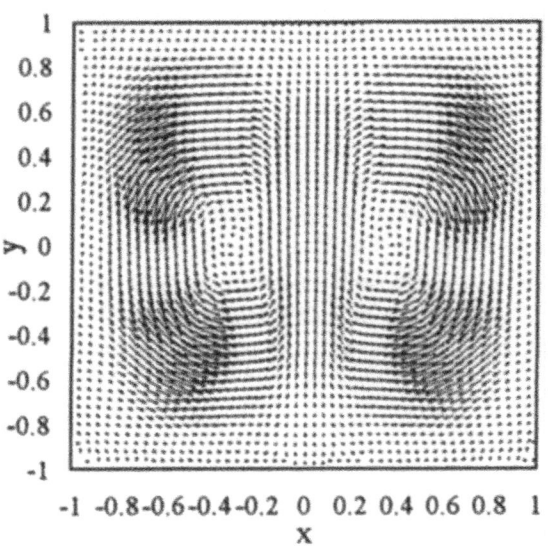

Figure 8.   The initial velocity perturbation. Two counter-rotating vortices.

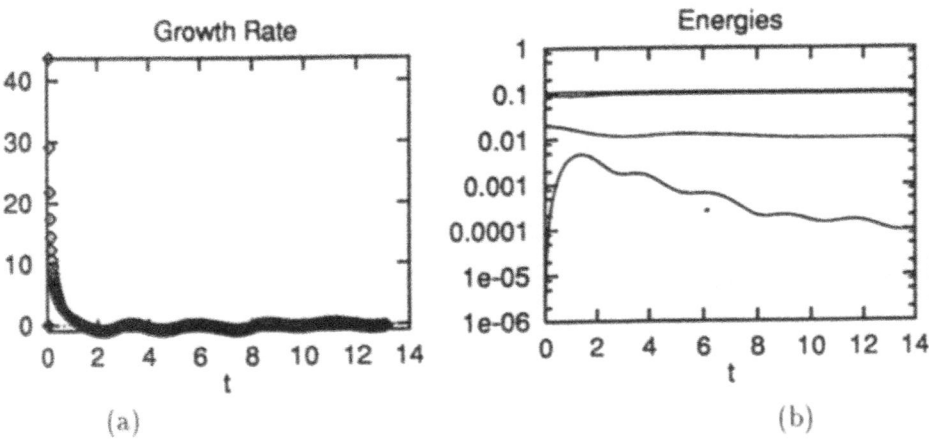

(a)   (b)

Figure 9.   (a)The decay in the growth rate, (b) The time history of total, magnetic, and kinetic energies.

plane. The other figures were obtained at the time levels: 50,100,200,400,700 respectively (safety factor was set to 0.4). The initial velocity perturbation leads to a pressure build-up near the center as seen in Fig.10b,c. This central pressure forces the plasma particles mainly towards the lower and upper boundaries leading to a triple humped pressure profile as shown in Fig.10d.

As the time proceeds, the pressure (and hence all other physical quantities) reach the equilibrium state as shown for the pressure in Figs.10e,f. These results show that the code developed here can be used to simulate Tokamak plasmas. This requires the use of the cylindrical coordinates along with triangular grids which is the subject of the subsequent papers.

5. Conclusion

The solution of hyperbolic equations with simple first order upwind differencing produces errors that accumulate in time an it produces unphysical expansion shocks which slow down the convergence. To fix this problem, a new sonic fix that utilizes a physically correct dissipation term which is embedded in the fluxes is described in this paper. Along with the new sonic fix, the inclusion of the source term, related to the divergence of magnetic field, into the fluxes was also discussed. A wide variety of one or two dimensional test problems resulting from the solutions of Euler's and MHD equations are presented. The results show that the scheme with the new sonic fix along with the new divergence source strength is robust, accurate, and entropy satisfying and that it produces rather sharp contact discontinuities and shocks without spurious oscillations. The scheme is currently being improved to include curvilinear geometries and flux rotation.

6. Acknowledgements

The partial supports from NATO and NSF are acknowledged.

References

Aslan, N., "Computational investigations of ideal MHD plasmas with discontinuities," Ph.D. Thesis, University of Michigan, Nuclear Eng. Dept., 1993.

Aslan, N., "Numerical solutions of one-dimensional MHD equations by a fluctuation approach," *Int. J. for Num. Meth. in Fluids*, Vol. 22, 1996, pp. 569-580.

Aslan, N., "Numerical solutions of 2-D MHD equations by finite volume method with quadrilateral cells," *Proc. of Ninth Int. Conf. in Numerical Meth. in Laminar and Turbulent Flow*, Vol. 9-2, 1995, p. 1633.

Aslan, N., "Two dimensional solutions of MHD equations with a modified Roe's Method," *Int. J. Numer. Meth. in Fluids*, in print, 1996.

Aslan, N. and Kammash, T., "Developing Numerical Fluxes with New Sonic Fix for MHD Equations," submitted to *J. Comp. Phys.*, 1996.

Brio, M. and Wu, C.C., "An upwind differencing scheme for the equations of ideal magnetohydrodynamics," *J. Comp. Phys.*, Vol. 75, 1988, pp. 400-422.

Glaister, "An approximate Riemann solver for compressible flows with axial symmetry," *Num. Anal. Report*, Vol. 2/87, Dept. of Mathematics, University of Reading, 1987.

Gombosi, T.I., Powell, K.G., and De Zeeuw, D.L., "Axisymmetric modelling of cometary mass loading on an adaptively refined grid," *J. Geophys. Res.*, Vol. 99, A11, 1994, pp. 21525-21539.

Harten, A., "High-resolution schemes for hyperbolic conservation laws," *J. Comp. Phys.*, Vol. 49, 1983, pp. 235-269,

LeVeque, R.J., "High-resolution finite volume methods on arbitrary grids via wave propagation," *J. Comp. Phys.*, Vol. 78, 1988, pp. 36-63.

Niu, K., "Nuclear Fusion," Cambridge University Press, ISBN: 0 521 32994 9, 1988.

Powell, K.G., "An approximate Riemann solver for magnetohydrodynamics (that works in more than one dimension)," *J. Geophys. Res.*, ICASE Report No. 94-24, NASA Langley Research Center, Hampton, VA, 1994.

Roe, P.L., "Sonic Flux Formulae," *SIAM J. Sci. Stat. Comput.*, Vol. 13, 1992, pp. 611-630.

Sweby, P.K., "High-resolution schemes using flux limiters for hyperbolic conservation laws," *SIAM J. Numer. Anal.*, Vol. 21, 1984, pp. 995-1011.

van Leer, B., Lee, W.T., and Powell, K.G., "Sonic point capturing," AIAA Paper 89-1945, Vol. 49, 1983, pp. 235-269.

Zachary, A. and Colella, P., "A higher-order Godunov method for the equations of ideal magnetohydrodynamics," *J. Comp. Phys.*, Vol. 99, 1992, pp. 341-347.

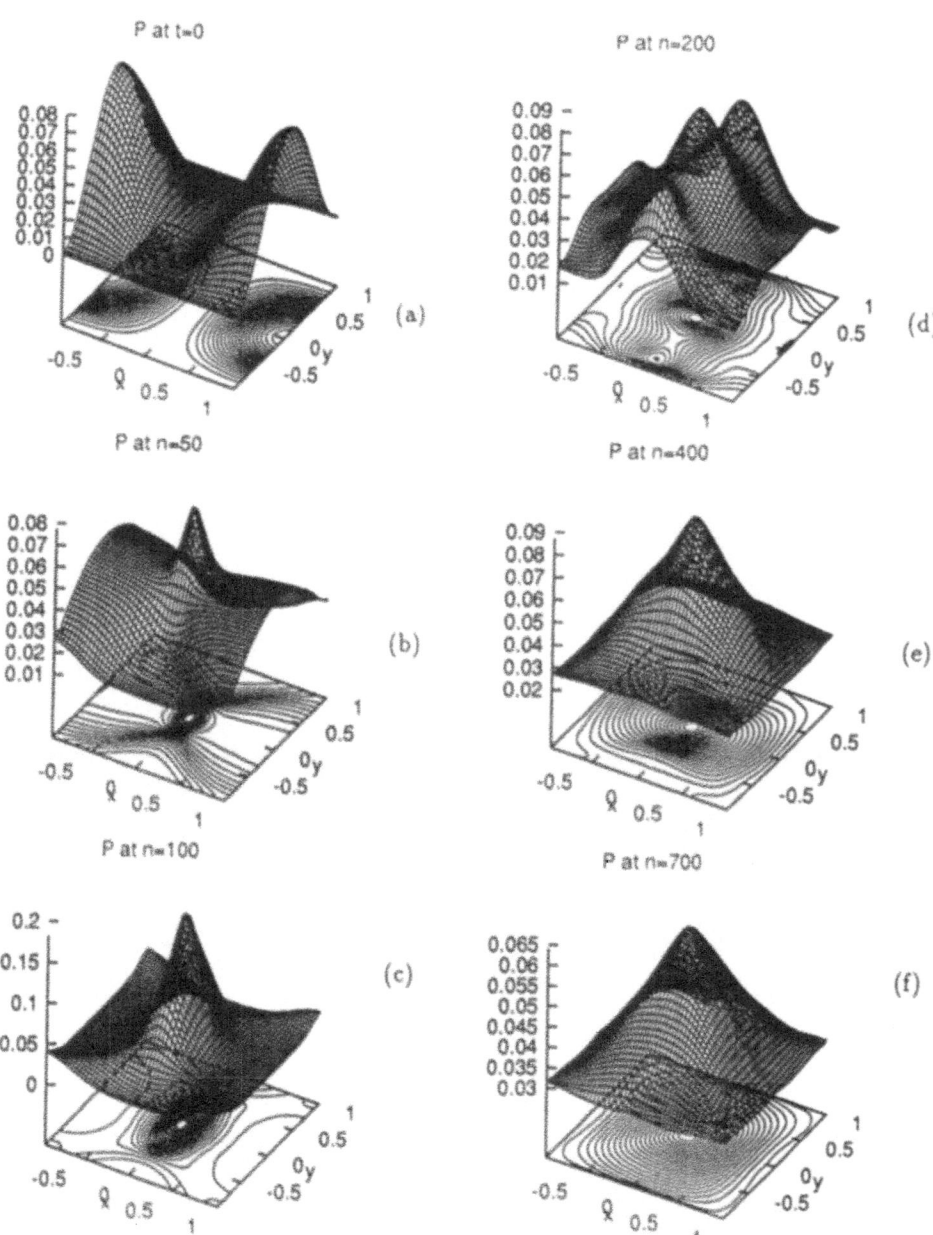

Figure 10. 2D version of the Kink perturbation. The surface plots denote the pressure profiles in [x,y].

LOCAL PRECONDITIONING: WHO NEEDS IT?

BRAM VAN LEER
Department of Aerospace Engineering
The University of Michigan
Ann Arbor, MI 48109-2118

Abstract.
Benefits of local preconditioning for the Euler equations are reviewed, and illustrated with numerical examples. Progress in the fight against the feared stagnation-point instability is discussed, and hope-giving numerical evidence is presented in support of this development.

1. Introduction

Over the past dozen years, local preconditioning has evolved from a gimmick used by some isolated individuals (Choi & Merkle, 1993; Turkel, 1987) for treating low-speed compressible flow, to an indispensible technique for convergence acceleration and accuracy preservation in flow computations, at all Mach and Reynolds numbers, and including turbulence modeling. I say "indispensible" because I believe this is the correct qualifier, but I must admit that the necessity of local preconditioning is not yet fully appreciated by the larger CFD community. Hence the title question: "Local Preconditioning: Who Needs It?" and the correct answer: "You." Let me try to convince you in what follows.

2. Convergence Acceleration: A Personal History

When I made my first steps into the realm of computational aeronautics during an extended sabbatical leave at ICASE, 1979-81, I had no concept whatsoever of steady-flow calculations. I was educated and employed in astrophysics, had specialized in cosmic gas dynamics, and the only flows I had computed were violently unsteady. In my experience any compressive velocity difference within a gas would create a shock wave, weak or strong,

335

V. Venkatakrishnan et al (eds.), Barriers and Challenges in Computational Fluid Dynamics, 335-349.

and I had great difficulty grasping that *steady* subsonic flows contain no shocks. My final product, an upwind transonic code, commissioned by Jerry South, then head of NASA Langley's Theoretical Aerodynamics Branch, was second-order accurate in space *and* time, and took tens of thousands of steps to converge to a steady state.

The next summer (1992) I came back to ICASE and worked with Eli Turkel. Eli taught me the magic trick of local time-stepping. When I finally had digested this daring concept, wide vistas of research opportunities suddenly unfolded before my eyes. I realized how for years I had been torturing myself, trying to maintain solution accuracy in unsteady calculations, and how much freedom the steady-state guys had in comparison. When marching to a steady state, anything goes; all that matters is to get there fast, and the quality of the final solution is dictated merely by the quality of the spatial discretization. My favorite spatial operators, the upwind-biased ones of the κ-family, worked very well for transonic flows, so I decided to indulge in the newly found freedom.

I also realized how glamorous the life of a successful marcher-to-a-steady-state could be. The whole fluid-dynamics community would be forever grateful and obliged to me, were I to come up with an acceleration method cutting convergence times by orders of magnitude. Fame and riches would befall to me, people would revere me wherever I would appear, love-struck women would crowd at my feet, in short, my life would be like that of a movie star. Somewhat disturbing was, though, that Eli Turkel did not appear to lead this kind of life, yet. Convergence acceleration might be a harder game than I had expected, after all. (And it was.)

Immediately after my return to Holland a unique opportunity arose ... in astrophysics. One of Leiden Observatory's all-time top students, Wim Mulder, came to me with a *steady* cosmic flow problem: to determine the density and velocity distributions in a disk galaxy with spiral structure induced by a non-axisymmetric gravitational potential. We set out to develop systems versions of the classical elliptic relaxation schemes, which seemed to keep much of their value when used with upwind spatial operators, because upwinding leads to diagonal dominance. We also developed a marching strategy called Switched Evolution/Relaxation (SER), in which the time step starts out small, yielding a close-to-physical evolution, but is allowed to grow when the residual drops, yielding Newton-type iterations. At the 1983 AIAA CFD conference in Danvers, MA, we presented Euler results for all of these schemes, and even some early multigrid results.

In 1985, after three fruitful years of working with Mulder, I felt confident enough to embark on a long journey through *uncharted* provinces of the land of convergence acceleration. My guidance was the firm belief that explicit relaxation schemes, *if properly constructed*, would be all you needed to make multigrid relaxation work for the Euler and Navier-Stokes equations. In essence I still believe this.

This journey would lead me toward multistage marching schemes with guaranteed maximal damping of all high-frequency error modes, that is, maximal for the number of stages used, and to semicoarsened multigrid relaxation..

I knew there were two major parts to this adventure:

1. Optimizing high-frequency damping in scalar advection-diffusion schemes;
2. Developing a technique of local preconditioning to make the system of equations (Euler, Navier-Stokes) behave like a scalar equation. This, by itself, would be a powerful single-grid acceleration method, since it would remove the stiffness due to the spread in physical time scales. For the Euler equations it means equalization of the characteristic speeds.

I also knew the second task definitely was the harder one, but I could not have foreseen that today, in 1996, there still are some serious barriers and challenges in the development of local preconditioning matrices.

But first, the good news.

3. The Good News

3.1. THREE PROVEN BENEFITS OF EULER PRECONDITIONING

If properly designed, local preconditioning for the Euler equations may have all of the following well-documented benefits.

1. It removes from the Euler equations the stiffness caused by the range of characteristic speeds, thus improving the convergence rate of any marching scheme (van Leer et al., 1991; Lee, 1991; Godfrey et al., 1993; Turkel et al., 1996);
2. It makes the system behave more like a scalar equation, thus facilitating the design of effective auxiliary techniques, such as multi-grid smoothers (Lynn & van Leer, 1993; Lynn & van Leer, 1995; Lynn, 1995; Lynn et al., 1996), residual smoothing (Lynn, 1995) and approximate factorization (Lee, 1996);
3. It produces discretizations that remain accurate in the limit of vanishing Mach number (van Leer et al., 1991; Lee, 1991; Godfrey et al., 1993; Turkel et al., 1993; Turkel et al., 1996; Reed, 1995; Lee, 1996).

In short, there are single-grid-relaxation, multi-grid-relaxation and accuracy benefits to the simple explicit technique of local preconditioning. Those who are not taking advantage of these are selling themselves short.

I shall not attempt to reiterate all the numerical evidence in support of these desirable properties, but will restrict myself to showing some less known numerical results on multigrid convergence in Section 3.3. Accuracy preservation is further discussed in Section 3.4.

An additional benefit of local Euler preconditioning is that it casts the equations in a form that is pre-eminently suited for treatment by fluctuation splitting. This requires a lengthier explanation, given in the next subsection.

3.2. ... AND ANOTHER BENEFIT

During the 1990s our understanding of the Euler convergence process has significantly increased. There is mounting evidence that the treatment of the true advection equations hidden in the Euler system (e.g. advection of entropy) has to be different from that of the coupled acoustic equations arising in subsonic flow. Whether starting from the equations in conservation form or primitive variables, this calls for a suitable decomposition of the system, in which the embedded acoustic equations are isolated from the advection equations.

Ta'asan (Ta'asan, 1993) suggests using a set of "canonical" equations based on the steady form of the Euler equations; this has its roots in vintage rules prescribed by Brandt in his 1984 Multi-Grid Guide (Brandt, 1982). In Brandt's work the emphasis is on achieving the theoretical limit of multigrid convergence, and not necessarily on the accuracy of the spatial operator.

In contrast, Roe and Mesaros (Mesaros & Roe, 1995; Mesaros, 1995) searched for a genuinely multi-dimensional splitting of the spatial Euler operator ("fluctuation splitting") that would match accuracy to compactness. They ended up discretizing of the system that results from applying the local preconditioning of Van Leer, Lee and Roe (van Leer *et al.*, 1991). The most advanced applications of these preconditioned equations are by Deconinck et al. (Deconinck & Degrez, 1996), and use the so-called matrix form of the PSI advection scheme, which automatically distinguishes between omnidirectional acoustic wave propagation and unidirectional advection.

The combination of the preconditioned equations and the fluctuation-split scheme, formulated on unstructured triangular or tetrahedral grids, differs in essence from standard Euler schemes. Such a numerical strategy will not easily be adopted by the larger CFD community, which has a considerable effort invested in higher-order Godunov-type methods with TVD limiters. Fortunately, it has been demonstrated that the use of the preconditioned equations alone, with a corresponding modification of the spatial discretization[1], already has a large pay-off.

---

[1]If the Euler equations are written as $U_t + AU_x + BU_y + CU_z = 0$, the preconditioned equations read $U_t + P(AU_x + BU_y + CU_z) = 0$ or $P^{-1}U_t + AU_x + BU_y + CU_z = 0$. When discretizing these equations, besides multiplying the residual by the precondition-

3.3. EFFECT ON SINGLE- AND MULTI-GRID RELAXATION

As mentioned above, the equalization of wave speeds, achieved by optimal Euler preconditioners, removes the stiffness encountered in steady-state calculations, especially explicit calculations, yielding convergence acceleration already on a single grid. Furthermore, the concentration of discrete eigenvalues, achieved by optimal Euler preconditioners, makes it possible to design multistage marching schemes with optimal high-frequency damping regardless of flow angle and Mach number; these are particularly suited for multigrid relaxation on semi-coarsened grids. The single-grid and multigrid convergence-acceleration mechanisms are independent, so their individual effects add up; this was demonstrated conclusively by Tai (Tai, 1990) for 1-D Euler schemes and Lynn (Lynn, 1995) for 2-D Euler schemes. Some of Lynn's results were presented at the 1995 AIAA CFD Conference in San Diego, CA; below I present a different sample, intended to drive the message of the double benefit home once more.

Another message worth repeating is that the local preconditioning matrix resulting from retaining only the main-diagonal block of an implicit upwind Euler discretization (point/Jacobi relaxation) *does not have any single-grid acceleration effect* (Lynn et al., 1996). Point/Jacobi-type preconditioning is currently enjoying a renewed interest (Allmaras, 1995; Pierce & Giles, 1996) because of its smoothing effect on high-frequency error combinations, useful in multigrid relaxation with semi-coarsening; for this reason it was already recommended by Mulder (Mulder, 1989).

Table 1 illustrates the compound single/multigrid benefit for a set of second-order-accurate calculations of inviscid subsonic flow ($M_{inflow} = 0.35$) in a 2D channel with a circular bump on one wall (1979 GAMM Workshop parameters: thickness 4.2% of chord, height 2 chords). A high-resolution spatial discretization ($\kappa = 0$, Van Albada limiter, Roe's Riemann solver) was used; time marching was done with a 4-stage scheme optimized for best damping of all high x- and y-frequency combinations, as needed for semi-coarsened multigrid relaxation. A "design graph" for this scheme, i. e. the optimally scaled Fourier footprint of the spatial discretization plotted on top of the scheme's stability domain and amplification-level lines, is shown in Figure 1.

The two things to be learned from Table 2 (work needed for residual reduction) are:

---

ing matrix, one must modify the artificial-dissipation matrices so that they become suited for the preconditioned equations. If the flux Jacobian is $\mathbf{A}$, the corresponding dissipation matrix is some average of $|\mathbf{A}|$, and the preconditioning matrix $\mathbf{P}$, then the dissipation matrix must be changed into $\mathbf{P}^{-1}|\mathbf{PA}|$. This means that after preconditioning the dissipation matrix becomes $|\mathbf{PA}|$, which matches the preconditioned coefficient matrix $\mathbf{PA}$.

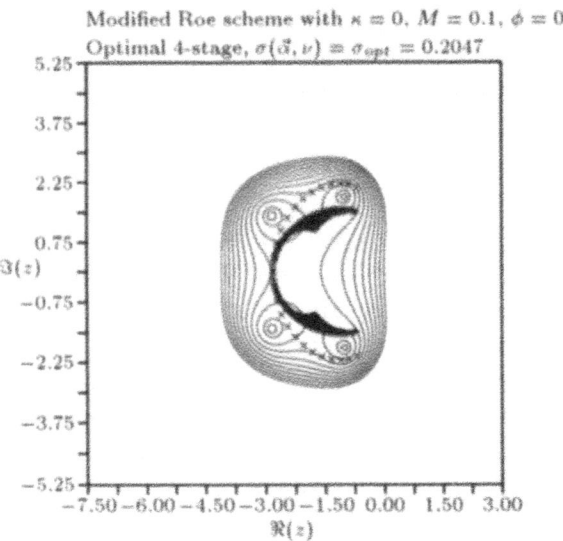

Figure 1. Design graph of optimal four-stage scheme for the $\kappa = 0$ preconditioned Euler operator; $M = 0.1$, $\phi = 0°$.

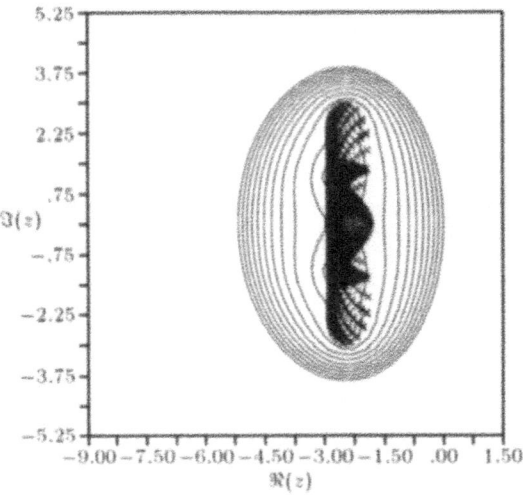

Figure 2. Design graph of optimal four-stage scheme for the first-order preconditioned Euler operator with explicit residual smoothing; $M = 0.1$, $\phi = 45°$, $\epsilon = 0.1$.

(a) Local matrix preconditioning yields faster convergence than local time-stepping ($\equiv$ scalar preconditioning) on a single grid. In this subsonic case the speed up is not so impressive, as the single-grid convergence process is dominated by acoustic waves bouncing between the walls; these are not affected by the preconditioning.

(b) The maximum gain from multigrid relaxation is greater for local matrix preconditioning than for local time-stepping.

A contrasting aspect is provided by Table 2, which contains the convergence data for a supersonic flow case ($M_{\text{inflow}} = 1.4$). This set clearly brings out the single-grid benefit: the single-grid acceleration by matrix preconditioning is so strong that adding more grid levels yields little additional speed-up.

In the next test problem the initial values consist of a uniform state (flow angle $= 10°$, range of M-values) with random perturbations on a 32×32 uniform grid; at the boundaries free-stream conditions are prescribed. A first-order upwind spatial operator with optimized four-stage time-stepping was used. Table 3 shows work needed till convergence for various relaxation strategies: single-grid with local time-stepping (SG-LTS), single-grid with matrix time-stepping (SG-MTS), regular multigrid with local time-stepping (MG-LTS), regular multigrid with matrix time-stepping (MG-MTS). Both local time-stepping methods are affected by the variation of condition number with Mach number, whereas the matrix time-stepping methods are virtually insensitive to Mach number.

The above acceleration can still be enhanced at little computational cost by introducing explicit residual smoothing, which significantly increases the allowable CFL number without compromising high-frequency damping. Owing to the preconditioning the smoothing is equally effective for all physical modes (acoustic and advective). Figure 2 shows a design graph for the preconditioned 4-stage first-order upwind scheme when explicit residual smoothing, with coefficient $\epsilon = 0.1$ is added. The effect of the residual smoothing is an increase of the optimized CFL number from 2.64 to 3.63, without sacrificing high-frequency damping. The improved convergence is clear upon comparing Table 4 (perturbation test, work needed using explicit residual smoothing) with Table 3.

3.4. EFFECT ON ACCURACY AT LOW MACH NUMBER

One unforeseen side effect of certain preconditioners, such as those of the Van Leer-Turkel family, is that they prevent the deterioration of accuracy encountered when computing flows at ever-decreasing Mach number on a fixed grid, a phenomenon first reported by Volpe (Volpe, 1991). This beneficial and very useful effect of local preconditioning is owing to the mod-

Number of stages

		Local time-stepping				Matrix preconditioning			
		3	4	5	6	3	4	5	6
No.	1	3702	3740	3685	3744	3126	2976	2380	2916
of	2	2241	1622	1284	1087	1366	1271	780	838
grid	3	692	**582**	**582**	583	388	301	632	615
levels	4	755	751	705	708	319	**191**	194	219

TABLE 1. Work units required to reduce $\|TE\|_1$ to 5×10^{-2}; channel flow, $M = 0.35$; 64×32 grid. Defect-correction cycles; nested iteration for initial guess.

Number of stages

		Local time-stepping				Matrix preconditioning			
		3	4	5	6	3	4	5	6
No.	1	1422	1384	1365	1344	420	420	425	414
of	2	895	933	781	753	356	350	332	**309**
grid	3	724	898	**722**	769	361	370	360	358
levels	4	809	1047	993	732	436	399	380	400

TABLE 2. Same as Table 1, but for $M = 1.4$.

ification of the artificial-dissipation matrices that has to be implemented anyway for stability reasons. This required modification was originally considered a nuisance, since it makes preconditioning more intrusive; now we know that codes for compressible flow actually *need* such a modification if they are to be applied to problems in which incompressible and compressible flow occur side by side. Such situations arise, for instance, in propulsion (Choi & Merkle, 1993), in low-speed high-lift flow (Turkel *et al.*, 1996) and in V/STOL maneuvering (Reed, 1995). Local preconditioning makes it possible to solve flow problems of this kind with a single code, overcoming both the loss of accuracy and the loss of convergence speed, not a small accomplishment for a modest explicit technique.

Since its discovery (van Leer *et al.*, 1991; Lee, 1991), the accuracy benefit has been demonstrated over and over in numerical tests (van Leer *et al.*, 1991; Lee, 1991; Godfrey *et al.*, 1993; Turkel *et al.*, 1993; Turkel *et al.*, 1996; Mesaros & Roe, 1995; Mesaros, 1995; Reed, 1995; Deconinck & Degrez, 1996; Lee, 1996); we need not repeat any of these here. An important development, still worth emphasizing, is that the effect has also been explained with mathematical analysis. The asymptotic analysis by Turkel (Turkel *et al.*, 1993) et al. provides the detailed structure a preconditioning matrix must have in order to properly balance the artificial dissipation terms with

Mach Number

	0.1	0.35	0.5	0.85	0.99	1.01	1.2	2.0	5.0
SG-LTS	1224	408	352	984	1512	1552	784	272	164
SG-MTS	280	272	272	272	224	188	160	148	128
MG-LTS	996	401	263	263	816	1757	290	252	228
MG-MTS	76	76	83	83	83	83	83	90	103

TABLE 3. Work required to reduce residual norm by 10^{-5}; perturbed uniform flow, $\phi = 10°$.

Mach Number

	0.1	0.35	0.5	0.85	0.99	1.01	1.2	2.0	5.0
SG-LTS	688	204	196	556	848	916	436	152	108
SG-MTS	176	172	168	168	136	116	96	88	72
MG-LTS	760	256	208	166	650	1438	180	180	187
MG-MTS	62	62	62	69	69	69	69	76	83

TABLE 4. Same as Table 3, but with the use of explicit residual smoothing ($\epsilon = 0.1$).

the inviscid flux terms for $M \to 0$ (lack of balance causes the loss of accuracy). Reed's (Reed, 1995) analysis is based on the modified equation and is less detailed; it shows that the truncation error of the discrete pressure equation is multiplied by M^2 owing to the preconditioning.

An important observation based on the analysis of Turkel et al. is that point/Jacobi-type preconditioning does not have the structure needed for accuracy preservation in the incompressible limit. This is clear from its common use, which does not include any change to the artificial-dissipation matrices. However, even if it would be used to modify the latter, this would not properly balance the terms in the scheme when M approaches 0.

The analysis in (Turkel *et al.*, 1993) does not answer the question if a local loss of accuracy can result in a very limited region of low Mach number, such as a stagnation region, if no preconditioning is used. We suspect there indeed is a local loss of accuracy, but there has been no study to confirm this. If proven correct, it might have far-reaching consequences for our trust in standard discretizations.

4. The Bad News

One serious problem associated with the use of local preconditioning, even if it does the right thing in the limit of incompressibility, is that it commonly breaks down *locally* when the Mach number vanishes, i. e. in a stagnation point. This break-down manifests itself in non-convergence, if not instability.

Three independent factors are known to cause this problem; they are, in order of increasing severity:

1. Sensitivity of the preconditioned equations to strong initial transients in velocities and pressures, such as created by impulsive starting values (Darmofal & Schmid, 1996);

2. Sensitivity of the preconditioning matrix to the flow angle for $M \to 0$ (van Leer *et al.*, 1995), an point of concern particularly for the Van Leer-Lee-Roe matrix;

3. Degeneration of the eigenvector system of the preconditioned equations for $M \to 0$(Darmofal & Schmid, 1996), a problem plaguing most preconditioners currently in use.

The first factor can easily be eliminated by starting with smoother initial values, for instance, those created by taking the first few steps with a nonpreconditioned scheme.

To reduce the flow-angle sensitivity, Van Leer et al. (van Leer *et al.*, 1995) chose to allow an increase of the condition number from 1 to 2; this at least made a nonconservative flow code converge to an accurate low-speed solution, which previously had been unattainable (Mesaros & Roe, 1995).

As a "fix" for eigenvectors becoming parallel, Darmofal and Schmid (Darmofal & Schmid, 1996) suggested to limit the value of M in the preconditioner from below, for instance,

$$M_{\text{lim}} = \max(M, \epsilon M_\infty), \tag{1}$$

where ϵ is a small fraction; this was shown to cure even conservative schemes. The necessity of such limiting had already been found in practice by Turkel (Turkel, 1993), although Turkel attributed it to the loss of symmetrizability of the system after preconditioning. Unfortunately, the appearance of two different Mach numbers considerably complicates the analysis and design of any but the simplest preconditioners, as well as the construction of the corresponding artificial-dissipation matrices; the powers of symbolic manipulation are easily exhausted. In addition, the choice of the cut-off value for M is problem-dependent; a "safe" larger value of M_{lim} may noticeably slow down convergence.

Recently, a group of preconditioning matrices has been found (Lee, 1996; Lee *et al.*, 1997) that suffer much less from the above eigenvector degeneration. These matrices have their root in a matrix for the 1-D Euler equations that produces an orthogonal eigenvector structure for the entire Mach-number range, just as the original equations have. In the 2-D extension,

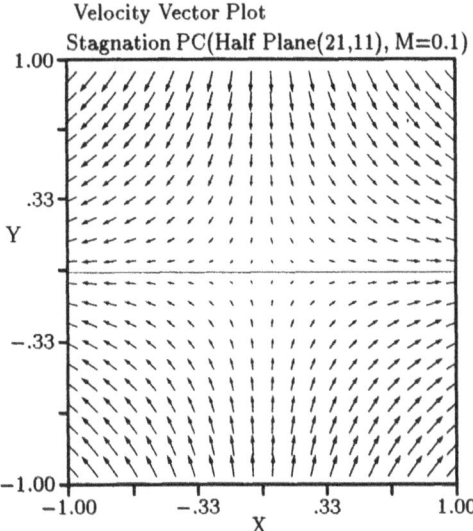

Figure 3. Flow field of half-plane stagnation flow, computed with the "stagnation preconditioner." The upper half is for $b_0 = 0$, the lower half for $b_0 = 1.5$

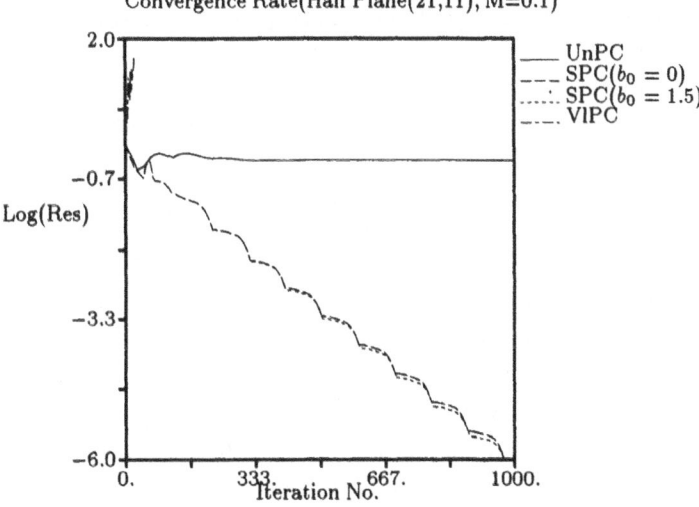

Figure 4. Residual history for half-plane stagnation-flow calculation; UnPC = Unpreconditioned; SPC = Stagnation preconditioner; VlPC = Van Leer preconditioner.

eigenvectors related to plane waves moving in the flow direction are orthogonal in the limit of $M \to 0$, and non-parallel for waves moving normal
to the flow.

These matrices can be linked to the family that includes the Van Leer-
Lee-Roe and Turkel matrices. To fix our thoughts, let us assume that the
Euler equations are expressed in terms of the symmetrizing variables $\mathbf{U}$,
defined by $d\mathbf{U} = (dp/\rho a, du, dv, dw, dS)$, and that the flow is aligned with
the positive x-axis. Using the notation $\beta = \sqrt{1 - M^2}$ for $M < 1$, the Van
Leer/Turkel family with free parameter f has the generic two-dimensional
form

$$
P_{VL,96} = \begin{pmatrix}
\frac{M^2}{\beta} & -\frac{M}{\beta}f & 0 & 0 \\
-\frac{M}{\beta} & 1 + \frac{f}{\beta} & 0 & 0 \\
0 & 0 & \beta & 0 \\
0 & 0 & 0 & 1
\end{pmatrix}
\tag{2}
$$

For $f = 1$ this is the Van Leer-Lee-Roe matrix, for $f = 0$ a version of the
Turkel matrix, optimized for the transonic flow regime. If, for small M, we
choose f close to -1, i. e.

$$
f \approx -1 + rM^2,
\tag{3}
$$

it is seen that the element P_{22} becomes small:

$$
P_{22} \approx (r - \frac{1}{2})M^2.
\tag{4}
$$

This form, with $P_{22} \sim M^2$ can be made positive-definite, and has the
improved eigenvector structure mentioned earlier.

The first matrix of this type, which has $r = -\frac{1}{2}$, was developed by D. Lee
(Lee, 1996) with the purpose of removing any flow-angle dependence. It
later was found that this matrix produced a favorable eigenvector structure
as well.

The performance of this "staganation preconditioner" can be judged
from the converged velocity fields shown in Figure 3; with this matrix one
can successfully compute stagnating flow in the half-plane, *without using the
Darmofal-Schmid fix*. The corresponding convergence histories are shown
in Figure 4. Note the absence of convergence without preconditioning, and
the instability caused by Van Leer reconditioning. Turkel preconditioning
(not shown) produces a similar instability at half the growth rate.

The new preconditioner, though, is not perfect: when dropping the solid-
wall reflection condition in an attempt to compute a full-plane stagnation-
flow, the new preconditioner generated large amounts of vorticity, preventing convergence. The creation of spurious vorticity is a serious remaining
problem in Euler preconditioning. It appears that avoiding vorticity sources

in the preconditioned equations puts a constraint on the preconditioning matrix that conflicts with other natural restrictions (Lee, 1996).

One version of the stagnation preconditioner, with $r = \frac{7}{2}$, can be smoothly linked to the Van Leer-Lee-Roe preconditioner, which is the best performer at $M = 1$. The new stagnation-friendly all-purpose preconditioner follows the matrix structure of (2), with f defined by

$$f = -\frac{\beta - 2M^2}{1 + M^2}. \tag{5}$$

This matrix produces orthogonal eigenvectors for waves moving in the streamwise direction *for all Mach numbers*, as well as satisfaction of symmetrizability and positivity conditions. In spite of the complicated form of (5), the numerical implemenation of this matrix is quite simple, because any matrix of the form (2) produces a simply structured artificial-viscosity matrix $\mathbf{P}^{-1}|\mathbf{PA}|$, while $\mathbf{P}^{-1}|\mathbf{PB}|$ is independent of f. We repeat that the Darmofal-Schmid fix (1) does complicate the construction of artificial-viscosity matrices.

5. From Euler to Navier-Stokes

The extension of local preconditioners from the Euler to the Navier-Stokes equations is not trivial. The wave speeds now become complex as a result of the dissipative terms, which greatly complicates any analysis, even when carried out with symbolic manipulation. A new parameter, the Reynolds number, appears in the equations; in their discretized form the key quantity is the cell Reynolds number Re_h. For medium (≈ 1) to high values of Re_h there is little need to deviate from the Euler preconditioner (Lee, 1996; Turkel *et al.*, 1996), but for lower values of Re_h this "Euler approach" fails.

Im my judgment it is still too early to give a satisfactory account of Navier-Stokes preconditioners linked to the optimal Euler family (2). Instead, I will refer to two forthcoming publications (Lee, 1996; Lee *et al.*, 1997), in which a design analysis is presented of local Navier-Stokes preconditioners, effective at all Mach and Reynolds numbers, and even for turbulent flow, along with some illustrative numerical results.

6. Conclusions

The systematic study of local preconditioning has provided a wealth of results regarding the structure of the Euler equations and their discretizations. Properly designed preconditioning matrices can equalize physical time scales in the continuum equations, boost the significant terms in the discretized equations so that these are not drowned by the truncation error, and thus go far beyond the single benefit shared with Jacobi preconditioning, i. e., effective high-frequency damping. The issue of robustness remains, but considerable progress has been made in fighting the stagnation-point instability without introducing an arbitrary cut-off parameter in the matrices. Local matrix preconditioning is bound to become a standard technique for Euler and Navier-Stokes calculations, a worthy successor to its scalar little sister, local time-stepping.

References

Allmaras, S.R., "Analysis of semi-implicit preconditioners for multigrid solution of the 2-d compressible Navier-Stokes equations," AIAA Paper 95-1651-CP, 1995.

Brandt, A., "Guide to multigrid development," *Multigrid Methods*, W. Hackbush and U. Trottenberg, eds., No. 960, Lecture Notes in Mathematics, Springer Verlag, 1982.

Choi, Y.-H. and Merkle, L., "The application of preconditioning in viscous flows," *Journal of Computational Physics*, Vol. 195, 1993, pp. 207-223.

Darmofal, D.L. and Schmid, P.J., "The importance of eigenvectors for local preconditioning of the Euler equations," *Journal of Computational Physics*, Vol. 127, 1996, pp. 346-362.

Deconinck, H. and Degrez, G., "Monotone shock-capturing cell vertex schemes for the Euler and Navier-Stokes equations on unstructured grids", *Fifteenth International Conference on Numerial Methods in Fluid Mechanics*, Springer, 1996.

Godfrey, A.G., Walters, R.W., and van Leer, B., "Preconditioning for the Navier-Stokes equations with finite-rate chemistry," AIAA Paper 93-0535, 1993.

Lee, D., "Local preconditioning of the Euler and the Navier-Stokes equations," PhD thesis, University of Michigan, 1996.

Lee, D., van Leer, B., and Lynn, J.F., "A local Navier-Stokes preconditioner for all Mach and cell Reynolds numbers," AIAA Paper 97-2024-CP, 1997.

Lee, W.-T., "Local preconditioning of the Euler equations," PhD thesis, University of Michigan, 1991.

Lynn, J.F., "Multigrid solution of the Euler equations with local preconditioning," PhD thesis, University of Michigan, 1995.

Lynn, J.F. and van Leer, B., "Multi-stage schemes for the Euler and Navier-Stokes equations with optimal smoothing," AIAA Paper 93-3355-CP, 1993.

Lynn, J.F. and van Leer, B., "A semi-coarsened multigrid solver for the Euler and Navier-Stokes equations with local preconditioning," AIAA Paper 95-1667-CP, 1995.

Lynn, J.F., van Leer, B., and Lee, D., "Multigrid solution of the Euler equations with local preconditioning," *Fifteenth International Conference on Numerical Methods in Fluid Mechanics*, Springer 1996.

Mesaros, L.M., "Multi-dimensional fluctuation splitting schemes for the Euler Equations on unstructured grids," PhD thesis, University of Michigan, 1995.

Mesaros, L.M. and Roe, P.L., "Multidimensional fluctuation-splitting schemes based on decomposition methods," AIAA Paper 95-1699-CP, 1995.

Mulder, W.A., "A new multigrid approach to convection problems," *Journal of Computational Physics*, Vol. 83, 1989, pp. 303-323.

Pierce, N.A. and Giles, M.B., "Preconditioning compressible flow calculatons on stretched meshes," AIAA Paper 96-0889, 1996.

Reed, C.L., "Low speed preconditioning applied to the compressible Navier-Stokes equations," PhD thesis, University of Texas at Arlington, 1995.

Ta'asan, S., "Canonical forms of multidimensional inviscid flows," ICASE Report No. 93-34, 1993.

Tai, C.-H., "Acceleration techniques for explicit Euler codes," PhD thesis, University of

Michigan, 1990.

Turkel, E., "Preconditioned methods for solving the incompressible and low speed compressible equations," *Journal of Computational Physics*, Vol. 72, 1987, pp. 277-298.

Turkel, E., "Review of preconditioning methods for fluid dynamics," ICASE Report No. 92-47, 1993.

Turkel, E., Fiterman, A., and van Leer, B., "Preconditioning and the limit to the incompressible flow equations," ICASE Report No. 93-42, 1993.

Turkel, E., Vatsa, V.N., and Radespiel, R., "Preconditioning methods for low-speed flows," 1996.

van Leer, B., Lee, W.T., and Roe, P.L., "Characteristic time-stepping or local preconditioning of the Euler equations," *AIAA 10th Computational Fluid Dynamics Conference*, 1991, pp 260-282; also AIAA-91-1552-CP.

van Leer, B., Mesaros, L., Tai, C.-H., and Turkel, E., "Local preconditioning in a stagnation point," *12th AIAA Computational Fluid Dynamics Conference*, 1995, pp. 88-101. Also AIAA-95-1654-CP.

Volpe, G., "On the use and accuracy of compressible flow codes at low Mach numbers," AIAA Paper 91-1662, 1991.

THE QUEST FOR DIAGONALIZATION OF DIFFERENTIAL SYSTEMS

PHIL ROE

W.M. Foundation Laboratory for Computational Fluid Dynamics
Department of Aerospace Engineering
University of Michigan, Ann Arbor Michigan
and ICASE, NASA Langley, Hampton, VA

AND

ELI TURKEL

School of Mathematical Sciences
Sackler Faculty of Exact Sciences
Tel-Aviv University, Tel-Aviv, Israel
and ICASE, NASA Langley, Hampton, VA

Abstract.

We consider ways of reducing systems of constant coefficient differential equations to a set of scalar equations. For two dimensional systems in (x, y) this is done in general. When one introduces a time variable in addition to the two space variables this can be analyzed by introducing a preconditioned system. This is equivalent to using a congruence transformation to diagonalize the steady state equations.

In three dimensions we only consider the specific example of the linearized Euler equations. We demonstrate that these can be diagonalized by replacing the first order set of equations by scalar third order equations. If one insists on not increasing the order of the derivatives then one must distinguish between subsonic and supersonic flows.

1. Introduction

The concept of diagonalization plays an important role in many *CFD* algorithms. Conventional upwinding techniques, as applied to systems of equations, rest on the diagonalization of one-dimensional operators, which fol-

V. Venkatakrishnan et al (eds.), Barriers and Challenges in Computational Fluid Dynamics, 351-369.
© 1998 *Kluwer Academic Publishers.*

lows from strong hyperbolicity. These methods are successful in enhancing the robustness of the resultant codes especially at hypersonic speeds. Preconditioning can be viewed as modifying the transient behavior of the solution at the continuum level. This was first used to allow larger time steps for low Mach number regions (Turkel, 1987). Later it was found that preconditioning is also necessary to get the correct convergence of a compressible code to an incompressible code (Turkel et $al.$, 1994). Here we view preconditioning as a technique to allow (block) diagonalization. This attempts to isolate solution modes whose convergence can be enhanced by special treatments. This has a substantial side effect on improving the accuracy of codes at locally low Mach numbers. Strongly related to this concept are recent proposals to apply independent spatial discretizations to the elliptic and hyperbolic components of the steady state equations. In all these cases the object is to identify groups of variables as operators that would benefit from special numerical treatment. Therefore, it is timely to review the mathematics of diagonalization, especially as it relates to the Euler equations of compressible flow.

We shall consider general systems, at least in two dimensions. However, our analysis is always restricted to the linear case with constant coefficients. Our typical system is either a steady state problem $\mathcal{L}u = 0$ or the unsteady problem $u_t + \mathcal{L}u = 0$. When diagonalizing such a system we distinguish between three types of diagonalization.

1. A similarity transformation is introduced so that $Q\mathcal{L}Q^{-1}$ is diagonal. In this case if we define $\mathbf{v} = Q\mathbf{u}$ then we have a diagonal system for the variables $\mathbf{v}$. Hence, this is equivalent to a change of variables.

2. We precondition the unsteady problem to get $\mathbf{P}^{-1}u_t + \mathcal{L}u = 0$ where $\mathbf{P}$ is positive definite and $\mathcal{L}$ is symmetric. We now assume a decomposition $\mathbf{P} = TS$ and define $v = T^{-1}u$. Then the equation becomes $v_t + (S\mathcal{L}T)\,v = 0$. A particular case is when $S = T^*$. This is convenient since every positive matrix $\mathbf{P}$ has the decomposition $\mathbf{P} = TT^*$. In this case we have a congruence transformation of $\mathcal{L}$. Hence, preconditioning is equivalent to a congruence transformation plus a similarity transformation.

3. In the above two cases we have implicitly assumed that the transformation Q contains only constant coefficients and not differential operators. Hence, we are not changing the order of the differential system. A different approach is to allow the transformations to depend on derivatives. This is equivalent to replacing the first order system by a set of higher order equations. This again can be done via a change of variables, i.e. a similarity transform or else via a preconditioning i.e., a congruence transform.

In all these cases we demand that the steady state solution be independent of the transformations used so that either a similarity or congruence transformation may be allowed.

We first consider equations in one space direction plus time so that $\mathcal{L} = A\partial_x$. In this case any strongly hyperbolic $(m \times m)$ system of the form

$$\partial_t \mathbf{u} + A\partial_x \mathbf{u} = 0$$

is reducible to m scalar convection equations of the form

$$\partial_t(\ell_k \cdot \mathbf{u}) + \lambda_k \partial_x(\ell_k \cdot \mathbf{u}) = 0,$$

where ℓ_k is a left eigenvector of A, and λ_k is the corresponding eigenvalue. This follows since for a strongly hyperbolic system A is diagonalizable. Thus, the canonical form is simply the familiar linear advection equation

$$\partial_t u + a\partial_x u = 0.$$

In two dimensions we consider first the steady-state equations

$$A\partial_x u + B\partial_y u = 0. \tag{1}$$

When A and B are symmetric it is known (Au-Yeung, 1969) that this equation can be diagonalised by a congruence transformation if and only if some real linear combination of A and B is positive definite. In that case the canonical form is again linear advection

$$a\partial_x u + b\partial_y u = 0. \tag{2}$$

If the above condition is not met, then we shall show that some of the eigenvalues of $A^{-1}B$ appear in conjugate pairs, and a second canonical form appears which is a (2×2) system, equivalent under scaling and axis transformations to the Cauchy-Riemann equations

$$\begin{pmatrix} 1 & 0 \\ 0 & -1 \end{pmatrix} \partial_x \begin{pmatrix} u \\ v \end{pmatrix} + \begin{pmatrix} 0 & -1 \\ -1 & 0 \end{pmatrix} \partial_y \begin{pmatrix} u \\ v \end{pmatrix} = 0. \tag{3}$$

The general two-dimensional system can then be written in a form that is block diagonal, with a mixture of scalar and (2×2) blocks. When the steady solution is sought by marching in a time-like variable, we show that a preconditioning matrix P can always be found, such that this block structure is preserved in the transient solution of

$$\mathbf{P}^{-1}\partial_t \mathbf{u} + A\partial_x \mathbf{u} + B\partial_y \mathbf{u} = 0$$

The scalar blocks have then the canonical form

$$\partial_t u + a\partial_x u + b\partial_y u = 0. \tag{4}$$

The (2×2) blocks, although they behave elliptically in the steady state, behave hyperbolically in the transient phase, and the associated variables follow a wave equation

$$\frac{\partial^2 \psi}{\partial t^2} = \frac{\partial^2 \psi}{\partial x^2} + \frac{\partial^2 \psi}{\partial y^2} \tag{5}$$

We consider (4) and (5) as our basic canonical forms. Although both of these equations are hyperbolic, considered as time-dependent problems, only (4) is associated with any distinct direction (namely $dy/dx = b/a$). By contrast, (5), during the transient phase, is associated with characteristic surfaces, $(x - x_0)^2 + (y - y_0)^2 - (t - t_0)^2 = const$, rather than with individual directions. In the steady state, (4) is a hyperbolic equation while (5) reduces to an elliptic equation, Laplace's equation, and so has no uniquely distinguished directions.

In three dimensions, we have found three canonical forms, which are

$$\frac{\partial \psi}{\partial t} + a\frac{\partial \psi}{\partial x} + b\frac{\partial \psi}{\partial y} + c\frac{\partial \psi}{\partial z} = 0, \tag{6}$$

and (3×3) systems equivalent to either

$$\frac{\partial^2 \psi}{\partial t^2} = \frac{\partial^2 \psi}{\partial x^2} + \frac{\partial^2 \psi}{\partial y^2} + \frac{\partial^2 \psi}{\partial z^2}, \tag{7}$$

or

$$\left(\frac{\partial}{\partial t} - \frac{\partial}{\partial x}\right)^2 \psi = \frac{\partial^2 \psi}{\partial y^2} + \frac{\partial^2 \psi}{\partial z^2}, \tag{8}$$

In this paper we will only present the decomposition and not discuss any applications.

2. The Two-dimensional Case

2.1. GENERAL STEADY-STATE PROBLEMS

We consider

$$A\partial_x u + B\partial_y u = 0 \tag{9}$$

with A and B symmetric. This is a reasonable assumption since most physical systems can be described by symmetric hyperbolic equations.

According to a result of Au-Yeung (Au-Yeung, 1969) this equation can be diagonalized by a congruence transformation if a real linear combination of A and B is positive definite. In the present context, this condition has a physical interpretation. Suppose that we add a time-dependent term $\partial_t u$ to (9). Then the speeds with which waves propagate in a direction θ are the eigenvalues of $A \cos \theta + B \sin \theta$. If this matrix is positive definite for $\theta = \theta^*$, then the domain of influence of a disturbance at the origin lies in the half-plane $x \cos \theta^* + y \sin \theta^* > 0$. Therefore the domain of influence does not enclose the origin, and tangents to it from the origin may be drawn. These are the characteristic directions for the steady solution. This situation arises whenever the flow speed is greater than the speed of any signal relative to the medium (supersonic case). Thus, the two dimensional supersonic flow equations are always diagonalizable.

To deal with the subsonic case, we begin by attempting a formal diagonalization of (9). We premultiply the equation by a matrix S, and effect a change of variable $Qv = u$, so that

$$(SAQ)\partial_x v + (SBQ)\partial_y v = 0. \tag{10}$$

We require that SAQ, SBQ be representable as

$$SAQ = \Lambda_A^{\#},$$
$$SBQ = \Lambda_B^{\#},$$

where $\Lambda_A^{\#}, \Lambda_B^{\#}$ denote matrices having the same block diagonal structure. (Note that matrices sharing such a structure form a closed group under multiplication and inversion.) Then

$$(\Lambda_A^{\#})^{-1}\Lambda_B^{\#} = (Q^{-1}A^{-1}S^{-1})(SBQ) = Q^{-1}A^{-1}BQ$$

so that Q must represent a similarity transformation that block diagonalises $A^{-1}B$. (If A happens to be singular we may diagonalize $B^{-1}A$ instead). At the present stage, the matrix S is still arbitrary. We first attempt a complete diagonalization.

$$A^{-1}B = R\Lambda L \qquad \text{where} \qquad R = L^{-1}$$

If this succeeds, we can take $S = LA^{-1}$ and $Q = R$, to derive

$$\partial_x(Lu) + \Lambda\partial_y(Lu) = 0.$$

representing a set of characteristic equations for the characteristic variables $\mathbf{v}_k = \ell_k \cdot \mathbf{u}$ propagating along directions $dy/dx = \lambda_k$. This is a similarity transformation that reduces A to the identity matrix and diagonalizes B.

Suppose this fails in that

$$\det(A^{-1}B - \lambda I) = 0$$

has a pair of conjugate roots

$$\lambda_{1,2} = \lambda_R \pm i\lambda_I.$$

Then we will also find conjugate eigenvectors

$$\mathbf{r}_{1,2} = \mathbf{r}_R \pm i\mathbf{r}_I,$$
$$\ell_{1,2} = \ell_R \pm i\ell_I.$$

That is,

$$A^{-1}B = \begin{bmatrix} \cdot & \cdot & & & \\ \cdot & \cdot & & & \\ r_1 & r_2, & \cdot & \cdot & \cdot \\ \cdot & \cdot & & & \\ \cdot & \cdot & & & \end{bmatrix} \begin{bmatrix} \lambda_1 & & & \\ & \lambda_2 & & \\ & & \cdot & \\ & & & \cdot \end{bmatrix} \begin{bmatrix} \cdot & \cdot & \ell_1 & \cdot & \cdot \\ \cdot & \cdot & \ell_2 & \cdot & \cdot \\ & & \cdot & & \\ & & \cdot & & \end{bmatrix}$$

Now introduce the complex matrices

$$J = \frac{1}{\sqrt{2}} \begin{bmatrix} 1 & 1 & & \\ i & -i & & \\ & & \sqrt{2} & \\ & & & \cdot \end{bmatrix}, \quad J^T = \frac{1}{\sqrt{2}} \begin{bmatrix} 1 & i & & \\ 1 & -i & & \\ & & \sqrt{2} & \\ & & & \cdot \end{bmatrix}$$

and rewrite the factorization

$$A^{-1}B = (RJ^T)((J^T)^{-1}\Lambda J^{-1})(JL) = R^{\#}\Lambda^{\#}L^{\#}$$

Then it can be verified that $R^{\#}, L^{\#}$ are real matrices, and that $\Lambda^{\#}$ is the real symmetric matrix

$$\Lambda^{\#} = (J^T)^{-1}\Lambda J^{-1} = \begin{bmatrix} \lambda_R & \lambda_I & & \\ \lambda_I & -\lambda_R & & \\ & & \lambda_3. & \\ & & & \cdot \end{bmatrix}$$

Now we write the governing equations in the form (10) with $S = (R^{\#})^{-1}A^{-1}$ and $Q = (L^{\#})^{-1}$; the result, after partial simplification, is

$$(R^{\#})^{-1}(L^{\#})^{-1}\partial_x v + \Lambda^{\#}\partial_y v = 0$$

Here,

$$(R^{\#})^{-1}(L^{\#})^{-1} = (L^{\#}R^{\#})^{-1}$$
$$= (JLRJ^T)^{-1}$$

$$= (JJ^T)^{-1}$$

$$= \begin{bmatrix} 1 & 0 & \\ 0 & -1 & \\ & & 1 \\ & & \end{bmatrix}.$$

Thus, when expressed in terms of variables $\mathbf{v} = L^{\#}\mathbf{u}$, the equations decouple into $(m-2)$ advection equations, and the 2×2 symmetric elliptic system

$$\begin{pmatrix} 1 & 0 \\ 0 & -1 \end{pmatrix} \partial_x \begin{pmatrix} u \\ v \end{pmatrix} + \begin{pmatrix} \lambda_R & \lambda_I \\ \lambda_I & -\lambda_R \end{pmatrix} \partial_y \begin{pmatrix} u \\ v \end{pmatrix} = 0. \qquad (11)$$

If there is more than one conjugate pair of eigenvalues, the treatment above can easily be extended by adding additional (2×2) blocks to the matrix J, so we have

THEOREM Any $(m \times m)$ first order constant coefficient hyperbolic system in two dimensions decomposes into $(m - 2k)$ convection equations and k independent (2×2) symmetric elliptic systems.

REMARK The $(m \times 2)$ blocks that appear in $R^{\#}$ provide basis vectors for the subspaces inhabited by the elliptic subproblems; the $(2 \times m)$ blocks that appear in $L^{\#}$ are operators that project an arbitrary vector into these subspaces. We note that Binding (Binding, 1990) has also connected the diagonalization problem with that of the generalized eigenvalues.

2.2. THE STEADY EULER EQUATIONS

Take the vector of unknowns to be $d\mathbf{u} = (\frac{dp}{\rho c}, du, dv, dp - c^2 d\rho)^T$, and adopt coordinates aligned with the flow direction. Then A, B are the symmetric matrices

$$A = \begin{bmatrix} M & 1 & 0 & 0 \\ 1 & M & 0 & 0 \\ 0 & 0 & M & 0 \\ 0 & 0 & 0 & M \end{bmatrix}, \qquad B = \begin{bmatrix} 0 & 0 & 1 & 0 \\ 0 & 0 & 0 & 0 \\ 1 & 0 & 0 & 0 \\ 0 & 0 & 0 & 0 \end{bmatrix}$$

We then find that the equation $A^{-1}B = R\Lambda L$ becomes

$$A^{-1}B = \begin{bmatrix} 0 & 0 & \frac{M}{M^2-1} & 0 \\ 0 & 0 & \frac{-1}{M^2-1} & 0 \\ \frac{1}{M} & 0 & 0 & 0 \\ 0 & 0 & 0 & 0 \end{bmatrix} =$$

$$\begin{bmatrix} M & M & 0 & 0 \\ -1 & -1 & 1 & 0 \\ \sqrt{M^2-1} & -\sqrt{M^2-1} & 0 & 0 \\ 0 & 0 & 0 & 1 \end{bmatrix} \begin{bmatrix} \frac{1}{\sqrt{M^2-1}} & 0 & 0 & 0 \\ 0 & \frac{-1}{\sqrt{M^2-1}} & 0 & 0 \\ 0 & 0 & 0 & 0 \\ 0 & 0 & 0 & 0 \end{bmatrix} \begin{bmatrix} \frac{1}{2M} & 0 & \frac{1}{2\sqrt{M^2-1}} & 0 \\ \frac{1}{2M} & 0 & \frac{-1}{2\sqrt{M^2-1}} & 0 \\ \frac{1}{M} & 1 & 0 & 0 \\ 0 & 0 & 0 & 1 \end{bmatrix}$$

For supersonic flow, the transformation $\mathbf{v} = L\mathbf{u}$ produces four real characteristic equations, which are, in primitive variables,

$$\begin{aligned} \partial_x S &= 0 & dS &= dp - a^2 d\rho \\ \partial_x H &= 0 & dH &= dp + \frac{\rho}{2}(u^2) = dp + \rho q\, dq \end{aligned}$$

$$\begin{aligned} (\beta \partial_x + \partial_y)(\beta dp + \rho u\, dv) &= 0 \\ (\beta \partial_x - \partial_y)(\beta dp - \rho u\, dv) &= 0 \end{aligned}$$

where $\beta^2 = M^2 - 1$.

If $M^2 < 1$, two of these characteristic equations break down. We may then have recourse to the alternative factorization $R^\# \Lambda^\# L^\#$, which is

$$A^{-1}B = \begin{bmatrix} 0 & 0 & \frac{M}{M^2-1} & 0 \\ 0 & 0 & -\frac{1}{M^2-1} & 0 \\ \frac{1}{M} & 0 & 0 & 0 \\ 0 & 0 & 0 & 0 \end{bmatrix} =$$

$$\begin{bmatrix} \sqrt{2}M & 0 & 0 & 0 \\ -\sqrt{2} & 0 & 1 & 0 \\ 0 & -\sqrt{2(1-M^2)} & 0 & 0 \\ 0 & 0 & 0 & 1 \end{bmatrix} \begin{bmatrix} 0 & \frac{-1}{\sqrt{1-M^2}} & 0 & 0 \\ \frac{-1}{\sqrt{1-M^2}} & 0 & 0 & 0 \\ 0 & 0 & 0 & 0 \\ 0 & 0 & 0 & 0 \end{bmatrix} \begin{bmatrix} \frac{1}{\sqrt{2}M} & 0 & 0 & 0 \\ 0 & 0 & \frac{1}{\sqrt{2(1-M^2)}} & 0 \\ \frac{1}{M} & 1 & 0 & 0 \\ 0 & 0 & 0 & 1 \end{bmatrix}$$

From this factorization we recover the same streamwise characteristic equations as in the supersonic case. It is interesting to note that the entropy characteristic corresponds to a shared eigenvector of the matrices A, B, whereas the enthalpy characteristic does not. The two acoustic characteristics collapse into the elliptic system deriving from (11), which is

$$\begin{pmatrix} 1 & 0 \\ 0 & -1 \end{pmatrix} \partial_x \begin{pmatrix} \frac{\beta_*}{\rho u} dp \\ dv \end{pmatrix} + \begin{pmatrix} 0 & \frac{-1}{\beta_*} \\ \frac{-1}{\beta_*} & 0 \end{pmatrix} \partial_y \begin{pmatrix} \frac{\beta_*}{\rho u} dp \\ dv \end{pmatrix} = 0. \qquad (12)$$

where $\beta_*^2 = 1 - M^2$.

These are the Cauchy-Riemann equations written in Prandtl-Glauert coordinates $(x, y/\beta_*)$. Hence, we can express this as two scalar second order elliptic equations.

This decomposition differs from that found by Ta'asan He employed, as canonical variables, (S, H, u, v) and his decomposition was not completely block diagonal, containing some additional coupling terms. The variables which emerge from the present treatment are (S, H, p, v) and there is no coupling between blocks.

2.3. PRECONDITIONING

Suppose that the time-dependent two-dimensional problem obeys

$$\partial_t u + A\partial_x u + B\partial_y u = 0 \tag{13}$$

where A and B are symmetric. To find a decomposition of this equation as it stands would require a three-dimensional treatment. However, there are occasions when only the steady state is of interest, but it is convenient to compute it by following some time-like evolution. In such cases (13) may be replaced (sometimes advantageously (Turkel, 1987; Turkel, 1993; van Leer *et al.*, 1991)) by

$$\mathbf{P}^{-1}\partial_t u + A\partial_x u + B\partial_y u = 0 \tag{14}$$

where P is a 'preconditioning matrix'. To retain well-posedness, P must be positive definite. In this section we point out that P can always be chosen to preserve the canonical block-diagonal form of the steady state equations.

As in the steady case, premultiply (14) by $(R^\#)^{-1}A^{-1}$ and make the change of variable $\mathbf{v} = L^\#\mathbf{u}$; this gives

$$(R^\#)^{-1}A^{-1}\mathbf{P}^{-1}(L^\#)^{-1}\partial_t v + (JJ^T)^{-1}\partial_x v + \Lambda^\#\partial_y v = 0.$$

For the decomposition to be preserved, it is neccessary and sufficient that

$$(R^\#)^{-1}A^{-1}\mathbf{P}^{-1}(L^\#)^{-1} = D^\#$$

where $D^\#$ is block diagonal of the same form as the matrices that multiply the space derivatives. Solving for $\mathbf{P}$ gives

$$\mathbf{P} = (R^\# ADL^\#)^{-1}.$$

where the only constraint on $D^\#$ is that it shall be positive definite to ensure well-posedness.

To apply this analysis to the Euler equations we take

$$D^\# = \begin{bmatrix} P & Q & 0 & 0 \\ R & S & 0 & 0 \\ 0 & 0 & \frac{1}{M} & 0 \\ 0 & 0 & 0 & \frac{1}{M} \end{bmatrix}.$$

This assigns the fluid velocity u to each of the scalar waves. We require $\Delta = PS - QR > 0$ for positive definiteness. After carrying through the required operations we find that

$$\mathbf{P}^{-1} = \begin{bmatrix} \frac{1}{M^2} - P\frac{\beta_*^2}{M} & \frac{1}{M} & Q\beta_* & 0 \\ \frac{1}{M} & 1 & 0 & 0 \\ R\beta_* & 0 & -SM & 0 \\ 0 & 0 & 0 & 1 \end{bmatrix}.$$

and also

$$\mathbf{P} = \begin{bmatrix} -\frac{S}{\Delta}\frac{M}{\beta_*^2} & \frac{S}{\Delta}\frac{1}{\beta_*^2} & -\frac{Q}{\Delta}\frac{1}{\beta_*} & 0 \\ \frac{S}{\Delta}\frac{1}{\beta_*^2} & 1 - \frac{S}{\Delta}\frac{1}{M\beta_*^2} & \frac{Q}{\Delta}\frac{1}{M\beta_*} & 0 \\ -\frac{R}{\Delta}\frac{1}{\beta_*} & \frac{R}{\Delta}\frac{1}{M\beta_*} & -\frac{P}{\Delta}\frac{1}{M} & 0 \\ 0 & 0 & 0 & 1 \end{bmatrix}.$$

This is the most general preconditioner that preserves the canonical decomposition. It may be noted that $\det P = (\beta_*^2\Delta)^{-1}$. The preconditioner of van Leer et al. (van Leer et al., 1991) corresponds to the special case $P = S = -1/(M\beta_*)$, $Q = R = 0$, $\Delta = 1/(M^2\beta_*^2)$.

It is still an open question which choice of $\mathbf{P}$ leads to the fastest and most robust convergence to a steady state for a numerical code. This is principally because it is not easy to translate these qualities into mathematical terms. Here, we will identify robustness with the behavior of the solutions as $M \to 0$. We identify the rate of convergence with the condition number i.e. the ratio of the largest and smallest propagation speeds in the transient solution. We wish this condition number to be chosen as small as possible. [1] We will also insist on symmetry requirements for well-posedness.

When the preconditioned equations are written in Cartesian coordinates not aligned with the flow, terms may appear which depend on the flow direction as the flow velocity vanishes. A necessary and sufficient condition to avoid this phenomenon is that $P_{22} = P_{33}$, which implies

$$\beta_*^2(M\Delta + P) = S. \tag{15}$$

The wavespeeds in the direction θ are given by

$$\det[A\cos\theta + B\sin\theta - P^{-1}\lambda]$$

---

[1] In practice life is not this simple. Choosing different preconditioners also changes the "artificial" viscosity associated with the scheme which in turn affects the rate of convergence. Hence, in applications it is not always the lowest condition number that gives the quickest convergence (Turkel et al., 1994; Turkel et al., 1996).

which gives a twice repeated root $\lambda = M \cos\theta$, together with the quadratic term

$$\Delta\beta^2\lambda^2 + [(P-S)\beta^2\cos\theta + (Q+R)\beta\sin\theta]\lambda = \beta^2\cos^2\theta + \sin^2\theta. \quad (16)$$

To keep this symmetrical with respect to the streamline, we must take $Q + R = 0$, and we consider first the case $Q = R = 0 \Rightarrow \Delta = AD$. The wavespeeds in the x-direction are then $1/P$ and $-1/S$, and in the y-direction $\pm 1/(\beta_* \sqrt{PS})$. It can be shown that the best condition number is obtained from the preconditioner of van Leer *et al.*, but unfortunately this does not satisfy the condition (15) at low Mach numbers. We require, in fact, that as $M \to 0$, then

$$(S + 1/M)(P - 1/M) = -1/M^2.$$

This is the graph of a hyperbola in the SP plane. We are only concerned with the branch that passes through the origin (the other branch leads to imaginary wavespeeds), and only with the half of that branch for which P, S are both negative (the other half does not make P positive definite). The condition number based solely on the 'acoustic' waves is

$$K = \max(S/P, P/S)$$

which is unity at the origin and degrades away from it. Taking into account that the flow velocity is M, and that $|P| > |S|$, we actually have

$$K = \max\left(\frac{S}{P}, \frac{P}{S}, \frac{1}{MS}, MP\right).$$

The point on the hyperbola that minimizes this is

$$P = \frac{-1}{M} \qquad S = \frac{-1}{2M}$$

leading to K=2.

Of course, the preconditioner does not have to satisfy (15) away from $M = 0$, and can be blended in with the van Leer *et al.* preconditioner, by some such choice as

$$P = \frac{-1}{M\beta_*}, \qquad S = \frac{-1}{M * \beta_*(1 + \beta_*)}$$

which leads to the simple preconditioner

$$\mathbf{P} = \begin{bmatrix} \frac{M^2}{\beta_*} & \frac{-M}{\beta_*} & 0 & 0 \\ \frac{-M}{\beta_*} & \frac{1+\beta_*}{\beta_*} & 0 & 0 \\ 0 & 0 & \beta_*(1+\beta_*) & 0 \\ 0 & 0 & 0 & 1 \end{bmatrix}, \quad \mathbf{P}^{-1} = \begin{bmatrix} \frac{1+\beta_*}{M^2} & \frac{1}{M} & 0 & 0 \\ \frac{1}{M} & 1 & 0 & 0 \\ 0 & 0 & \frac{1}{\beta_*(1+\beta_*)} & 0 \\ 0 & 0 & 0 & 1 \end{bmatrix}.$$

$$(17)$$

Rather strikingly, these are identical with those given by van Leer *et al.*, except for the factors $(1 + \beta_*)$ in the $()_{33}$ elements.

3. The Three-dimensional Case

3.1. GENERAL STEADY-STATE PROBLEMS

The most common block diagonal forms in three-dimensional problems appear to have blocks that are either scalar or else (3×3). The possibility of (2×2) blocks is ruled out, apart from exceptional cases, by an anlysis of Gilquin, Laurens and Rosier (Gilquin *et al.*, 1993), who considered the general (2×2) symmetric system in multi-dimensions. They only require that the system be strongly hyperbolic (i.e. combinations of the coefficient matrices can always be diagonalized). They then prove that such systems can always be decomposed into a convective part (possibly with speed zero) and a *two dimensional* elliptic part. Therefore, (2×2) blocks are incapable of representing three-dimensional elliptic systems.

Due to the complexity of the general three dimensional case we shall only consider the special case of the linearized Euler equations.

3.2. EULER EQUATIONS

We write the linearized Euler equations in the operator-matrix notation

$$\mathbf{w}_t + \mathcal{L}\mathbf{w} = 0 \tag{18}$$

with $d\mathbf{w} = (\frac{dp}{\rho c}, du, dv, dw, dS)^t$. Define

$$Q = u\partial_x + v\partial_y + w\partial_z.$$

Since all coefficients are assumed constant Q commutes with ∂_x, ∂_y and ∂_z. Hence,

$$\mathcal{L} = QI + c \begin{pmatrix} 0 & \partial_x & \partial_y & \partial_z & 0 \\ \partial_x & 0 & 0 & 0 & 0 \\ \partial_y & 0 & 0 & 0 & 0 \\ \partial_z & 0 & 0 & 0 & 0 \\ 0 & 0 & 0 & 0 & 0 \end{pmatrix}.$$

If we Fourier transform this system (with Fourier variables ξ, η, ζ) we get

$$\hat{\mathcal{L}} = QI + ic \begin{pmatrix} 0 & \xi & \eta & \zeta & 0 \\ \xi & 0 & 0 & 0 & 0 \\ \eta & 0 & 0 & 0 & 0 \\ \zeta & 0 & 0 & 0 & 0 \\ 0 & 0 & 0 & 0 & 0 \end{pmatrix}.$$

Since $\hat{\mathcal{L}}$ depends on the identity matrix and a scalar times a symmetric matrix it can always be diagonalized by a similarity transformation even though $\hat{\mathcal{L}}$ is not self-adjoint. $\hat{\mathcal{L}}$ has two 'acoustic' eigenvalues, $Q \pm \sqrt{\xi^2 + \eta^2 + \zeta^2}$ and a triple eigenvalue of Q.

<u>Definition:</u> The transformation is called algebraic if Q and Q^{-1} are independent of the Fourier variables ξ, η, ζ. The transformation is differential if Q and Q^{-1} depend on polynomials of the Fourier variables. In this case we are effectively increasing the order of the differential system of equations. Finally the transformation is pseudo-differential if it depends on more general functions of the Fourier variables.

Straightforward calculations show that $\hat{\mathcal{L}}$ can be diagonalized by a similarity transformation with a unitary matrix

$$
\mathbf{Q}^{-1} = \begin{pmatrix}
0 & 0 & \frac{1}{\sqrt{2}} & \frac{1}{\sqrt{2}} & 0 \\
-\frac{\eta}{\sqrt{\xi^2+\eta^2}} & \frac{\xi\zeta}{N} & \frac{\xi}{\lambda_+\sqrt{2}} & \frac{\xi}{\lambda_-\sqrt{2}} & 0 \\
\frac{\xi}{\sqrt{\xi^2+\eta^2}} & \frac{\eta\zeta}{N} & \frac{\eta}{\lambda_+\sqrt{2}} & \frac{\eta}{\lambda_-\sqrt{2}} & 0 \\
0 & -\frac{\xi^2+\eta^2}{N} & \frac{\zeta}{\lambda_+\sqrt{2}} & \frac{\zeta}{\lambda_-\sqrt{2}} & 0 \\
0 & 0 & 0 & 0 & 1
\end{pmatrix}
$$

where $\lambda_\pm = \pm\sqrt{\xi^2 + \eta^2 + \zeta^2}$, and $N = |\lambda|\sqrt{\xi^2 + \eta^2}$. $\mathbf{Q}^{-1}$ is essentially unique. The eigenvectors corresponding to the acoustic eigenvalues are unique except for the order. For the triple convective eigenvalue one can choose a different orthogonal basis for the eigenspace but this can not change the basic form.

From this we see that one can only diagonalize the Euler equations (either time dependent or steady state) by a similarity transformation if pseudo-differential operators are used. These are not directly useful for any numerical methods. Their only direct use would be to approximate one of the square roots by a rational approximation in some appropriate limit. For example, $\sqrt{\xi^2 + \eta^2 + \zeta^2} = \xi\sqrt{1 + \left(\frac{\eta}{\xi}\right)^2 + \left(\frac{\zeta}{\xi}\right)^2}$. If the fractions are small we can then expand this in a Taylor series. Since we cannot diagonalize the Euler equations by a similarity transformation without using pseudo-differential operators we shall instead consider congruence transformations.

We now demonstrate that one can diagonalize the three dimensional system by a differential congruence transformation. Define

$$
D = Q^2 - c^2(\partial_x^2 + \partial_y^2 + \partial_z^2). \tag{19}
$$

We now replace (18) by the preconditioned system

$$
\mathbf{w}_t + \mathbf{P}_D \mathcal{L} \mathbf{w} = 0
$$

with

$$\mathbf{P}_D = \begin{pmatrix} Q^2 & -c\partial_x Q & -c\partial_y Q & -c\partial_z Q & 0 \\ -c\partial_x Q & Q^2 - c^2(\partial_y^2 + \partial_z^2) & c^2\partial_x\partial_y & c^2\partial_x\partial_z & 0 \\ -c\partial_y Q & c^2\partial_x\partial_y & Q^2 - c^2(\partial_x^2 + \partial_z^2) & c^2\partial_y\partial_z & 0 \\ -c\partial_z Q & c^2\partial_x\partial_z & c^2\partial_y\partial_z & Q^2 - c^2(\partial_x^2 + \partial_y^2) & 0 \\ 0 & 0 & 0 & 0 & I \end{pmatrix}$$

One can then verify that

$$\mathbf{P}_D\mathcal{L} = \begin{pmatrix} QD & 0 & 0 & 0 & 0 \\ 0 & QD & 0 & 0 & 0 \\ 0 & 0 & QD & 0 & 0 \\ 0 & 0 & 0 & QD & 0 \\ 0 & 0 & 0 & 0 & 1 \end{pmatrix}.$$

These equations imply that we can diagonalize the three dimensional Euler equations (by a congruence transformation) by converting them to scalar equations of at most third order. This is certainly not the minimal degree that can be achieved. In particular the enthalpy equations can be diagonalized while maintaining a first order equation. We can use simpler matrices than $\mathbf{P}_D$ by considering congruence transformations. Let

$$\mathbf{P}_E = \begin{pmatrix} Q & -c\partial_x & -c\partial_y & -c\partial_z & 0 \\ 0 & 1 & 0 & 0 & 0 \\ 0 & 0 & 1 & 0 & 0 \\ 0 & 0 & 0 & 1 & 0 \\ 0 & 0 & 0 & 0 & 1 \end{pmatrix} \qquad \mathbf{P}_E^* = \begin{pmatrix} Q & 0 & 0 & 0 & 0 \\ -c\partial_x & 1 & 0 & 0 & 0 \\ -c\partial_y & 0 & 1 & 0 & 0 \\ -c\partial_z & 0 & 0 & 1 & 0 \\ 0 & 0 & 0 & 0 & 1 \end{pmatrix}$$

then

$$\mathbf{P}_E\mathcal{L}\mathbf{P}_E^* = \begin{pmatrix} DQ & 0 & 0 & 0 & 0 \\ 0 & Q & 0 & 0 & 0 \\ 0 & 0 & Q & 0 & 0 \\ 0 & 0 & 0 & Q & 0 \\ 0 & 0 & 0 & 0 & Q \end{pmatrix}$$

and so we have diagonalized $\mathcal{L}$ by a congruence transformation (Turkel, 1993).

The operator D is just the full potential operator. For subsonic flow this is an elliptic operator and so invertible. For supersonic flow D is a hyperbolic operator. Similarly, Q is a hyperbolic operator denoting convection along a streamline. Thus, given appropriate boundary conditions this too is invertible. Hence, $\mathbf{P}$ is invertible except at stagnation points and along the sonic line.

3.2.1. *SUBSONIC - THREE DIMENSIONS*

We consider the preconditioned subsonic equations for three dimensions in streamwise coordinates. We shall analyze the specific preconditioning suggested by van Leer, Lee and Roe (van Leer *et al.*, 1991). By our previous discussion the decomposition also applies to the steady state equations without any preconditioning. Let $\beta_* = \sqrt{1 - M^2}$.

$$\frac{\beta_* + 1}{M^2} p_t + \frac{\rho u}{M^2} u_t + u p_x + \rho c^2 (u_x + v_y + w_z) = 0$$

$$\frac{1}{\rho u} p_t + u_t + u u_x + \frac{p_x}{\rho} = 0$$

$$\frac{1}{\beta_*} v_t + u v_x + \frac{p_y}{\rho} = 0$$

$$\frac{1}{\beta_*} w_t + u w_x + \frac{p_z}{\rho} = 0$$

$$S_t + u S_x = 0$$

S and H are convected along the stream line. We now have three equations for the unknowns (p,v,w). Let $(dq, dr, ds) = (\frac{\beta_*}{\rho u} dp, du, dv)$.

$$\begin{pmatrix} q \\ r \\ s \end{pmatrix}_t + \begin{pmatrix} -\beta_* u & 0 & 0 \\ 0 & \beta_* u & 0 \\ 0 & 0 & \beta_* u \end{pmatrix} \begin{pmatrix} q \\ r \\ s \end{pmatrix}_x + \begin{pmatrix} 0 & u & 0 \\ u & 0 & 0 \\ 0 & 0 & 0 \end{pmatrix} \begin{pmatrix} q \\ r \\ s \end{pmatrix}_y + \begin{pmatrix} 0 & 0 & u \\ 0 & 0 & 0 \\ u & 0 & 0 \end{pmatrix} \begin{pmatrix} q \\ r \\ s \end{pmatrix}_z = 0$$

We next rescale x by defining $x = \beta_* x'$. We rewrite this equation in terms of x' and then drop the prime. The effect is to eliminate the β_* term from the x derivative. Then in this modified coordinate system we get

$$\begin{pmatrix} q \\ r \\ s \end{pmatrix}_t + \begin{pmatrix} -u & 0 & 0 \\ 0 & u & 0 \\ 0 & 0 & u \end{pmatrix} \begin{pmatrix} q \\ r \\ s \end{pmatrix}_x + \begin{pmatrix} 0 & u & 0 \\ u & 0 & 0 \\ 0 & 0 & 0 \end{pmatrix} \begin{pmatrix} q \\ r \\ s \end{pmatrix}_y + \begin{pmatrix} 0 & 0 & u \\ 0 & 0 & 0 \\ u & 0 & 0 \end{pmatrix} \begin{pmatrix} q \\ r \\ s \end{pmatrix}_z = 0 \quad (20)$$

We wish to know when this first order system is equivalent to the Laplace equation. To do this we study the equivalence of the three dimensional Laplace equation with a first order system. Let

$$\frac{\partial^2 \psi}{\partial x^2} + \frac{\partial^2 \psi}{\partial y^2} + \frac{\partial^2 \psi}{\partial z^2} = 0.$$

Define $u = \psi_x$, $v = -\psi_y$ and $w = -\psi_z$. Then

$$\begin{pmatrix} -1 & 0 & 0 \\ 0 & 1 & 0 \\ 0 & 0 & 1 \end{pmatrix} \begin{pmatrix} u \\ v \\ w \end{pmatrix}_x + \begin{pmatrix} 0 & 1 & 0 \\ 1 & 0 & 0 \\ 0 & 0 & 0 \end{pmatrix} \begin{pmatrix} u \\ v \\ w \end{pmatrix}_y + \begin{pmatrix} 0 & 0 & 1 \\ 0 & 0 & 0 \\ 1 & 0 & 0 \end{pmatrix} \begin{pmatrix} u \\ v \\ w \end{pmatrix}_z = 0.$$

Hence, it is trivial that the second order three dimensional Laplace equation yields the first order system. However, one can not, in general, start with the (u, v, w) first order system and obtain a Laplace equation for each of the variables. By the existence of the potential ψ we also have $v_z - w_y = 0$. If this is added as an additional equation to the system then it is easily shown that (u, v, w) do satisfy a Laplace equation for each variable. Hence, to go from the system back to the second order Laplace equation for a potential ψ we have a compatibility requirement on the solution.

For the three dimensional Euler equations the "acoustic" portion corresponds to the Laplace equation (or time dependent wave equation) only if we satisfy a compatibilty constraint corresponding to the vanishing of the streamwise component of the vorticity. If this constraint is not met then we can only diagonalize the system by including higher derivatives.

Starting with (21) and differentiating we find that q satisfies the wave equation $\frac{\partial^2 q}{\partial t^2} = u^2 \Delta q$. However r and s do not obey a wave equation. Instead we get

$$\frac{\partial^2 r}{\partial t^2} = u^2 \left(\frac{\partial^2 r}{\partial x^2} + \frac{\partial^2 r}{\partial y^2} + \frac{\partial^2 s}{\partial y \partial z} \right) = u^2 \left(\Delta r + \frac{\partial}{\partial z} \left(\frac{\partial s}{\partial y} - \frac{\partial r}{\partial z} \right) \right)$$

$$\frac{\partial^2 s}{\partial t^2} = u^2 \left(\frac{\partial^2 s}{\partial x^2} + \frac{\partial^2 s}{\partial z^2} + \frac{\partial^2 r}{\partial y \partial z} \right) = u^2 \left(\Delta s - \frac{\partial}{\partial y} \left(\frac{\partial s}{\partial y} - \frac{\partial r}{\partial z} \right) \right)$$

Differentiate the equation for r by y and the equation for s by z and add we find that $\frac{\partial^2 (r_y + s_z)}{\partial t^2} = u^2 \Delta (r_y + s_z)$, i.e.$(r_y + s_z)$ also satisfies the wave equation. Since q satisfies a wave equation so does q_x and hence so does the divergence $q_x + r_y + s_z$. Define $\omega = r_z - s_y$ i.e. the z component of the vorticity. Differentiating the equation for r by z and the equation for s by y and subtracting we find that $\omega_t + u \omega_x = 0$. Thus, we have two "elliptic" variables, q and div(q,r,s) and one "hyperbolic" variable ω in addition to the previous two hyperbolic variables S and H. Hence, if we solve this three dimensional system with an initial condition that the vorticity is zero then the vorticity remains zero for all time. In this case we can go from the first order system back to the second order wave equation. In two dimensions the system is purely elliptic since ω is always identically zero.

3.2.2. *SUPERSONIC - THREE DIMENSIONS*

In three dimensions we consider the van Leer et al. preconditioned super-sonic equations in streamwise coordinates. As before this also applies to the non-preconditioned steady state equations. Let $\beta = \sqrt{M^2 - 1}$.

$$\frac{\beta M + 1}{M^2}p_t + \frac{\rho c}{M}u_t + up_x + \rho c^2(u_x + v_y + w_z) = 0$$

$$\frac{1}{\rho u}p_t + u_t + uu_x + \frac{p_x}{\rho} = 0$$

$$\frac{M}{\beta}v_t + uv_x + \frac{p_y}{\rho} = 0$$

$$\frac{M}{\beta}w_t + uw_x + \frac{p_z}{\rho} = 0$$

$$S_t + uS_x = 0$$

Solving for (p, u, v, w, S) we get

$$p_t + \frac{\beta u}{M}p_x + \frac{\rho c^2 M}{\beta}(v_y + w_z) = 0$$

$$u_t + uu_x - \frac{1}{\rho}\left(\frac{\beta}{M} - 1\right)p_x - \frac{c}{\beta}(v_y + w_z) = 0$$

$$v_t + \frac{\beta}{M}(uv_x + \frac{p_y}{\rho}) = 0$$

$$w_t + \frac{\beta}{M}(uw_x + \frac{p_z}{\rho}) = 0$$

$$S_t + uS_x = 0$$

As before the system for (p,v,w) decouples from the rest. Let $\mathbf{w} = (\frac{\beta}{\rho u}dp, dvdw)^t$. Then

$$\mathbf{w}_t + \frac{\beta u}{M}\begin{pmatrix} 1 & 0 & 0 \\ 0 & 1 & 0 \\ 0 & 0 & 1 \end{pmatrix}\mathbf{w}_x + \begin{pmatrix} 0 & c & 0 \\ c & 0 & 0 \\ 0 & 0 & 0 \end{pmatrix}\mathbf{w}_y + \begin{pmatrix} 0 & 0 & c \\ 0 & 0 & 0 \\ c & 0 & 0 \end{pmatrix}\mathbf{w}_z = 0 \quad (21)$$

Normalizing (21) we have

$$\mathbf{w}_t + \mathbf{w}_x + B\mathbf{w}_y + C\mathbf{w}_z = 0$$

with

$$B = \begin{pmatrix} 0 & 1 & 0 \\ 1 & 0 & 0 \\ 0 & 0 & 0 \end{pmatrix} \qquad C = \begin{pmatrix} 0 & 0 & 1 \\ 0 & 0 & 0 \\ 1 & 0 & 0 \end{pmatrix}$$

This is a new canonical form which cannot be reduced to either a hyperbolic (convective) equation or to an elliptic (acoustic) equation. Instead since the coefficient matrix for the x direction reduces to the identity matrix we can introduce new variables $\tau = t$ and $\xi = x - t$. Then

$$\mathbf{w}_\tau + B\mathbf{w}_y + C\mathbf{w}_z = 0 \qquad \mathbf{w} = (q, r, s)^t$$

This is identical to the canonical form for the subsonic case considered above. Hence, q satisfies a wave equation and r, s satisfy

$$\frac{\partial^2 r}{\partial t^2} = \Delta r + \frac{\partial}{\partial z}\left(\frac{\partial s}{\partial y} - \frac{\partial r}{\partial z}\right)$$

$$\frac{\partial^2 s}{\partial t^2} = \Delta s - \frac{\partial}{\partial y}\left(\frac{\partial s}{\partial y} - \frac{\partial r}{\partial z}\right)$$

So, $r_y + s_z$ satisfies a two dimensional wave equation while $(s_y - r_x)_\tau = 0$ in the moving coordinate system. Hence, the vorticity component $s_y - r_x$ satisfies a one dimensional convection equation, in the steamwise direction, as we already have found.

Thus, the supersonic system consists of three convective equations for the entropy, enthalpy and vorticity that decouple. In addition we have a system of two equations that is a combined hyperbolic-elliptic system in the sense that it is an elliptic system in the y and z directions in a system moving in the x direction.

References

Au-Yeung, Y.H., *Proceedings Amer. Math. Soc.*, Vol. 20, 1969, pp. 545-548.

Binding, P., "Simultaneous Diagonalisation of Several Hermitian Matrices," *SIAM J. Matrix Analysis and Applic.*, Vol. 11, 1990, pp. 531-536.

Gilquin, H., Laurens, J., and Rosier, C., "Multi-Dimensional Riemann Problems for Linear Hyperbolic Systems: Part I," in *Nonlinear Hyperbolic Problems: Theoretical, Applied and Computational Aspects*, Donato, A. and Oliveri, F., eds., Vieweg, 1993, pp. 276-279.

Mesaros, L. and Roe, P., "Multidimensional Fluctuation Splitting Schemes Based on Decomposition Methods," *AIAA 12th Computational Fluid Dynamics Conference*, 1995, pp. 582-591.

Roe, P. and Mesaros, L., "Solving Steady Mixed Conservation Laws by Elliptic/Hyperbolic Decomposition," *15th International Conference Numerical Methods Fluid Dynamics*, Springer Lecture Notes in Physics, to appear, 1997.

Ta'asan, S., "Canonical-Variables Multigrid Method for Steady-State Euler Equations," *Fourteenth Inter. Conf. Numer. Meth. Fluid Dynamics*, Lecture Notes in Physics, Deshpande et al., eds., Springer, 1994.

Ta'asan, S., "Essentially Optimal Multigrid Method for Steady State Euler Equations," *AIAA Paper 95-0209*, 1995.

Turkel, E., "Preconditioned Methods for Solving the Incompressible and Low Speed Compressible Equations," *Journal of Computational Physics*, Vol. 72, 1987, pp. 277-298.

Turkel, E., "A Review of Preconditioning Methods for Fluid Dynamics," *Applied Numerical Mathematics*, Vol. 12, 1993, pp. 257-284.

Turkel, E., Fiterman, A., and van Leer, B., "Preconditioning and the Limit of the Compressible to the Incompressible Flow Equations for Finite Difference Schemes," *Frontiers of Computational Fluid Dynamics 1994*, Caughey, D.A. and Hafez, M.M., eds., John Wiley and Sons, 1994, pp. 215-234.

Turkel, E., Vatsa, V.N, and Radespiel, R., "Preconditioning Methods for Low Speed Flow," *AIAA Paper 96-2460*, 1996.

van Leer, B., Lee, W.T., and Roe, P.L., "Characteristic Time-Stepping or Local Preconditioning of the Euler Equations," *AIAA Paper 91-1552*, 1991.

MULTIDIMENSIONAL UPWINDING: UNFOLDING THE MYSTERY

DAVID SIDILKOVER
ICASE
Mail Stop 403
NASA Langley Research Center
Hampton, VA 23681

Abstract.

Some important advances took place during the last several years in the development of genuinely multidimensional upwind schemes for the compressible Euler equations. In particular, a robust, high-resolution genuinely multidimensional scheme which can be used for any of the flow regimes computations was constructed (see (Sidilkover, 1994b; Sidilkover, 1994a; Sidilkover, 1995)). This paper summarizes briefly these developments and outlines the *fundamental* advantages of this approach.

1. Introduction and brief review

The efficiency of the existing steady-state multigrid solvers routinely used for compressible flow problems in engineering practice is still very poor. There clearly exists a pressing need for more efficient algorithms.

In order to obtain a truly efficient steady-state solver, some *fundamental* issues concerning different aspects of an algorithm need to be addressed. The recently proposed genuinely multidimensional approach towards the construction of the discrete schemes for the compressible flow resolves some of these issues. A discussion in this regard is the main subject of this paper.

[0]This research was supported by the National Aeronautics and Space Administration under NASA Contract No. NAS1-19480 while the author was in residence at the Institute for Computer Applications in Science and Engineering (ICASE), NASA Langley Research Center, Hampton, VA 23681-0001.

V. Venkatakrishnan et al (eds.), Barriers and Challenges in Computational Fluid Dynamics, 371-386.
© 1998 *Kluwer Academic Publishers.*

1.1. MULTIDIMENSIONAL UPWINDING

The quest for a genuinely multidimensional upwind scheme began more than a decade ago. Initially it was motivated chiefly by the expectation that, once such a scheme is developed, it will imitate the fluid flow phenomena more accurately than the standard dimension-by-dimension approach. It was, however, suggested later in (Sidilkover, 1989),(Sidilkover & Brandt, 1993) that improving the efficiency of the the steady-state solvers may be the most important reason for developing the genuinely multidimensional approach.

One of the main difficulties in the numerical treatment of compressible flow is the possible presence of shocks in the solution. It is well known that a scheme that is both second order accurate and avoids under- and overshoots (which may trigger a nonlinear instability) near discontinuities has to be nonlinear. Such a scheme has to incorporate the so-called *high-resolution* mechanism, i.e. a smoothness monitor, that is usually implemented in the form of a flux-limiter. Initially, such schemes were developed for the one-dimensional case. Then, extending this approach to multidimensions was done on a dimension-by-dimension basis. The well known fact, however, is that the Gauss-Seidel relaxation is unstable when applied in conjunction with such schemes (Spekreijse, 1987). Therefore, the standard multi-grid solvers have to resort to the defect-correction technique or multistage Runge-Kutta relaxation and the efficiency of such solvers may be poor. A closer look reveals that the standard high-resolution discretizations suffer from the following deficiency: the high-frequency error components may be (nearly) invisible to the residuals of the discrete equations, i.e. the discrete scheme is (nearly) unstable. In turn this means that it may be inherently impossible to construct a good smoother (an important ingredient of a multigrid solver) using these discrete schemes.

A genuinely multidimensional advection scheme was constructed in (Sidilkover, 1989; Sidilkover & Brandt, 1993). The scheme was named "genuinely multidimensional" since it imitates well the anisotropy of the advection phenomena in two dimensions: the artificial dissipation is added only along the streamline, while the high-resolution mechanism affects significantly the cross-flow direction only. The key feature of this scheme is the two-dimensional limiter, i.e. the argument of a limiter-function is the ratio of finite differences in two different coordinate directions. The scheme was formulated in the control-volume context for Cartesian grids and relied on the compact 9-point-box stencil. The *fundamental* advantage of this approach is that the two-dimensional high-resolution mechanism does not damage the stability properties of the discretization.

The so-called "residual distribution" (or "fluctuation-splitting") schemes for scalar advection equation on unstructured triangular grids were presented in (Deconinck *et al.*, 1991). It was found later that these schemes have some links to the aforementioned genuinely two-dimensional control-volume approach for the advection equation. Exploration of these links led to the unification of the two approaches and resulted in a scheme that incorporated two-dimensional limiters and was formulated for unstructured triangular grids. This scheme (like that presented in (Sidilkover, 1989),(Sidilkover & Brandt, 1993)) can be given a purely *algebraic* interpretation. However, the task of extending these ideas to systems of equations appeared to be a complicated one.

Consider a hyperbolic system of partial differential equations in two dimensions

$$\mathbf{u}_t + A\mathbf{u}_x + B\mathbf{u}_y = 0, \tag{1}$$

where $\mathbf{u}$ is the vector of size N and A, B are $N \times N$ matrices. The matrices A and B in general do not commute. This means that they cannot be diagonalized simultaneously, i.e. the system cannot be written as N advection equations.

A prolonged effort was to represent locally the physics of compressible flow by finite number of simple waves using the local gradients of the solution (in the spirit of (Roe, 1986)) with intention to apply a genuinely two-dimensional advection scheme to each one of the simple waves. However, the discretizations for the Euler equations constructed in this way suffered from a severe lack of robustness.

The breakthrough approach that resulted in a robust genuinely multidimensional scheme, suitable for the computations of the entire range of the flow regimes was presented in (Sidilkover, 1994b). Then it was described in more detail including the implications for multigrid and extension to 3D in (Sidilkover, 1994a) and (Sidilkover, 1995). The key idea was not to try to apply the multidimensional advection to systems, but rather the same strategy that was used to construct the scalar scheme. The *algebraic* interpretation of the advection scheme played an important role at this point. It was crucial to recognize that a certain linear first order scheme based on standard upwind methodology can be used as a basic building block for the hyperbolic systems as well as for the scalar advection. The multidimensional high-resolution correction is then applied in a formal way similarly to the scalar case. The resulting scheme for the Euler equations was demonstrated to produce a very good quality solution for subsonic, transonic and supersonic regimes. The approach was called "genuinely-multidimensional" since it can be argued that it leads to a discrete scheme whose artificial dissipation is a rotationally-invariant differential operator. However, it was not clear if this property is of any direct practical importance.

The constructed high-resolution scheme for the Euler equations relies on a compact stencil. The result of this property is a smaller error in smooth regions and better resolution of discontinuities, compared to the standard dimension-by-dimension approach. However, in our view, these are only *marginal* improvements. The *fundamental* advantage of this approach is that its high-resolution mechanism does not damage the stability properties of the scheme. It was demonstrated in (Sidilkover, 1994a; Sidilkover, 1995) that the Collective Gauss-Seidel relaxation is stable when applied directly to the resulting high-resolution discretization of the hyperbolic systems. This results in a very simple, efficient and robust multigrid solver for the compressible Euler equations, suitable for the entire range of flow regimes.

Some researchers who were previously pursuing other directions adopted the genuinely-multidimensional approach proposed in (Sidilkover, 1994b; Sidilkover, 1994a; Sidilkover, 1995). A modification of the underlying first order linear scheme (the system N-scheme) aimed at improving the discontinuity resolution was proposed by van der Weide and Deconinck ("Positive Matrix Distribution Schemes" (van der Weide & Deconinck, 1996)).

It should be mentioned that important steps towards the construction of a genuinely multidimensional schemes for the Euler equations were made by Colella (Colella, 1984; Colella, 1990), LeVeque (LeVeque, 1988) and Radvogin (Radvogin, 1991). However, the nonlinear high-resolution corrections in these schemes rely on one-dimensional limiters, which introduces some of the dimension-by-dimension flavor.

Another very interesting related approach was proposed in (Giles *et al.*, 1989). A discretization for the triangular unstructured grids for the Euler equations was developed. The problem was that the scheme was linear, i.e. it did not incorporate any high-resolution mechanism.

1.2. MULTIGRID FOR ADVECTION DOMINATED PROBLEMS

One of the major reasons for the poor efficiency of the standard flow solvers (see (Brandt, 1984)) is the fact that for advection dominated problems the coarse grid provides only a fraction of the needed correction for certain error components. It is well known that the steady Euler equations contain two different factors: the advection and the Full-Potential type operators. The latter is either of elliptic or hyperbolic type depending on the flow regime (subsonic or supersonic). The difficulty mentioned above can be avoided ((Brandt, 1984)) by constructing a solver that distinguishes between different factors of the system and treats each one appropriately. In the subsonic case, for instance, the advection factor can be treated by marching and the elliptic factor – multigrid. The efficiency of such an algorithm will be essentially the same as that of the multigrid solver for the elliptic part only. Such algorithms are referred to as "essentially optimal". An approach to achieve a separation between the co-factors – the so-called Distributive Gauss-Seidel relaxation – was proposed in (Brandt, 1984). It was demonstrated in (Brandt & Yavneh, 1993) that using this approach one can obtain the essentially optimal multigrid efficiency for a staggered-grid discretization of the incompressible Navier-Stokes equations.

A related approach was proposed by Ta'asan in (Ta'asan, 1994) for the incompressible and compressible subsonic Euler equations. The staggered-grid discretization is based on the *canonical* variables formulation (see (Ta'asan, 1993)), that expresses the partitioning of the steady Euler equation into elliptic and advection factors. The essentially optimal multigrid efficiency was demonstrated using this approach for subsonic flow and body-fitted grids. A possible limitation of this approach may be that it is not directly generalizable for the viscous flow.

Another approach towards achieving the optimal multigrid efficiency is based on the pressure equation formulation of the Euler equations (see (Sidilkover & Ascher, 1995),(Roberts *et al.*, 1996)). Its main virtue is simplicity. Its limitation, however, is that it may be not generalizable to *viscous* compressible flow.

Following the work of Ta'asan, some researchers attempted to apply the idea of partitioning the Euler equations towards the construction of discrete schemes (see (Mesaros & Roe, 1995) and (Paillère, 1995)). It is well-known that the two-dimensional Euler equations in supersonic case can be written as four locally decoupled advection equations (see (Hirsch, 1990)). This property was used as a basis for applying the advection schemes to discretize the system in this case. In subsonic case, however, the distinction was made between the advection and the elliptic ("acoustic subsystem") partitions. The treatment of transonic flow, however, was problematic since it required matching of two different discretizations accross the sonic lines. Another drawback of these approaches was that they are cannot be generalized to three dimensions (see (van der Weide & Deconinck, 1996)). To conclude, the discrete schemes constructed in this way suffered from a lack of robustness and generality. No multigrid efficiency in subsonic regime was demonstrated either. Finally, some researchers who previously followed this direction adopted the genuinely-multidimensional approach proposed in (Sidilkover, 1994b; Sidilkover, 1994a; Sidilkover, 1995) (see (van der Weide & Deconinck, 1996)).

1.3. WHAT THIS PAPER IS ABOUT

In this paper, we first present a brief review of the construction of genuinely multidimensional schemes for the scalar advection and the compressible

Euler equations. We summarize the basic properties of the discretizations, emphasizing those that are unique to this approach and are of *fundamental* importance for practical purposes.

As it was mentioned earlier, it is not clear, what the direct practical implications of the *genuine multidimensionality* (or the rotational invariance of the artificial dissipation) property of the approach are. However, we present in this paper a heuristic argument suggesting that the *genuine multidimensionality* is closely related to the *factorizability* property of the discretization. The latter is of *fundamental* practical importance. It is necessary in order to construct a Distributive Gauss-Seidel relaxation that will allow one to decouple the advection and Full-Potential co-factors of the Euler system and thus to obtain an optimally efficient multigrid solver.

2. Genuinely multidimensional upwind approach: summary

In this section we briefly review the construction of the high-resolution genuinely multidimensional upwind schemes for the scalar advection equation and for the Euler system. Consider a general triangular grid covering the domain. Assume the discrete unknowns are defined at the grid nodes thus defining a linear function and allowing us to evaluate the gradients of the current solution approximation on each triangle. The approach is to construct discrete equations which are to be solved at each node from the portions of the residuals of the equations computed on the triangles having this node as a common vertex. In other words, portions of the residual of the equation evaluated on a particular triangle contribute to the construction of the discrete equations at the nodes of this triangle. The problem is to find the exact rules for this construction, so that the resulting discrete equations will have certain desirable properties.

2.1. ADVECTION SCHEME

We consider triangular element T, and choose two out of the three faces. Then we write the advection equation we wish to solve in the local coordinate system aligned with these two faces

$$u_t + au_x + bu_y = 0 \tag{2}$$

Without loss of generality we can consider a linear constant coefficient equation, since in general non-linear case we can linearize the equation on each triangle according to the conservative linearization procedure suggested in (Deconinck *et al.*, 1993).

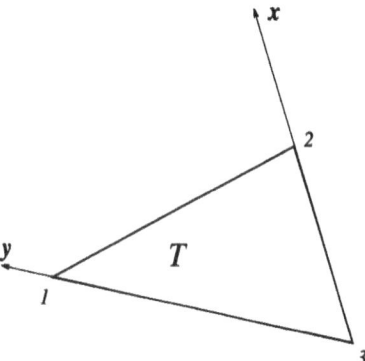

Figure 1. Triangle.

We can write the discrete equation at the grid node i in the following form

$$\sum_{j=1}^{n} c_{i,j}(u_j - u_i) = 0, \qquad (3)$$

Definition 1 *The discrete scheme* (3) *is said to be of the positive type if* $c_{i,j} \geq 0$ *for all* $j = 1, ..., n$.

Numerical soltuions obtained using a positive scheme satisfy a discrete *maximimum principle*. This property is useful to ensure that the discrete solution will be non-oscillatory near discontinuities.

We shall outline here the construction of a positive advection scheme. Residual of the equation (2) can be represented as a sum of two portions

$$r = r^x + r^y, \qquad (4)$$

where

$$r^x = -S_T a u_x^h; \qquad r^y = -S_T b u_y^h. \qquad (5)$$

Residual of the equation (2) on the triangle T contributes to the construction of the discrete equations to be solved at each of the three nodes of the triangle according to the following residual distribution formulae

$$
\begin{aligned}
&\text{node } 1 \leftarrow r^x(1 - \text{sign}(a)) \\
&\text{node } 2 \leftarrow r^x(1 + \text{sign}(a)) + r^y(1 - \text{sign}(b)) \\
&\text{node } 3 \leftarrow r^y(1 + \text{sign}(b))
\end{aligned}
\qquad (6)
$$

It easy to see that this construction results in a positive scheme since for any real (positive or negative) number z we have the following inequality $\pm z(1 \pm \text{sign}(z)) \geq 0$. The accuracy of such a scheme, though, is only first order.

Definition 2 *A discrete scheme is called <u>linearity preserving</u> if whenever the residual r on the triangle T vanishes, the contributions due to this residual lead to a zero update of the solution at each of the three nodes of the triangle.*

A *linearity preserving* scheme is second order accurate.

Define the following quantities

$$r^{x^*} = r^x + r^y \Psi(q); \qquad r^{y^*} = r^y + r^x \Psi(q)/q \qquad (7)$$

where $q = -r^x/r^y$. In this paper we assume that Ψ is the *minmod* limiter. Substituting r^{x^*}, r^{y^*} instead of r^x, r^y into (6) we obtain a high-resolution scheme. The important feature here is the two-dimensionality of the limiter, i.e. the fact that the argument of the limiter-function is a ratio of numerical derivatives in two different coordinate direction. ((Sidilkover & Brandt, 1993),(Sidilkover, 1989)).

Using the following limiter identity

$$r^y \Psi(q) \equiv -r^x \Psi(q)/q, \qquad (8)$$

it is easy to verify that the constructed nonlinear scheme is indeed both *positive* and *linearity preserving*.

2.2. EXTENSION TO THE EULER SYSTEM

Consider a hyperbolic system of partial differential equations

$$\mathbf{u}_t + A\mathbf{u}_x + B\mathbf{u}_y = 0. \qquad (9)$$

The discrete equation approximating (9) at node i can be written as follows

$$\sum_{j=1}^{n} C_{i,j}(\mathbf{u}_j - \mathbf{u}_i) = 0, \qquad (10)$$

Property of positivity can be formally extended to the system case.

Definition 3 *The discrete scheme* (10) *is said to be of the <u>positive type</u> if the matrices $C_{i,j}$ $(j = 1, ..., n)$ have non-negative eigenvalues.*

It is not clear, however, how to generalize the *maximum principle* for systems. No conclusions can be derived from the positivity property unless the additional assumption about the symmetry of the matrices $C_{i,j}$ $(j = 1, ..., n)$ is made. In this case the energy stability property of the scheme can be demonstrated. However, in case of the Euler system, this would require the use of the *symmetrizing* variables formulation, which is *non-conservative*.

This makes the energy stability property too restrictive to be of substantial practical importance for the Euler equations.

It is interesting to note though that the standard high-resolution schemes, if carefully implemented, are of the *positive* type. Therefore, we aimed at constructing a genuinely multidimensional high-resolution scheme ((Sidilkover, 1994b; Sidilkover, 1994a; Sidilkover, 1995)), which is of the *positive* type as well.

Assume that the hyperbolic system of equations (9) is written in the non-orthogonal coordinate frame aligned with the two of the faces of triangle T (Fig.1). Residuals of the system on triangle T can be represented as a sum of two portions

$$\mathbf{R} = \mathbf{R}^x + \mathbf{R}^y, \quad \text{where } \mathbf{R}^x = -S_T A \mathbf{u}_x^h; \quad \mathbf{R}^y = -S_T B \mathbf{u}_y^h \quad (11)$$

Consider the following residual distribution formula

$$\begin{aligned} &\text{node } 1 \leftarrow \mathbf{R}^x(I - \text{sign}(A)) \\ &\text{node } 2 \leftarrow \mathbf{R}^x(I + \text{sign}(A)) + \mathbf{R}^y(I - \text{sign}(B)) \\ &\text{node } 3 \leftarrow \mathbf{R}^y(I + \text{sign}(B)) \end{aligned} \quad (12)$$

Assuming that matrix M has a complete set of real eigenvalues (definition of the hyperbolicity of a system) it is easy to see that matrix $\pm M(I \pm \text{sign}(M))$ will is non-negative definite. This means that the scheme defined by (12) is of the positive type by construction (as an upwind scheme is expected to be).

In order to obtain a *positive* high-resolution genuinely multidimensional scheme for a hyperbolic system we may have first to rewrite system (9) (as it was done in (Sidilkover, 1994b; Sidilkover, 1994a; Sidilkover, 1995)) in a different set of variables. The *auxiliary* variables $\mathbf{v} = (s, u, v, p)^T$ (see (Sidilkover, 1994b; Sidilkover, 1994a; Sidilkover, 1995)) are a good choice for the Euler system (here s is the entropy, u, v - the velocity components and p is the pressure). System (9) rewritten in varibales $\mathbf{v}$ takes the following form

$$\mathbf{v}_t + \mathcal{A}\mathbf{v}_x + \mathcal{B}\mathbf{v}_y = 0 \quad (13)$$

where

$$\mathbf{u} = T\mathbf{v} \quad (14)$$

As before, we can compute residual $\mathbf{r}$ of system (13) on triangle T and represent it as a sum of two portions:

$$\mathbf{r} = \mathbf{r}^x + \mathbf{r}^y, \quad \text{where } \mathbf{r}^x = -S_T \mathcal{A}\mathbf{v}_x^h; \quad \mathbf{r}^y = -S_T \mathcal{B}\mathbf{v}_y^h. \quad (15)$$

Considering the following residual distribution formula

$$\begin{aligned} &\text{node } 1 \leftarrow T\mathbf{r}^x(I - \text{sign}(\mathcal{A})) \\ &\text{node } 2 \leftarrow T[\mathbf{r}^x(I + \text{sign}(\mathcal{A})) + \mathbf{r}^y(I - \text{sign}(\mathcal{B}))] \\ &\text{node } 3 \leftarrow T\mathbf{r}^y(I + \text{sign}(\mathcal{B})), \end{aligned} \quad (16)$$

we arrive at a positive first order accurate scheme that is *identical* to (12).

Introduce the following quantities

$$r_i^{x^*} = r_i^x + r_i^y \Psi(q_i); \qquad r_i^{y^*} = r_i^y + r_i^x \Psi(q_i)/q_i \tag{17}$$

with $q_i = -r_i^x/r_i^y$, and $i = 1, ..., N$ Denote by $\mathbf{r}^{x^*}$ and $\mathbf{r}^{y^*}$ vectors whose components are $r_i^{x^*}$ and $r_i^{y^*}$ $(i = 1, \ldots, N)$ respectively. The high-resolution genuinely two-dimensional scheme can be obtained by substituting $\mathbf{r}^{x^*}$ and $\mathbf{r}^{y^*}$ instead of the $\mathbf{r}^x$ and $\mathbf{r}^y$ into (16).

Using the limiter identity

$$r_i^y \Psi(q_i) \equiv -r_i^x \Psi(q_i)/q_i, \text{ for } i = 1, ..., N \tag{18}$$

it is possible to show that the genuinely two-dimensional high-resolution scheme is both positive and linearity preserving. We emphasize here again, that in order to achieve this property for the Euler equations, it was necessary to use another (non-conservative) form of the equations when introducing the high-resolution mechanism ((Sidilkover, 1994b; Sidilkover, 1994a; Sidilkover, 1995)). The *auxiliary* variables formulation was suitable for this purpose. The only justification for the desirability of the positivity property for systems of equations is that the standard high-resolution schemes, if properly implemented, satisfy this property.

In addition to the smaller error in smooth regions and sharper resolution of discontinuities (compared to the standard dimension-by-dimension methods), the constructed scheme also offers a possibility to optimize the stencil in order to resolve a particular discontinuity layer (by choosing two faces of a triangular element that are the closest to the direction of this layer). The important advantage of this approach, however, is that the genuinely multidimensional high-resolution mechanism constructed in this section does not damage the stability properties of the scheme.

3. Separation of co-factors

Compressible steady Euler equations in two dimensions can be written in the following matrix form

$$L\mathbf{u} = 0, \tag{19}$$

where

$$\mathbf{u} = (s, u, v, p)^T, \tag{20}$$

$$L = \begin{pmatrix} Q & 0 & 0 & 0 \\ 0 & \rho Q & 0 & \partial_x \\ 0 & 0 & \rho Q & \partial_y \\ 0 & \rho \partial_x & \rho \partial_y & \frac{1}{c^2} Q \end{pmatrix}, \tag{21}$$

and $Q = \vec{U} \cdot \nabla$ is the advection operator. In order to find out what is the type of the system of equations we can look at the determinant of matrix L

$$det(L) = \rho^2 (Q)^2 [\frac{1}{c^2} Q^2 - \Delta] \qquad (22)$$

The determinant has two distinct co-factors: one of advection and another Full-Potential types.

In the rest of this paper we shall consider the subsonic case only. The Full-Potential factor in this case is of the elliptic type.

It was suggested in (Brandt, 1984) that the different co-factors should also be treated differently (each one in the appropriate way). One way to do this is to cast the equations into such a form that the co-factors can be discretized separately. Canonical variables formulation (Ta'asan, 1994) as well as pressure equation based schemes (see (Roberts et al., 1996)) belong to this category. A more general approach (as proposed by Brandt in (Brandt, 1984)) is to discretize the equations in some primitive form, but to design relaxation of Distributive Gauss-Seidel type (DGS) such that it will separate the treatment of co-factors. However, the discrete scheme suitable for this purpose should be *factorizable*, i.e. satisfy the discrete analog of the property (22).

We would like to construct a linear first order upwind "positive" scheme such that it is factorizable and is also upgradable to second order using the genuinely two-dimensional high-resolution mechanism.

First, we shall take a closer look at the one-dimensional case.

3.1. ONE-DIMENSIONAL CASE

Consider a first order upwind scheme for the one-dimensional Euler equations. Without loss of generality we consider the primitive variable formulation

$$L^h \mathbf{u}^h = 0 \qquad (23)$$

where

$$\mathbf{u}^h = (s^h, u^h, p^h)^T \qquad (24)$$

and

$$L^h = \begin{pmatrix} -\frac{h}{2}|u|\partial_{xx}^h + Q^{2h} & 0 & 0 \\ 0 & \rho(-\frac{h}{2}c\partial_{xx}^h + Q^{2h}) & \partial_x^{2h} - \frac{h}{2}\frac{u}{c}\partial_{xx}^h \\ 0 & \rho(\partial_x^{2h} - \frac{h}{2}\frac{u}{c}\partial_{xx}^h) & -\frac{h}{2c}\partial_{xx} + \frac{1}{c^2}Q^{2h} \end{pmatrix}, \qquad (25)$$

where h is a meshsize, ∂_{xx}^h is a central approximation of the second derivative, ∂_x^{2h} is a central approximation of the first derivative and $Q^{2h} = u\partial_x^{2h}$ is the advection operator.

3.1.1. *Factorization*

The determinant of L^h:

$$det(L^h) = \rho(-\frac{h}{2}|u|\partial_{xx}^h + Q^{2h})[(1 - M^2)\partial_{xx}^h]$$ (26)

The first factor is the upwind scheme approximating an advection operator corresponding to the entropy equations. The Full-Potential factor is approximated by a "short" central difference. The issue of factorization appears to be trivial in this case, since the momentum and the pressure equations correspond solely to the elliptic factor.

3.1.2. *Distributive Gauss-Seidel relaxation*

Introducing new variables

$$(s^h, u^h, p^h)^T = \mathbf{u}^h = M^h \mathbf{w}^h = M^h(s^h, w^h, \phi^h)^T$$ (27)

the Euler system will take the following form

$$L^h M^h \mathbf{w}^h = 0$$ (28)

Assuming that that

$$M^h = \begin{pmatrix} 1 & 0 & 0 \\ 0 & -\frac{1}{2}\frac{1}{c}\partial_{xx}^h + \frac{1}{c^2}Q^{2h} & \partial_x^{2h} - \frac{h}{2}\frac{u}{c}\partial_{xx}^h \\ 0 & -\rho(\partial_x^{2h} - \frac{h}{2}\frac{u}{c}\partial_{xx}^h) & \rho(\frac{h}{2}c\partial_{xx}^h - Q^{2h}) \end{pmatrix}$$ (29)

we obtain

$$L^h M^h = \begin{pmatrix} -\frac{h}{2}|u|\partial_{xx}^h + Q^{2h} & 0 & 0 \\ 0 & \rho(1 - M^2)\partial_{xx}^h & 0 \\ 0 & 0 & \rho(1 - M^2)\partial_{xx}^h \end{pmatrix}$$ (30)

The philosophy of the Distributive Gauss-Seidel relaxation is as follows: We would like to store at the gridpoints the primitive variables $\mathbf{u}$ and not the auxiliary variables $\mathbf{w}$. For this purpose the updates in w and ϕ corresponding to the relaxing the elliptic factor at point i should be translated into the updates in u, p at points $(i - 1), (i), (i + 1)$ according to M^h.

3.1.3. *Links between DGS relaxation and the Riemann solver*

It is well known that the one-dimensional Euler system can be diagonalized, i.e. rewritten as a set of (locally) decoupled advection equations for the *characteristic* variables $(s, \alpha^+, \alpha^-)^T$, where

$$\alpha_x^+ = \rho c u_x + p_x, \quad \text{and} \quad \alpha_x^- = \rho c u_x - p_x.$$ (31)

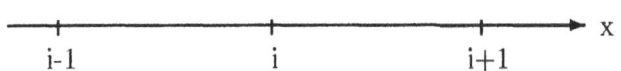

Figure 2.   One-dimensional grid

Some algebra reveals that relaxing the Full-Potential (elliptic) factor corresponds to:

 - at point $(i-1)$ - update α^+, keep α^- the same;
 - at point $(i+1)$ - update α^-, keep α^+ the same;
 - at point i - update both α^+ and α^-.

We can conclude that there are some links between the characteristic variables formulation (approximate Riemann solver) and the design of the DGS.

3.2. TWO-DIMENSIONAL CASE

We shall look now for a *factorizable* upwind scheme in two dimensions.

3.2.1. *Dimension-by-dimension approach*
Considering here the isentropic case (no loss of generality for the purpose of the discussion presented here), we can write such a scheme in the following symbolic form

$$L^h \mathbf{u}^h = 0 \tag{32}$$

The modified equations (or the First Differential Approximation – FDA) corresponding to the scheme dimension-by-dimension

$$FDA(L^h) =$$
$$\begin{pmatrix} \rho[-\frac{h}{2}(c\partial_{xx} + |v|\partial_{yy}) + Q] & 0 & \partial_x - \frac{h}{2}\frac{1}{c}u\partial_{xx} \\ 0 & \rho[-\frac{h}{2}(|u|\partial_{xx} + c\partial_{yy}) + Q] & \partial_y - \frac{h}{2}\frac{1}{c}v\partial_{yy} \\ \rho(\partial_x - \frac{h}{2}\frac{u}{c}\partial_{xx}) & \rho(\partial_y - \frac{h}{2}\frac{v}{c}\partial_{yy}) & -\frac{h}{2}\frac{1}{c}\Delta + Q \end{pmatrix} \tag{33}$$

It is easy to see that the matrix (33) cannot be factorized.

3.3. GENUINELY MULTIDIMENSIONAL APPROACH

The approach towards the construction of discrete schemes for the Euler equations ((Sidilkover, 1994b; Sidilkover, 1994a)) was called "genuinely multidimensional" since it leads to schemes that retain (to a certain extent) the rotational invariance property of the Euler equations. Namely, it can be argued that the artificial dissipation terms present in these schemes approximate a rotationally invariant differential operator. In its turn this may mean that the waves oblique to the grid are "properly upwinded" or, in other words, the same approximate Riemann solver can be "recovered" in an arbitrary direction.

It is not clear whether or not this property, though intuitively appealing, is of any direct practical importance for the steady-state computations. Therefore, we do not discuss it in detail. However, we have observed previously that there are some links between the approximate Riemann solver and the design of DGS in one dimensional case. Therefore, the following conjecture seems reasonable.

Conjecture *A genuinely multidimensional scheme is* <u>*factorizable*</u>.

It was pointed out in (Sidilkover, 1994a) that some of the multidimensional second order corrections can be added without limiters resulting in a linear 'positive' scheme with essential multidimensional character. Namely, for the u-momentum equation in subsonic case, those are the cross-derivative correction terms that compensate for the loss of accuracy due to the artificial dissipation in x direction. For the v-momentum equations those will be the terms that compensate for the loss of accuracy due to the artificial dissipation in y-direction.

Writing such a scheme in the symbolic form

$$L_{[2D]}^h \mathbf{u}^h = 0, \tag{34}$$

and considering the corresponding FDA

$$FDA(L_{[2D]}^h) =$$

$$\begin{pmatrix} \rho[-\frac{h}{2}(c\partial_{xx} + |v|\partial_{yy}) + Q] & \rho[-\frac{h}{2}(c - |v|)\partial_{xy}] & \partial_x - \frac{h}{2}\frac{1}{c}Q_x \\ \rho[-\frac{h}{2}(c - |u|)\partial_{xy}] & \rho[-\frac{h}{2}(|u|\partial_{xx} + c\partial_{yy}) + Q] & \partial_y - \frac{h}{2}\frac{1}{c}Q_y \\ \rho(\partial_x - \frac{h}{2}\frac{u}{c}\partial_{xx}) & \rho(\partial_y - \frac{h}{2}\frac{v}{c}\partial_{yy}) & -\frac{h}{2}\frac{1}{c}\Delta + Q \end{pmatrix} \tag{35}$$

we can easily verified that FDA matrix is *factorizable*. The added multidimensional correction played a crucial role in achieving this property.

However, the FDA does not uniquely define a discrete scheme. Moreover, not all the discrete schemes corresponding to a certain FDA are *factorizable*. A *factorizable* scheme corresponding the above mentioned FDA was constructed on a quad-type grid. The details concerning the scheme as well as the Distributive Gauss-Seidel relaxation will be given elsewhere.

4. Summary

An approach towards constructing a genuinely multidimensional upwind scheme was intoduced in (Sidilkover, 1994b; Sidilkover, 1994a; Sidilkover,

1995). The *fundamental* advantage of this approach for practical purposes is that the multidimensional high-resolution mechanism (unlike the standard one) does not damage the stability properties of the scheme. The conclusion made in this paper is that the genuinely multidimensional approach leads to a scheme that is also *factorizable*. The practical importance of this property is that an optimally efficient multigrid solver can be obtained through the construction of an appropriate Distributive Gauss-Seidel relaxation. Also, since the *factorizability* property can be easily verified, we suggest it is used as the definition of *genuine multidimensionality* of a scheme.

Due to its generality, the genuinely multidimensional approach for discretization of the Euler equations may play a crucial role in constructing an optimally efficient multigrid flow solver suitable for engineering computations. This is because

- it does not rely on casting the equations into any special form.
- extends to the compressible Navier-Stokes equations.

References

Brandt, A., "Multigrid techniques: 1984 guide with applications to fluid dynamics," The Weizmann Institute of Science, Rehovot, Israel, 1984.

Brandt, A. and Yavneh, I., "Accelerated multigrid convergence and high Reynolds recirculating flows," *SIAM J. Sci. Statist. Comput.*, Vol. 14, 1993, pp. 607-626.

Colella, P., "Multidimensional upwind methods for hyperbolic conservation laws," *Technical Report LBL-17023*, Lawrence Berkeley Report, 1984.

Colella, P., "Multidimensional upwind methods for hyperbolic conservation laws," *J. Comp. Phys.*, Vol. 87, 1990, p. 171.

Deconinck, H., Roe, P.L., and Struijs, R., "A multidimensional generalization of Roe's flux difference splitter for the Euler equations," *Comput. & Fluids*, Vol. 22, 1993, pp. 215-222.

Deconinck, H., Struijs, R., and Roe, P.L., "Fluctuation splitting for multidimensional convection problem: An alternative to finite volume and finite element methods," *VKI Lecture Series 1990-3 on Computational Fluid Dynamics*, Von Karman Institute, Brussels, Belgium, March 1991.

Giles, M., Anderson, W.K., and Roberts, T.W., "The upwind control volume scheme for unstructured triangular grids," *Technical Memorandum 101664*, NASA, 1989.

Hirsch, Ch., "Numerical computation of internal and external flows," Vol. 2, J. Wiley & Sons, 1990.

LeVeque, R.J., "High resolution finite volume methods on arbitrary grids via wave propagation," *J. Comp. Phys.*, Vol. 78, 1988.

Mesaros, L. and Roe, P.L., "Multidimensional fluctuation-splitting schemes based on decomposition methods," *AIAA 95-1699*, 12th AIAA CFD Meeting, San Diego, June 19-22, 1995.

Paillère, H., Deconinck, H., and Roe, P.L., "Conservative upwind residual-distribution schemes based on the steady characteristics of the Euler equations," *AIAA 95-1700*, 12th AIAA CFD Meeting, San Diego, June 19-22, 1995.

Radvogin, Yu. B., "Quasi-monotonous multidimensional difference schemes with second order accuracy," *Soviet Academy of Science Preprint*, 1991.

Roberts, T.W., Sidilkover, D., and Swanson, R.C., "Textbook multigrid efficiency for the steady Euler equations," submitted to the *13th CFD AIAA Meeting*, 1996.

Roe, P.L., "Discrete modles for the numerical analysis of time-dependent multidimensional gas dynamics," *J. Comp. Phys.*, Vol. 63, 1986, pp. 458-476.

Sidilkover, D., "Numerical solution to steady-state problems with discontinuities," PhD thesis, The Weizmann Institute of Science, Rehovot, Israel, 1989.

Sidilkover, D., "A genuinely multidimensional upwind scheme and efficient multigrid solver for the compressible Euler equations," *ICASE Report No. 94-84*, 1994.

Sidilkover, D., "A genuinely multidimensional upwind scheme for the compressible Euler equations," *Proceedings of the Fifth International Conference on Hyperbolic Problems: Theory, Numerics, Applications*, J. Glimm *et al.*, eds., World Scientific, June 1994.

Sidilkover, D., "Multidimensional upwinding and multigrid," *AIAA 95-1750*, 12th AIAA CFD Meeting, San Diego, June 19-22, 1995.

Sidilkover, D. and Ascher, U., "A multigrid solver for the steady-state Navier-Stokes equations using pressure-Poisson formulation," *Matematica Aplicada e Computational*, Vol. 14, 1995, pp. 21-35.

Sidilkover, D. and Brandt, A., "Multigrid solution to steady-state 2D conservation laws," *SIAM J. Numer. Anal.*, Vol. 30, 1993, pp.249-274.

Spekreijse, S., "Multigrid solution of monotone second-order discretization of hyperbolic conservation laws," *Math. Comp.*, Vol. 49, 1987, pp. 135-155.

Ta'asan, S., "Canonical forms of multidimensional steady inviscid flows," *ICASE Report No. 93-94*, 1993.

Ta'asan, S., "Canonical-variables multigrid method for steady-state Euler equations," *ICASE Report No. 94-14*, 1994.

van der Weide, E. and Deconinck, H., "Positive matrix distribution schemes for hyperbolic systems, with applications to the Euler equations," *Proceedings of the Third ECCOMAS CFD Conference*, J.A. Desideri *et al.*, eds., Paris, France, John Wiley & Sons, September 1996.

LIST OF ATTENDEES

Ramesh Agarwal*†
NIAR
Wichita State University
1845 Fairmount
Wichita, KS 67260-0093
(316) 978-6427
agarwal@niar.twsu.edu

W. Kyle Anderson†
Mail Stop 128
NASA Langley Research Center
Hampton, VA 23681-0001
(757) 864-2164
w.k.anderson@larc.nasa.gov

Eyal Arian
ICASE
Mail Stop 403
NASA Langley Research Center
Hampton, VA 23681-0001
(757) 864-2208
arian@icase.edu

Necdet Aslan*†
Marmara Universitesi
Fen-Edeb, Fizik Bolumu
Ziverbey, Istanbul
TURKEY
necdet@nem.nukleer.gov.tr

Harold Atkins†
Mail Stop 128
NASA Langley Research Center
Hampton, VA 23681-0001
(757) 864-2308
h.l.atkins@larc.nasa.gov

Abdelkader Baggag
ICASE
Mail Stop 403
NASA Langley Research Center
Hampton, VA 23681-0001
(757) 864-9817
baggag@icase.edu

Alvin Bayliss†
Department of Engineering
  Sciences
Northwestern University
Technological Institute
Evanston, IL 60208
(847) 491-5585
alvin@alvin.eecs.nwu.edu

Daryl Bonhaus
Mail Stop 128
NASA Langley Research Center
Hampton, VA 23681-0001
(757) 864-2293
d.l.bonhaus@larc.nasa.gov

---

*Speaker
†Panel Member

388

Achi Brandt[†]
Department of Applied Math.
The Weizmann Institute of
  Science
Rehovot, 76100
ISRAEL
(011) 972-8342345
mabrandt@weizmann.weizmann.ac.il

Mark Carpenter[*†]
Mail Stop 128
NASA Langley Research Center
Hampton, VA 23681-0001
(757) 864-2318
m.h.carpenter@larc.nasa.gov

Sukumar Chakravarthy[*†]
Metacomp Technologies, Inc.
650 S. Westlake Blvd., Suite 203
Westlake Village, CA 91362-3804
(805) 371-8750
sukumarcr@aol.com

Chau-Lyan Chang
High Technology Corporation
28 Research Drive
Hampton, VA 23666
(757) 865-0818
chang@htc-tech.com

Wenlong Dai[*†]
University of Minnesota
116 Church Street, SE
Minneapolis, MN 55455
(612) 626-0050
wenlong@lcse.umn.edu

Stephen Davis
U.S. Army Research Office
P.O. Box 12211
Research Triangle Park,
  NC 27709-2211
(919) 549-4284
sdavis@aro.ncren.net

Rosa Donat
University of Valencia
Dept. de Matematica Aplicada
c/Dr. Moliner, 50
Burjassot 46100
SPAIN
donat@godella.matapl.uv.es

Michael Giles[*†]
Oxford University Computing
  Laboratory
Wolfson Building
Parks Road
Oxford OX1 3QD
UNITED KINGDOM
(011) 44-1865-273862
giles@comlab.oxford.ac.uk

John Goodrich*†
Mail Stop 5-11
NASA Lewis Research Center
Cleveland, OH 44135
(216) 433-5922
fsgdrch@jgoodrich.lerc.nasa.gov

Wagdi G. Habashi*†
Department of Mechanical
  Engineering
Concordia University
1455 de Maisonneuve Blvd.
  West, ER 301
Montreal Quebec H3GIM8
CANADA
(514) 848-3165
habashiw@cfdlab.concordia.ca

Mohamed Hafez*†
Department of Mechanical
  Engineering
University of California, Davis
Davis, CA 95616
(916) 752-0212
mhafez@ucdavis.edu

Ehtesham Hayder
ICASE
Mail Stop 403
NASA Langley Research Center
Hampton, VA 23681-0001
(757) 864-4746
hayder@icase.edu

Jeffrey Hittinger
Department of Aerospace
  Engineering
The University of Michigan
2001 FXB Building
Ann Arbor, MI 48109-2118
(313) 764-7573
jhitt@engin.umich.edu

W. H. Hui
Department of Mathematics
The Hong Kong University of
  Science and Technology
Clear Water Bay
Kowloon, HONG KONG
(011) 852-2358-7415
whhui@uxmail.ust.hk

David Jacqmin*†
Mail Stop 5-11
NASA Lewis Research Center
Cleveland, OH 44122
(216) 433-5853
fsdavid@tess.lerc.nasa.gov

Leland Jameson
ICASE
Mail Stop 403
NASA Langley Research Center
Hampton, VA 23681-0001
(757) 864-2191
lmj@icase.edu

390

Jim Jones
Center for Applied Scientific
  Computing (CASC)
Lawrence Livermore National
  Laboratory
POB 808, L-561
Livermore, CA 94551
(510) 423-5194
jjjones@llnl.gov

Dinesh Kaushik
Department of Computer Science
Old Dominion University
Norfolk, VA 23529-0162
(757) 683-3266
kaushik@cs.odu.edu

David E. Keyes
Department of Computer Science
Old Dominion University
Norfolk, VA 23529-0162
(757) 683-4928
keyes@cs.odu.edu

Dennis Lankford
Sverdrup Tech., Inc.
1099 Avenue C
Arnold Air Force Base,
  TN 37389-9013
(615) 454-5825
lankford@hap.arnold.af.mil

Torbjorn Larsson
Prosolvia R&T, Inc.
2855 Coolidge Highway
Suite 104
Troy, MI 48084
(810) 649-1200
tobbe@prosolvia.com

Timur Linde*†
Department of Aerospace
  Engineering
University of Michigan
2004 FXB Bldg.
Ann Arbor, MI 48109
(313) 763-2397
linde@engin.umich.edu

Rainald Löhner*†
Institute for Computational
  Sciences
  and Informatics
George Mason University
Mail Stop 5C3
Fairfax, VA 22030-4444
(703) 993-4075
lohner@beethoven.gmu.edu

Robert Lowrie*†
Los Alamos National
  Laboratory
CIC-19, MS B256
Los Alamos, NM 87545
(505) 667-2121
lowrie@lanl.gov

Michele Macaraeg
Mail Stop 128
NASA Langley Research Center
Hampton, VA 23681-0001
(757) 864-2295
m.g.macaraeg@larc.nasa.gov

Robert MacCormack*†
Department of Aeronautics and
   Astronautics
Stanford University
Durand Building
Stanford, CA 94305
(415) 723-4627

Laurence Manzi
Applied Physics Laboratory
Johns Hopkins University
Laurel, MD 20723-6099
(301) 953-5000
larry_manzi@jhuapl.edu

Dimitri Mavriplis
ICASE
Mail Stop 403
NASA Langley Research Center
Hampton, VA 23681-0001
(757) 864-2213
dimitri@icase.edu

Duane Melson
Mail Stop 128
NASA Langley Research Center
Hampton, VA 23681-0001
(757) 864-2227
n.d.melson@larc.nasa.gov

Stanley Osher*†
Department of Mathematics
University of California, Los Angeles
Los Angeles, CA 90095
(310) 825-1758
sjo@math.ucla.edu

Sergei Pevchin
AOE Department
VPI & SU
P.O. Box 11705
Blacksburg, VA 24062-1705
(540) 231-6242
pevchin@aoe.vt.edu

Alex Pothen
Department of Computer Science
Old Dominion University
Norfolk, VA 23529-0162
(757) 683-4414
pothen@cs.odu.edu

Elbridge Gerry Puckett*†
Department of Mathematics
University of California, Davis
Davis, CA 95616
(916) 757-2839
puckett@math.ucdavis.edu

Philip Roe*†
Department of Aerospace
    Engineering
University of Michigan, Ann Arbor
Ann Arbor, MI 48109-2140
(313) 764-3394
philroe@caen.engin.umich.edu

Scott Rimbey
Department of Math Sciences
University of Delaware
Newark, DE 19711
(302) 831-1868
rimbey@math.udel.edu

Dave Rudy
Mail Stop 139
NASA Langley Research Center
Hampton, VA 23681-0001
(757) 864-2297
d.h.rudy@larc.nasa.gov

Thomas Roberts
Mail Stop 128
NASA Langley Research Center
Hampton, VA 23681-0001
(757) 864-6804
t.w.roberts@larc.nasa.gov

Yousef Saad
Department of Computer Science
University of Minnesota
4-192 Ee/CSci Building
200 Union Street
Minneapolis, MN 55455
(612) 624-7804
saad@cs.umn.edu

Kevin Roe
ICASE
Mail Stop 403
NASA Langley Research Center
Hampton, VA 23681-0001
(757) 864-7368
kproe@icase.edu

Manuel Salas
ICASE
Mail Stop 403
NASA Langley Research Center
Hampton, VA 23681-0001
(757) 864-2174
salas@icase.edu

David Sidilkover*†
ICASE
Mail Stop 403
NASA Langley Research Center
Hampton, VA 23681-0001
(757) 864-7312
sidilkov@icase.edu

Bjorn Sjogreen*†
Department of Scientific
  Computing
University of Uppsala
Box 120
Uppsala 751 04
SWEDEN
(011) 46-18-18-29-72
bjorns@tdb.uu.se

Jerry South
Mail Stop 285
NASA Langley Research Center
Hampton, VA 23681-0001
(757) 864-3520
j.c.south@larc.nasa.gov

Chao-Ho Sung
David Taylor Model Basin
Code 5030
Bethesda, MD 20084-5000
(301) 227-1865
sung@sigma.dt.navy.mil

R. Charles Swanson
Mail Stop 128
NASA Langley Research Center
Hampton, VA 23681-0001
(757) 864-2235
r.c.swanson@larc.nasa.gov

Shlomo Ta'asan*†
Department of Mathematics
Canegie Mellon University
Pittsburgh, PA 15213
(412) 268-5582
shlomo@andrew.cmu.edu

James Thomas
Mail Stop 128
NASA Langley Research Center
Hampton, VA 23681-0001
(757) 864-2146
j.l.thomas@larc.nasa.gov

Eli Turkel*†
Department of Mathematics
Tel-Aviv University
Ramat-Aviv
Tel-Aviv
ISRAEL
(011) 97236408038
turkel@math.tau.ac.il

Semyon Tsynkov*†
Mail Stop 128
NASA Langley Research Center
Hampton, VA 23681-0001
(757) 864-2150
s.v.tsynkov@larc.nasa.gov

Bram van Leer*†
Department of Aerospace
  Engineering
University of Michigan
Ann Arbor, MI 48109-2118
(313) 764-4305
bram@engin.umich.edu

Veer Vatsa
Mail Stop 128
NASA Langley Research Center
Hampton, VA 23681-0001
(757) 864-2236
v.n.vatsa@larc.nasa.gov

V. Venkatakrishnan
Boeing Commercial Airplane Group
POB 3707, M/S 67-LF
Seattle, WA 98124-2207
(206) 234-3124
venkat.venkatakrishnan@boeing.com

August Verhoff*†
McDonnell Douglas Corporation
MC 1067126
P.O. Box 516
St. Louis, MO 63166
(314) 233- 6343

Jeffery White
Taitech, Inc.
AMC P.O. Box 33830
Wright Patterson Air Force
  Base, OH 45433-0830
(513) 255-4141
whiteja@possum.appl.wpafb.af.mil

Kun Xu
Department of Mathematics
Hong Kong University of
  Science and Technology
Clear Water Bay, Kowloon
HONG KONG

Osman Yasar
Oak Ridge National
  Laboratory
P.O. Box 2008
Mail Stop 6203
Oak Ridge, TN 37831
(423) 241-5629
yasar@ccs.ornl.gov

Helen Yee
Mail Stop 202A-1
NASA Ames Research Center
Moffett Field, CA 94035
(415) 604-4769
yee@nas.nasa.gov

Hong Zhang-Sun
Department of Mathematical
  Sciences
Clemson University
Clemson, SC 29634
(504) 388-4966
hongsu@math.clemson.edu

Steven Zalesak
NASA Goddard Space Flight
  Center
Code 934
Greenbelt, MD 20771
(301) 286-8935
zalesak@gondor.gsfc.nasa.gov

ICASE/LaRC Interdisciplinary Series in Science and Engineering

KLUWER ACADEMIC PUBLISHERS – DORDRECHT / BOSTON / LONDON

The manufacturer's authorised representative in the EU is Springer
Nature Customer Service Centre GmbH, Europaplatz 3, 69115 Heidelberg,
Germany. If you have any concerns regarding our products, please
contact ProductSafety@springernature.com

Printed and bound by CPI Group (UK) Ltd, Croydon, CR0 4YY

23/04/2026

02095628-0005